HIGH-PRECISION METHODS IN EIGENVALUE PROBLEMS AND THEIR APPLICATIONS

Differential and Integral Equations and Their Applications

A series edited by:
A.D. Polyanin
Institute for Problems in Mechanics, Moscow, Russia

Volume 1
Handbook of First Order Partial Differential Equations
A.D. Polyanin, V.F. Zaitsev and A. Moussiaux

Volume 2
Group-Theoretic Methods in Mechanics and Applied Mathematics
D.M. Klimov and V. Ph. Zhuravlev

Volume 3
Quantization Methods in the Theory of Differential Equations
V.E. Nazaikinskii, B.-W. Schulze and B. Yu. Sternin

Volume 4
Hypersingular Integral Equations and Their Applications
I.K. Lifanov, L.N. Poltavskii and G.M. Vainikko

Volume 5
Equations of Mathematical Diffraction Theory
Mezhlum A. Sumbatyan and Antonio Scalia

Volume 6
High-Precision Methods in Eigenvalue Problems and Their Applications
Leonid D. Akulenko and Sergei V. Nesterov

HIGH-PRECISION METHODS IN EIGENVALUE PROBLEMS AND THEIR APPLICATIONS

Leonid D. Akulenko
Sergei V. Nesterov

CRC Press
Taylor & Francis Group
Boca Raton London New York

CRC Press is an imprint of the
Taylor & Francis Group, an **informa** business
A CHAPMAN & HALL BOOK

CRC Press
Taylor & Francis Group
6000 Broken Sound Parkway NW, Suite 300
Boca Raton, FL 33487-2742

First issued in hardback 2019

ISBN-13: 978-0-415-30993-6 (hbk)

Catalog record is available from the Library of Congress

Visit the CRC Press Web site at www.crcpress.com

**Visit the Taylor & Francis Web site at
http://www.taylorandfrancis.com**

**and the CRC Press Web site at
http://www.crcpress.com**

Contents

Preface

The problem of finding eigenvalues and eigenfunctions and studying their behavior plays a crucial role in modern mathematics. These investigations are of utmost importance for theoretical and applied problems in mechanics, physics, physical chemistry, biophysics, mathematical economics, theory of systems and their optimization, theory of random processes, and many other branches of natural science. The results of investigations of the Sturm–Liouville problem and its numerous modifications served as a basis for the creation and development of new branches of mathematics: theory of boundary value problems and integral equations, theory of operators and spectral theory, functional analysis and variational calculus, and also analytical and numerical methods.

Finding eigenvalues (frequencies) and eigenfunctions (vibration shapes) is the main topic in the theory of vibrations and wave processes. Vibration phenomena are observed in objects of various types, discrete or continuously distributed in space. Great achievements in the studies of finite-dimensional vibration systems are due to the fundamental works of Lyapunov, Poincaré, Mandelstam, Timoshenko, Andronov, Bogolyubov, and Kolmogorov.

Foundations of the theory of wave processes were laid at the end of the 19th century and the beginning of the 20th. These fundamental investigations are associated with the names of Maxwell, Kirchhoff, Lord Rayleigh, Helmholtz, Lord Kelvin, Lamb, Love, Timoshenko, Mandelstam, Papaleksi, Sommerfeld, Hilbert, Courant, Morse, Feshbach, Kochin, and Sretenskii.

Of everlasting importance is the treatise, "Theory of Sound", written by Lord Rayleigh more than a century ago. It still remains a reference book for many specialists in the field of vibration theory and wave processes. Owing to comprehensive and detailed investigations of various wave phenomena, the formal mathematical results obtained for the Sturm–Liouville problem and other purely theoretical conclusions have acquired a definite physical meaning. Certain technical difficulties of a mathematical or computational character have often forced the authors to consider only linear homogeneous models of systems. As noticed by Lord Rayleigh, such models have definite limitations and it is necessary to study nonhomogeneous and nonlinear systems.

The authors of this book have tried to follow the basic scientific principles set forth in Lord Rayleigh's treatise [62] and hope to fill some gaps in the investigations of eigenvalue problems for essentially nonhomogeneous linear systems that have important applications in natural and technological sciences. This book presents a new numerical–analytical method, called the method of accelerated convergence, for solving boundary value problems for differential equations of the second or a higher order, as well as for some systems of differential equations. The coefficients of the equations may be arbitrary functions with a fairly wide range, and this seems an important factor for applications to modern technologies. The theoretical basis of this approach is provided by a relation between the eigenvalues and the length of the interval on which the boundary value problem is considered. This fundamental property has never been used in theoretical or applied studies.

The method of accelerated convergence, together with the corresponding computational algorithms, is very efficient in the sense that it is easy to implement, spares computational resources, is stable with respect to computer failure, allows for testing the precision of calculations, entails no error accumulation, and possesses accelerated convergence. In contrast to the existing approaches, the refinements for eigenvalues and eigenfunctions obtained by this method are based on the construction and the examination of a particular eigenvalue corresponding to a fixed vibration mode.

Preliminary analysis of the system reduces to simple operations, and application of the recurrent computational algorithm can be almost entirely formalized and computer-managed. The implementation of this method requires highly precise integration of a sequence of Cauchy problems of the same type. The estimates involved in the preliminary stage of the investigation are obtained by standard calculations, which can be carried out relatively simply. Personal computers provide adequate means for on-the-fly highly precise calculations of eigenvalues and eigenfunctions. Such calculations are needed for problems of parametric synthesis of complex systems with a wide range of multi-dimensional structural parameters. Due to its outstanding precision, the method presented here can be used for identifying some important qualitative and analytical characteristics of eigenvalues and eigenfunctions depending on the said parameters, in particular, if these parameters are close to their limit values resulting in singularities in the equations.

Analytical and numerical advantages of the method developed here are demonstrated by the construction of approximate solutions for wide classes of problems, some of which are just model examples, while others are quite meaningful problems, interesting for their own sake or for various investigations in the theory of vibrations and wave processes. We consider problems formulated in standard terms for second- and fourth-order equations with boundary conditions of different types, and also for systems of second-order equations. Special attention is given to generalized eigenvalue problems in which the spectral parameter is nonlinearly involved in the equation or the boundary conditions. The importance of studying generalized eigenvalue problems is due to the fact such problems arise as a result of refinement of vibration models of distributed systems. Our investigations of vibration systems in multi-dimensional domains, in some important cases, result in generalized boundary value problems with nonseparable parameters whose eigenvalues can be effectively constructed by the method of accelerated convergence.

It should be mentioned that it was not the aim of the authors to present the methods and algorithms developed by them in most general formal terms, but it seems more important to describe the basic concepts of these methods and illustrate their potential by solving meaningful problems. Some of these problems are just test models, while others have important applications in mechanics, vibration theory, hydrodynamics, elasticity, and other fields. The interested reader may find that in order to solve some specific eigenvalue problems, it is much easier to adapt the methods and algorithms of this book to his purposes than to take a common approach based on the Rayleigh–Ritz method, the Bubnov–Galiorkin method, or the finite element method. With growing computational experience, the reader may extend the scope of applications of the method of accelerated convergence and study other classes of vibration systems. Some problems for further investigations are formulated at the end of several chapters.

This book consists of 15 chapters divided into sections and subsections. The enumeration of the formulas is separate for each section; thus, (2.3.15) indicates formula 15 in Section 2.3.

Chapters 1–5 are dedicated to the classical Sturm–Liouville problem and its generalizations. Chapter 1 is of an introductory character and contains statements of the eigenvalue problem in analytical and variational terms, the general scheme of its solution for different types of boundary conditions, reduction of the problem to the standard form, and its reduction to Fredholm and Volterra integral equations of the second kind, a summary of analytical and functional-numerical methods for the solution of the eigenvalue problem. Chapter 2 is crucial for the entire book and contains an exposition of the method of accelerated convergence based on an original procedure of the introduction of a small parameter and a relation between the eigenvalue and the length of the interval in integral and differential forms. Modifications of that method are also suggested for different types of boundary conditions occurring in applications. In Chapter 3, we describe the results of analytical solution of the Sturm–Liouville problem for a system with slowly varying coefficients, and some attention is given to the exposition and the justification of a perturbation method for the construction of approximations of the spectrum and an orthonormal basis of eigenfunctions. Chapter 4 contains the results of our investigations of generalized Sturm–Liouville problems with coefficients depending on the spectral parameter in an arbitrary nonlinear manner.

This class of problems essentially differs from the classical case, and its investigation requires a specific approach based on the method of accelerated convergence. Asymptotic solutions of the generalized Sturm–Liouville problem for higher vibration modes are obtained in Chapter 5. This approach is very effective in the case of smooth coefficients.

Chapters 6–9 are aimed at the development of effective numerical-analytical methods for solving self-conjugate boundary value problems for fourth-order equations, as well as systems of second-order equations. In Chapter 6, we establish oscillation properties of eigenvalues and eigenfunctions and formulate constructive statements similar to the Sturm oscillation theorem, comparison theorems and their corollaries. The notion of sagittary function is used for the construction of recurrent algorithms, which allow us to obtain effective two-sided estimates for eigenvalues and the corresponding approximations of eigenfunctions. A highly precise numerical-analytical method similar to the Newton method is developed in Chapter 7 for fourth-order boundary value problems. An analytical perturbation method for systems with slowly varying parameters is expounded in Chapter 8. The results of Chapters 6 and 7 are applied to the problem of parametric synthesis for strongly inhomogeneous beams with different types of boundary conditions for which the problem is self-conjugate, in particular, the conditions of elastic fixation for angular and linear displacements at the end-points. Some generalizations of the method of accelerated convergence for vector problems of Strum–Liouville type are studied in Chapter 9. These results can also be used for constructing approximations of eigenvalues and eigenfunctions of some problems for partial differential equations.

Approximate solutions of some specific eigenvalue problems for one-dimensional distributed systems are considered in Chapters 10–12. In Chapter 10, we study vibrations and stability of some elastic systems for a wide range of their determining parameters. We consider a heavy inhomogeneous rotating thread with additional tension, an inhomogeneous rod subjected to longitudinal loading (Euler's buckling problem), and a long narrow loaded beam (lateral buckling). Surface waves on a rotating spherical layer of heavy fluid and internal waves in a strongly stratified heavy fluid with unperturbed horizontal boundary are studied in Chapter 11. In Chapter 12, we examine parametric vibrations of Hill-type systems and apply these results to problems of stability of vibrations of elastic mechanical systems and satellites.

Chapters 13–15 contain our results regarding frequencies and shapes of free vibrations of systems described by partial differential equations (membranes and plates). In Chapter 13, we consider a rectangular membrane of variable density described by a smooth function of spatial coordinates. Vibrations of such a membrane are studied with the help of a modified algorithm of accelerated convergence. In Chapter 14, we construct and study approximations of frequencies and shapes of free vibrations of an inhomogeneous membrane whose surface density varies sharply, almost stepwise (cross-shaped inhomogeneities). And finally, Chapter 15 contains results of analytical and numerical-analytical investigations of free vibrations of elliptic membranes and plates clamped at the edge for a wide range of eccentricity values. A distinctive feature of such problems is that the independent variables cannot be separated completely and one has to solve two or more coupled boundary value problems of the Sturm–Liouville type.

The authors express their gratitude to R. P. Soldatova and S. A. Kumakshev for their valuable assistance in the preparation of the manuscript for publication.

The authors hope that this book will be helpful for a wide range of scientists, engineers, and postgraduate students with interests in applied mathematics, mechanics, physics, and engineering sciences.

Leonid D. Akulenko
Sergei V. Nesterov

BASIC NOTATIONS

Latin Characters

$\operatorname{Arg} f(x,a)$	the set of all roots of the equation $f(x,a)=0$;
$\arg f(x,a)$	a root of the equation $f(x,a)=0$;
$C_1, C_2, \ldots$	arbitrary constants;
$r,\ \varphi,\ z$	cylindrical coordinates: $r=\sqrt{x^2+y^2}$, $x=r\cos\varphi$, $y=r\sin\varphi$;
$r,\ \theta,\ \varphi$	spherical coordinates: $r=\sqrt{x^2+y^2+z^2}$, $x=r\sin\theta\cos\varphi$, $y=\sin\theta\sin\varphi$, $z=r\cos\theta$;
t	time ($t\geq 0$);
u	unknown function (dependent variable);
$u_n(x)$	the n-th eigenfunction (mode);
x	an independent variable;
x, y, z	spatial (Cartesian) coordinates;
x_1, x_2, x_3	spatial (Cartesian) coordinates.

Greek Characters

Δ	Laplace operator; in the two-dimensional case, $\Delta=\dfrac{\partial^2}{\partial x^2}+\dfrac{\partial^2}{\partial y^2}$; in the three-dimensional case, $\Delta=\dfrac{\partial^2}{\partial x^2}+\dfrac{\partial^2}{\partial y^2}+\dfrac{\partial^2}{\partial z^2}$;
$\Delta\Delta$	biharmonic operator; in the two-dimensional case, $\Delta\Delta=\dfrac{\partial^4}{\partial x^4}+2\dfrac{\partial^4}{\partial x^2\partial y^2}+\dfrac{\partial^4}{\partial y^4}$;
λ_n	the n-th eigenvalue;
λ_n^*	an upper bound for the n-th eigenvalue;
λ_{n*}	a lower bound for the n-th eigenvalue.

Derivatives

partial derivatives of $u=u(x,t)$: $\quad u_x=\dfrac{\partial u}{\partial x},\ u_t=\dfrac{\partial u}{\partial t},\ u_{xx}=\dfrac{\partial^2 u}{\partial x^2},\ u_{tt}=\dfrac{\partial^2 u}{\partial t^2},\ \ldots,$

derivatives of $u=u(x)$: $\quad u'=\dfrac{du}{dx},\ u''=\dfrac{d^2u}{dx^2},\ u'''=\dfrac{d^3u}{dx^3},\ u''''=\dfrac{d^4u}{dx^4};$

derivatives of $u=u(t)$: $\quad \dot u=\dfrac{du}{dt},\ \ddot u=\dfrac{d^2u}{dt^2}.$

Authors

Leonid D. Akulenko, Professor, D.Sc., Ph.D., is a prominent scientist in the fields of applied mathematics, control theory and mechanics.

Leonid D. Akulenko graduated from the Department of Physics of the Moscow State University in 1965. He received his Ph.D. degree in 1968 and D.Sc. degree in 1983 at the Institute for Problems in Mechanics of the Russian Academy of Sciences. Since 1972, Professor Akulenko has been a member of the staff at the Institute for Problems in Mechanics of the Russian Academy of Sciences. He is also Professor at the Moscow Institute of Physics and Technology.

Professor Akulenko is the author of three books and more than 170 research papers. He has made important contributions in the fields of control theory, nonlinear oscillations, asymptotic methods, and mathematical physics. In 1998, he was awarded the State Prize of the Russian Federation for his research in control theory and in 1997, he was awarded the Andronov Prize of the Russian Academy of Sciences for his research in mechanics.

Address: Institute for Problems in Mechanics, Vernadsky Avenue 101 Bldg 1, 119526 Moscow, Russia.

Sergei V. Nesterov, Professor, D.Sc., Ph.D., is a prominent scientist in the fields of hydrodynamics, applied mathematics, and differential equations.

Sergei V. Nesterov graduated from the Department of Physics of the Moscow State University in 1960. He received his Ph.D. degree in 1966 and D.Sc. degree in 1973 at the Institute of Oceanology of the Russian Academy of Sciences. Since 1989, Professor Nesterov has been Deputy Director of the Institute for Problems in Mechanics of the Russian Academy of Sciences. He is also Professor at the Bauman Moscow State Technical University.

Professor Nesterov is the author of more than 140 research papers. He has made important contributions in the fields of linear and nonlinear waves, asymptotic methods, and mathematical physics. In 1997, he was awarded the Andronov Prize of the Russian Academy of Sciences for his research in mechanics. In 1999, the title of Honored Scientist of the Russian Federation was conferred on him.

Address: Institute for Problems in Mechanics, Vernadsky Avenue 101 Bldg 1, 119526 Moscow, Russia.

Chapter 1

Statement of Eigenvalue Problems. Basic Methods of Their Solution

This chapter has a traditional introductory character. Various approaches to the investigation and approximate solution of eigenvalue problems are demonstrated in the case of the classical Sturm–Liouville problem with boundary conditions of the first, the second, or the third kind, and also periodic boundary conditions (the fourth kind). One of the main approaches to these eigenvalue problems is based on the analysis of solutions of the corresponding differential equation. The other approaches discussed here are those based on the direct variational method and the methods of functional analysis widely used in theoretical investigations and numerical procedures.

1.1. Statement of the Sturm–Liouville Problem

Problems that require an investigation of eigenvalues and eigenfunctions arise in connection with numerous topics in mechanics, the theory of vibrations and stability, hydrodynamics, elasticity, acoustics, electrodynamics, quantum mechanics, etc. Such problems are formulated in many different ways. In order to identify the essential features of eigenvalue problems, we start with their two simplest classical statements – in differential form and in variational form.

1.1.1. Boundary value problem for eigenvalues and eigenfunctions. The Sturm–Liouville problem is formulated as follows: *Find the values of the parameter λ for which there exists a nontrivial solution $u(x, \lambda)$ of the linear second-order differential equation* [4, 20, 22–24, 29, 33, 46, 47, 55]

$$(p(x)u')' + (\lambda r(x) - q(x))u = 0, \quad 0 < x < 1, \tag{1.1.1}$$

with the boundary conditions of the first kind

$$u(0) = u(1) = 0. \tag{1.1.2}$$

Here, $p(x)$, $r(x)$, $q(x)$ are known functions of the argument x varying on the unit segment, $x \in [0, 1]$. These functions are assumed sufficiently smooth, so that the constructions below make sense. If required by the context, the regularity properties of these functions will be specified separately. Moreover, it is assumed that the following conditions of positive definiteness are satisfied:

$$0 < p^- \le p(x) \le p^+ < \infty, \quad 0 < r^- \le r(x) \le r^+ < \infty, \quad 0 \le q(x) \le q^+ < \infty, \tag{1.1.3}$$

where $p^\pm$, $r^\pm$, q^+ are constants.

Note that if the argument x varies on an arbitrary bounded segment $a \le x \le b$, we can pass to another variable x_1 on the unit segment $0 \le x_1 \le 1$ by letting

$$x = (b-a)x_1 + a, \quad x_1 = \frac{x-a}{b-a}. \tag{1.1.4}$$

If the interval $[a, b]$ is unbounded, i.e., $b = \infty$ or (and) $a = -\infty$, one can use a nonlinear transformation of the argument and thus pass to the unit segment, for instance,

$$\begin{aligned} &x = a + \tan(\tfrac{\pi}{2}x_1) && \text{if } |a| < \infty, \ b = \infty; \\ &x = b - \tan[\tfrac{\pi}{2}(1 - x_1)] && \text{if } a = -\infty, \ |b| < \infty; \\ &x = \tan[\tfrac{\pi}{2}(2x_1 - 1)] && \text{if } a = -\infty, \ b = \infty. \end{aligned} \tag{1.1.5}$$

Transformations of the argument similar to (1.1.5) may be realized with the help of other elementary algebraic or transcendental functions. However, it should be kept in mind that in contrast to the regular transformation (1.1.4) replacements like (1.1.5) may reduce equation (1.1.1) to an equation with singularities at the points $x_1 = 1$ or (and) $x_1 = 0$. Situations in which conditions (1.1.3) are violated require special consideration.

Next, we discuss boundary conditions (1.1.2), which most commonly occur in theoretical investigations. It should be mentioned that in many applied problems these boundary conditions can only be satisfied in a fairly rough approximation. The actual fixation of the end-points is more adequately described by the boundary conditions of the third kind, the so-called *conditions of elastic fixation* [6, 22, 24, 33]

$$p(0)u'(0) - k_0 u(0) = 0, \quad p(1)u'(1) + k_1 u(1) = 0. \tag{1.1.6}$$

The coefficients k_0, k_1 characterize the stiffness of the fixation. Conditions (1.1.2) correspond to the limit case k_0, $k_1 \to \infty$, which can be realized in rough approximation. Boundary conditions of the second kind correspond to $k_0 = k_1 = 0$ and have the form

$$p(0)u'(0) = 0, \quad p(1)u'(1) = 0. \tag{1.1.7}$$

These boundary conditions are of great importance for theoretical and applied studies and have a definite mechanical or physical meaning (for instance, of free ends). A remarkable feature of boundary conditions (1.1.2), (1.1.6), (1.1.7) is that they ensure the self-conjugate character of the corresponding problem.

Together with the above three types of boundary conditions which establish a relation between the values of u, u' separately at each end-point $x = 0, 1$, an important role is played by the boundary condition of the fourth kind, the so-called *mixed condition* [19, 24, 42, 43, 68]

$$u(0) = u(1), \quad p(0)u'(0) = p(1)u'(1). \tag{1.1.8}$$

If the functions $p(x), r(x), q(x)$ are periodically extended outside the interval $0 < x < 1$ (1.1.4), then conditions (1.1.8) determine periodic solutions (called parametric vibrations, if $x = t$ is time). Note that for different types of boundary conditions, the behavior of solutions is different and the corresponding problems have to be approached in different ways.

As to the form of the original state equation (1.1.1), it is useful to keep in mind the following methods of its simplification. By a nonsingular linear transformation of the unknown function u (the Liouville transformation $u_* = u\sqrt{p}$), one obtains an equation with the coefficient $p \equiv 1$ by the highest-order derivative,

$$u_*'' + (\lambda r_*(x) - q_*(x))u_* = 0, \quad 0 < x < 1, \tag{1.1.9}$$

where

$$u_* = u\sqrt{p}, \quad r_* = \frac{r}{p}, \quad q_* = \frac{q}{p} + \frac{(\sqrt{p})''}{\sqrt{p}}.$$

Of course, this change of the variables is admissible, if $p(x)$ is twice differentiable. Boundary conditions (1.1.6)–(1.1.8) should be changed accordingly.

For a smooth function r_*, equation (1.1.9) admits further simplification by changing the argument x to z and the function $u_*(x)$ to $v(z)$,

$$\begin{gathered} v'' + (\lambda - s(z))v = 0, \quad v = v(z) = r_*^{1/4}(x)u_*(x), \quad z = \int_0^x \sqrt{r_*(\xi)}d\xi, \\ s = \frac{q_*}{r_*} + \frac{1}{4r_*^{3/4}}\left(\frac{r_*'}{r_*^{5/4}}\right)', \quad x = x(z), \quad 0 \le z \le \ell = \int_0^1 \sqrt{r_*(x)}\,dx. \end{gathered} \tag{1.1.10}$$

The structure of equation (1.1.10) allows us to go back to the variable $x = x(z)$, since r_*, q_* are functions of the argument x. Note that the new argument z can be normalized according to (1.1.4). Equations of the form (1.1.10) are widely used in theoretical physics. However, from the standpoint of applications, the transformations (1.1.9), (1.1.10) entail excessive computational difficulties.

The original form of the state equation (1.1.1) presumes that the function $p(x)$ is continuously differentiable, which is not always the case. However, this obstacle can be easily overcome by the introduction of a suitable "impulse" variable $\theta = pu'$ and the Cauchy representation

$$u' = \frac{\theta}{p(x)}, \qquad \theta' = -(\lambda r(x) - q(x))u. \tag{1.1.11}$$

System (1.1.11) is more convenient for analytical and, especially, for numerical analysis. Boundary conditions (1.1.2), (1.1.6)–(1.1.8) can be easily written in terms of the variables u, θ.

Problem (1.1.1)–(1.1.3) (the Sturm–Liouville problem) is an example of eigenvalue problems. The values of the parameter λ for which there exist its nontrivial solutions $u(x, \lambda)$ are called *eigenvalues* and the solutions themselves are called *eigenfunctions* (or *vibration modes*). In the theory of vibrations of various distributed systems, the values $\lambda = \omega^2$ determine *frequencies of free vibrations* (or *eigenfrequencies*) ω, and the functions $u(x, \lambda)$ determine *shapes of free vibrations*.

1.1.2. Variational statement of the eigenvalue problem. It is well known [4, 22, 24, 28, 45, 46] that the Sturm–Liouville problem (1.1.1), (1.1.2) is equivalent to the following variational isoperimetric problem: *Find a continuously differentiable function* $u(x)$, $0 \le x \le 1$, *satisfying boundary conditions* (1.1.2) *and realizing the minimum of the functional*

$$J[u] = \int_0^1 [p(x)u'^2 + q(x)u^2]\,dx \to \min_u, \quad u(0) = u(1) = 0, \tag{1.1.12}$$

under the additional isoperimetric condition (condition of normalization with the weight $r(x)$*)*

$$\Phi[u] = \int_0^1 r(x)u^2(x)\,dx \equiv \|u\|_r^2 = 1. \tag{1.1.13}$$

The minimum of the functional $J[u]$ (1.1.12) on the entire set of admissible functions satisfying the said conditions (the so-called global minimum) is equal to the first eigenvalue $\lambda_1 > 0$ of the Sturm–Liouville problem. The function $u_1(x)$ realizing that minimum is the first eigenfunction of the problem, i.e., $J[u_1(x)] = \lambda_1$.

It is also well known that subsequent eigenvalues $\lambda_2, \lambda_3, \ldots, \lambda_n, \ldots$ and eigenfunctions $u_1(x), u_2(x), \ldots, u_n(x), \ldots$ are obtained by narrowing the class of admissible functions $u(x)$; more precisely, the minimum of the functional $J[u]$ is sought on the subspace of all admissible functions orthogonal with the weight $r(x)$ to all eigenfunctions constructed on previous stages. In particular, the additional condition in the variational problem for $\lambda_2, u_2(x)$ has the form

$$\Psi_1[u] = \int_0^1 r(x)u_1(x)u\,dx \equiv (u, u_1)_r = 0. \tag{1.1.14}$$

In order to find λ_n, $u_n(x)$, one has to find a differentiable function $u(x)$ realizing the minimum of the functional $J[u]$ under the boundary conditions (1.1.2), the normalization conditions (1.1.13), and $(n-1)$ orthogonality conditions of the type (1.1.14),

$$\Psi_k[u] = (u, u_k)_r = 0, \quad k = 1, 2, \ldots, n-1. \tag{1.1.15}$$

The original idea to find eigenvalues and eigenfunctions on the basis of the variational principle should be credited to Lord Rayleigh [62]. An essential contribution to the variational approach is due to Courant and Hilbert [24], as well as Krylov [37]. Numerous investigations in this direction contributed to the development of functional analysis, the theory of linear operators, the spectral theory, and other branches of modern mathematics and mathematical physics [17, 25, 30, 45, 54, 66].

The equivalence of the variational problem (1.1.12)–(1.1.15) and the Sturm–Liouville problem (1.1.1), (1.1.2) may be established on the basis of the Euler–Lagrange equation for the functional $I[u]$,

$$I[u] = J[u] - \lambda\Phi[u] = \int_0^1 \left[p(x)u'^2 + q(x)u^2 - \lambda r(x)u^2\right] dx, \tag{1.1.16}$$

where λ is a Lagrange multiplier and $u(x)$ satisfies conditions (1.1.2). The Euler–Lagrange equation for (1.1.16) coincides with equation (1.1.1), in which λ is the parameter to be determined, together with the corresponding function $u(x, \lambda)$. One has to find the values λ_n and the functions $u_n(x) = u(x, \lambda_n)$, $n = 1, 2, \ldots$

Conversely, let λ_n, $u_n(x) = u(x, \lambda_n)$ be a solution of the Sturm–Liouville problem (1.1.1), (1.1.2). Then, equation (1.1.1) holds as identity for this solution. Let us multiply this identity by $u_n(x)$ and integrate the result in x from 0 to 1. Integrating by parts the first term with $(pu_n')'u_n$, we obtain

$$pu_n'u_n|_0^1 - \int_0^1 [p(x)u_n'^2 + q(x)u_n^2]\,dx - \lambda_n \int_0^1 r(x)u_n^2\,dx = 0. \tag{1.1.17}$$

From (1.1.17), taking into account the boundary conditions (1.1.1), and also (1.1.12), (1.1.13), (1.1.6), we find that

$$I[u_n] = J[u_n] - \lambda_n\Phi[u_n] = 0, \quad \lambda_n = \frac{J[u_n]}{\Phi[u_n]}. \tag{1.1.18}$$

Since the function $u_n(x)$ is defined to within a nonzero constant coefficient, it can be required that $\Phi[u_n] = \|u_n\|_r^2 = 1$, i.e., (1.1.13) holds. Then, (1.1.18) implies that $J[u_n] = \lambda_n$.

The orthogonality (with the weight $r(x)$) of two solutions $u_n(x) = u(x, \lambda_n)$ and $u_m(x) = u(x, \lambda_m)$ corresponding to different eigenvalues $\lambda_n \neq \lambda_m$ of the Sturm–Liouville problem is established in a standard manner [24, 36, 37, 46, 66]. Let us multiply the identities (1.1.1) for u_n and u_m by u_m and u_n, respectively, subtract the second relation from the first, and integrate the result in x. Integrating by parts, we get

$$(\lambda_n - \lambda_m) \int_0^1 r(x)u_n(x)u_m(x)\,dx \equiv (\lambda_n - \lambda_m)(u_n, u_m)_r = 0. \tag{1.1.19}$$

Thus, we have established the equivalence of the differential and the variational statements of the eigenvalue problem. From the standpoint of analytical and numerical studies, these two approaches complement one another.

Next, we describe some analytical and numerical methods of solving the Sturm–Liouville problem. This problem bears the names of those who created the theory and has a fairly long history of about 250 years, during which many analytical, numerical-analytical, and numerical methods have been developed for its investigation.

1.2. Analytical Methods of Solving the Sturm–Liouville Problem

The main difficulties of analytical solution of the problem are due to the fact that there are no general methods of integrating equation (1.1.1) for arbitrary $\lambda > 0$ and finding the values of λ for which there exist solutions satisfying conditions (1.1.2).

1.2.1. General scheme of analytical solution. Here, we outline the standard formal procedure for solving the classical problem (1.1.1), (1.1.2). Suppose that for any fixed $\lambda > 0$ we know two linearly independent solutions $U_1(x,\lambda)$, $U_2(x,\lambda)$ of equation (1.1.1),

$$c_1U_1(x,\lambda) + c_2U_2(x,\lambda) \not\equiv 0, \quad c_1^2 + c_2^2 > 0, \quad 0 \le x \le 1. \tag{1.2.1}$$

The general solution of this equation can be represented as a linear combination of U_1 and U_2, i.e.,

$$u(x,\lambda) = c_1U_1(x,\lambda) + c_2U_2(x,\lambda). \tag{1.2.2}$$

The constants c_1, c_2 are to be determined, together with λ. For this purpose, we use boundary conditions (1.1.2),

$$\begin{aligned} c_1U_1(0,\lambda) + c_2U_2(0,\lambda) &= 0, \\ c_1U_1(1,\lambda) + c_2U_2(1,\lambda) &= 0. \end{aligned} \tag{1.2.3}$$

The necessary and sufficient condition for the existence of a nontrivial solution of system (1.2.3) gives us the characteristic equation for the determination of λ, i.e., the determinant $\varDelta$ of system (1.2.3) should be equal to zero,

$$\varDelta(\lambda) = U_1(0,\lambda)U_2(1,\lambda) - U_2(0,\lambda)U_1(1,\lambda) = 0. \tag{1.2.4}$$

Equation (1.2.4) determines the values of the parameter $\lambda > 0$ for which the Sturm–Liouville problem admits a solution $u(x,\lambda)$.

As shown in [24, 29, 33, 46, 51, 52, 66], equation (1.2.4) has countably many simple roots $\lambda_1, \lambda_2, \ldots, \lambda_n, \ldots$. The set of these roots has no finite limit points, and the roots can be enumerated in increasing order,

$$0 < \lambda_1 < \lambda_2 < \ldots < \lambda_n < \lambda_{n+1} < \ldots, \quad n \ge 1, \tag{1.2.5}$$

$$\lambda_n \sim n^2, \quad \lambda_{n+1} - \lambda_n \sim n, \quad n \to \infty.$$

Every eigenvalue λ_n corresponds to a single eigenfunction

$$u_n(x) = u(x,\lambda_n), \quad n \ge 1, \tag{1.2.6}$$

defined to within a constant coefficient. The countable set of all functions $\{u_n(x)\}$ forms a complete orthogonal (with weight $r(x)$) system of linearly independent functions. Moreover, the functions $u_n(x)$ possess some important oscillation properties, whose investigation is the main topic of the Sturm theory [20, 24, 29, 33, 59]. The main facts of the oscillation theory, the Sturm comparison theorems, and their consequences can be found in various courses and handbooks on ordinary differential equations [20, 31, 36, 51, 52, 69, et al.] and are supposed to be known to the reader. A more general version of the Sturm theory developed by the authors for fourth-order equations and systems of second-order equations is described below, in Chapters 4 and 5.

The complexity of the expression (1.2.4) and the difficulty of calculations strongly depend on the choice of the functions U_1, U_2. Consider some fixed value λ_n, $n = 1, 2, \ldots$, from the sequence (1.2.5). Usually, one considers smallest eigenvalues $\lambda_1, \lambda_2, \ldots$, which are of main importance for applications. From the first equation in (1.2.3) we find a relation between c_1, c_2, and from (1.2.2) we obtain the desired solution,

$$c_2 = -c_1 \frac{U_1(0, \lambda_n)}{U_2(0, \lambda_n)}, \qquad U_2(0, \lambda_n) \neq 0,$$
$$u = u(x, \lambda_n) = c_1 u_n(x), \qquad u_n(x) \equiv U_1(x, \lambda_n) - \frac{U_1(0, \lambda_n)}{U_2(0, \lambda_n)} U_2(x, \lambda_n). \tag{1.2.7}$$

If $U_2(0, \lambda_n) = 0$, then $c_1 = 0$ and the solution has the form $u(x, \lambda_n) = c_2 U_2(x, \lambda_n)$. Similarly, one can utilize the second equation in (1.2.3). Obviously, the functions U_1, U_2 cannot be both equal to zero at the same point of the interval $0 \le x \le 1$. The constant c_1 in (1.2.7) (or c_2) can be chosen from the additional normalization condition with weight $r(x)$ (see (1.1.13))

$$c_1^2 \int_0^1 r(x) u_n^2(x)\, dx = 1 \qquad \left(c_2^2 \int_0^1 r(x) U_2^2(x, \lambda_n)\, dx = 1 \right). \tag{1.2.8}$$

The property (1.2.8) is used in theoretical investigations for writing functional series in concise form [24, 25, 45, 46, 54, 66]. However, (1.2.8) is not always convenient for the numerical construction of the functions U_1, U_2. In numerical-analytical analysis, one usually takes

$$u(0, \lambda_n) = 0, \quad u'(0, \lambda_n) = 1, \tag{1.2.9}$$

i.e., one constructs a solution of the Cauchy problem (1.1.1), (1.2.9) for a fixed λ_n. This corresponds to the condition $U_2(0, \lambda_1) \neq 0$ in (1.2.3) and the expression

$$c_1 = \frac{W(0, \lambda_n)}{U_2(0, \lambda_n)}, \qquad W(x, \lambda_n) \equiv W(0, \lambda_n) \neq 0, \tag{1.2.10}$$

where W is the Wronskian determinant for U_1, U_2.

The analytical method described above can be realized for a fairly narrow class of equations (1.1.1) admitting integration in terms of elementary functions. However, in order to determine the roots λ_n of the characteristic equation (1.2.4), numerical methods are usually required, since $\Delta(x)$ belongs to the class of transcendental functions. For some problems of mathematical physics [21, 24, 39, 43, 46, 55, 57, 58, 62], equation (1.1.1) admits a solution in a class of special functions for which there are well-known efficient analytical and numerical methods [1, 32, 65, 67].

Let us consider some examples illustrating the above method of solving the Sturm–Liouville problem.

Example 1. In equation (1.1.1) with constant coefficients p, r, q we can change the parameter λ and obtain the simplest equation

$$u'' + \mu u = 0, \qquad \mu = (\lambda r - q) p^{-1}, \qquad \lambda = (\mu p + q) r^{-1}, \tag{1.2.11}$$

which corresponds to the case of $p = r \equiv 1$, $q \equiv 0$. For $\mu > 0$, we have $U_1 = \sin \sqrt{\mu} x$, $U_2 = \cos \sqrt{\mu} x$. According to (1.2.2)–(1.2.4), we get $c_2 = 0$, $\mu_n = (\pi n)^2$, i.e., $u_n(x) = c_n \sin \pi n x$, and the eigenvalues have the form $\lambda_n = (\mu_n p + q) r^{-1}$. The normalization conditions of the type (1.2.8)–(1.2.10) yield $c_n = \sqrt{2/r}$ or $c_n = (-1)^n (\pi n)^{-1}$. Note that for $\mu \le 0$, there are no nontrivial solutions, and we can consider only positive values of the index $n = 1, 2, \ldots$

Example 2. Consider equation (1.1.1) with piecewise constant coefficients p, r, q taking the values,

$$p(x) = \begin{cases} p_1 & \text{if } 0 \le x \le a, \\ p_2 & \text{if } a < x \le 1, \end{cases} \quad r(x) = \begin{cases} r_1 & \text{if } 0 \le x \le a, \\ r_2 & \text{if } a < x \le 1. \end{cases} \quad q(x) = \begin{cases} q_1 & \text{if } 0 \le x \le a, \\ q_2 & \text{if } a < x \le 1. \end{cases} \tag{1.2.12}$$

This problem has no classical solutions, since the first derivative of $p(x)$ exists only in the sense of distributions and is a type of the Dirac δ-function. This entails discontinuity of $u'(x)$ at $x = a$. However, there is a solution in the sense of the variational principle (1.1.4)–(1.1.7). To show this, we introduce the generalized impulse $\theta = pu'$ (1.1.11) instead of the equation (1.1.8), (1.1.9) of Lagrange type and represent equation (1.1.1) as the following system:

$$u' = \frac{\theta}{p(x)}, \quad \theta' = -(\lambda r(x) - q(x))u, \tag{1.2.13}$$

The coefficients of the linear system (1.2.13) have finite jumps determined by (1.2.12), and therefore, this system can be integrated by standard methods. The functions u, θ are continuous in x at the point $x = a$, and the solutions $u = u_{1,2}(x)$, $\theta = \theta_{1,2}(x)$ constructed on the intervals $0 \le x \le a$, $a < x \le 1$ satisfy the conditions

$$u_1(0) = 0, \quad u_2(1) = 0, \quad u_1(a) = u_2(a), \quad \theta_1(a) = \theta_2(a). \tag{1.2.14}$$

On the intervals where the coefficients are constant, it is convenient to represent the functions u_1, u_2 in the form

$$\begin{aligned} u_1 &= c_1 \sin(\sqrt{\mu_1}\, x), & \theta_1 &= c_1 p_1 \sqrt{\mu_1} \cos(\sqrt{\mu_1}\, x), & \mu_1 &= \frac{\lambda r_1 - q_1}{p_1}; \\ u_2 &= c_2 \sin[\sqrt{\mu_2}\,(1-x)], & \theta_2 &= c_2 p_2 \sqrt{\mu_2} \cos[\sqrt{\mu_2}\,(1-x)], & \mu_2 &= \frac{\lambda r_2 - q_2}{p_2}. \end{aligned} \tag{1.2.15}$$

The functions u_1 and u_2 automatically satisfy zero boundary conditions at $x = 0$ and $x = 1$, respectively. The last two conditions in (1.2.14) ensure continuity of the functions $u(x)$, $\theta(x)$ at the point $x = a$ and give us a relation between the constants c_1, c_2, λ, together with a characteristic equation of the form (1.2.4),

$$\begin{gathered} c_1 \sin(\sqrt{\mu_1}\, a) - c_2 \sin[\sqrt{\mu_2}\,(1-a)] = 0, \\ c_1 p_1 \sqrt{\mu_1} \cos(\sqrt{\mu_1}\, a) - c_2 p_2 \sqrt{\mu_2} \cos[\sqrt{\mu_2}\,(1-a)] = 0; \\ \Delta(\lambda) = -\frac{\tan(\sqrt{\mu_1}\, a)}{p_1\sqrt{\mu_1}} + \frac{\tan[\sqrt{\mu_2}\,(1-a)]}{p_2\sqrt{\mu_2}} = 0, \quad \mu_{1,2} = \frac{\lambda r_{1,2} - q_{1,2}}{p_{1,2}}. \end{gathered} \tag{1.2.16}$$

The transcendental equation (1.2.16) for λ has finitely many roots λ_n satisfying conditions (1.2.5). Using (1.2.15), we can construct a family of eigenfunctions $u_n(x)$, $\theta_n(x)$ in which the constants c_1 and c_2 are related by one of the first two equalities in (1.2.16). Further constructions are carried out similarly to those of Example 1. For $a \to 0$ or $a \to 1$, the case (1.2.11) takes place. In the general case, the eigenvalues λ_n regarded as functions of the parameters a, $p_{1,2}$, $r_{1,2}$, $q_{1,2}$ can be found by numerical methods.

A similar scheme can be implemented for the construction of a solution of the equation whose coefficients $p(x)$, $r(x)$, $q(x)$ are constant on three or more segments. In numerical analysis, one often utilizes approximations based on dividing the interval $0 \le x \le 1$ into sufficiently many segments, with p, r, q taking constant values on each of these segments. However, such approximations are too rough, and this approach is ineffective for highly precise mass calculations.

Example 3. Consider two problems of the form (1.1), (1.2) that can be reduced to analytically integrable equations of Euler's type [29, 31, 33, 59].

Problem 1. Let

$$p = p_0, \quad r = r_0(x+\alpha)^{-2}, \quad q = q_0(x+\alpha)^{-2}, \quad \alpha > 0, \tag{1.2.17}$$

where p_0, r_0, q_0, α are constants. Introducing the parameter μ defined in (1.2.11), we obtain an expression for $u(x)$ and relations for c_1, c_2, λ (after the cancellation of $\sqrt{\alpha}$, $\sqrt{1+\alpha}$),

$$u = c_1U_1(x,\lambda) + c_2U_2(x,\lambda) = c_1\sqrt{x+\alpha}\sin[\nu\ln(x+\alpha)] + c_2\sqrt{x+\alpha}\cos[\nu\ln(x+\alpha\nu)];$$

$$\Delta(\lambda) = \sin\left(\nu\ln\frac{\alpha}{1+\alpha}\right) = 0, \quad \nu_n = \mp\frac{\pi n}{\ln(1+1/\alpha)}, \quad \lambda_n = \frac{p_0}{r_0}\left(\frac{1}{4} + \frac{(\pi n)^2}{\ln^2(1+1/\alpha)}\right) + \frac{q_0}{r_0}. \tag{1.2.18}$$

The relation between the coefficients c_1 and c_2 for $\lambda = \lambda_n$ can be obtained from the corresponding two relations in (1.2.3).

Problem 2. Let

$$p = p_0(x+\alpha), \quad r = r_0(x+\alpha)^{-1}, \quad q = q_0(x+\alpha)^{-1}. \tag{1.2.19}$$

Then, elementary transformations show that c_1, c_2, λ are related by

$$u = c_1U_1(x,\lambda) + c_2U_2(x,\lambda) = c_1\sin[\sqrt{\mu}\ln(x+\alpha)] + c_2\cos[\sqrt{\mu}\ln(x+\alpha)]; \tag{1.2.20}$$

$$\Delta(\lambda) = \sin\left(\sqrt{\mu}\ln\frac{\alpha}{1+\alpha}\right) = 0, \quad \mu_n = \frac{(\pi n)^2}{\ln^2(1+1/\alpha)}, \quad \lambda_n = \frac{p_0}{r_0}\frac{(\pi n)^2}{\ln^2(1+1/\alpha)} + \frac{q_0}{r_0}.$$

The eigenvalues λ_n in (1.2.18) and (1.2.20) differ by the constant $p_0/(4r_0)$. For $\lambda = \lambda_n$, using relation (1.2.3) between the coefficients c_1 and c_2, we construct the desired functions $u_n(x)$ to within an arbitrary constant factor. In both cases, we may restrict ourselves to positive values of the index n.

And finally, consider an equation that can be solved in terms of special functions.

Example 4. When studying vibrations of distributed systems in annular domains, one encounters a problem which, after elementary transformations, can be written as the following Sturm–Liouville problem:

$$\frac{1}{x+a}((a+x)u')' + \lambda u = 0, \quad u(0) = u(1) = 0, \tag{1.2.21}$$

where a is the radius of the internal circle, $0 < a < 1$.

Writing problem (1.2.21) in variational form, as in Subsection 1.1.2, we see that all eigenvalues λ_n are positive. There are two linearly independent solutions

$$U_1 = J_0(\sqrt{\lambda}(x+a)), \quad U_2 = N_0(\sqrt{\lambda}(x+a)). \tag{1.2.22}$$

Here, $J_0(z)$ and $N_0(z)$ are zero-order Bessel and Neumann functions, respectively. Detailed information about these functions can be found in [1, 32, 65, 67]. The characteristic equation (1.2.4) has the form

$$\Delta(\lambda) = J_0(\sqrt{\lambda}a)N_0(\sqrt{\lambda}(1+a)) - N_0(\sqrt{\lambda}a)J_0(\sqrt{\lambda}(1+a)) = 0. \tag{1.2.23}$$

In order to find the eigenvalues $\lambda_n = \lambda_n(a)$, one has to calculate the roots of the intricate transcendental equation (1.2.23) with the functions J_0, N_0 either defined by tables (with low precision), or represented as a series, or constructed numerically. Routine calculations like those of the Newton method or the bisectional method allow us to find these eigenvalues for a given a, for instance, $\lambda_1(1) = 9.75332$. It should be mentioned that the characteristic equation (1.2.23) often occurs in mathematical physics, and for some values of a and $n = 1, 2, \ldots$, the roots $\lambda_n(a)$ can be found in handbooks [32].

Although the above simple examples are mostly meant to illustrate the basic methods, they suggest that computational methods may be required both on the final stage of finding the roots λ_n of equation (1.2.4) in rare cases of analytical integrability of equation (1.1.1) and on the initial stage of finding solutions of equation (1.1.1) and the construction of the function $\Delta(\lambda)$. On the other hand, the analytical approach is very important in connection with testing the algorithms. Moreover, the analytical approach can be used as a basis for the creation of approximate analytic methods, as well as some asymptotic methods which allow us to study the solutions as $n \to \infty$. Perturbation methods are widely utilized in the theory of vibrations and other branches of mathematical and theoretical physics [27, 33–35, 42, 46–49, 62]. Some original versions of these methods developed by the authors are described in Chapters 2–5.

Together with the theory of differential equations, among the mathematical tools for studying the Sturm–Liouville problem (1.1.1), (1.1.2), an important role belongs to the theory of Fredholm and Volterra integral equations of the second kind. The theory of integral equations possesses great generality and has led to a number of fundamental results (such as the Schmidt theory [24, 28, 46, 67]) pertaining to the original eigenvalue problem.

1.2.2. Reduction to a Fredholm integral equation of the second kind. The boundary value problem for eigenvalues and eigenfunctions can be reduced to a homogeneous Fredholm integral equation of the second kind with the parameter λ. This equation may be studied by the methods of the Fredholm theory or those of the Schmidt theory of symmetric integral equations [24, 28, 54, 67], or by numerical methods with the reduction of the equation to a system of linear algebraic equations [22, 23, 50].

Next, we briefly describe the procedure of reducing problem (1.1.1), (1.1.2) to an integral equation with fixed integration limits, namely, a Fredholm equation. For this purpose, one considers a nonhomogeneous boundary value problem whose solution determines the kernel $r(\xi)G(x,\xi)$ of the integral equation. This problem has the form

$$\begin{gathered}(p(x)G')' - q(x)G = -\delta(x-\xi), \quad G = G(x,\xi),\\ G(0,\xi) = G(1,\xi) = 0, \quad 0 < x,\ \xi < 1.\end{gathered} \tag{1.2.24}$$

It is easy to check that the function G is symmetric: $G(x,\xi) = G(\xi,x)$. Here, $\delta(x)$ is the Dirac function, i.e., a distribution satisfying the integral condition

$$\int_a^b \delta(x-\xi)f(x)\,dx = \begin{cases} f(\xi), & a < \xi < b,\\ 0, & \xi < a,\ \xi > b,\end{cases}$$

where $f(x)$ is an arbitrary continuous function on the interval $a \le x \le b$. In particular, one can take $a = 0$, $b = 1$. Since $q(x) \ge 0$, problem (1.2.24) admits one and only one solution $G(x,\xi)$, $0 \le x,\ \xi \le 1$, which is continuous in its arguments x, ξ jointly. The solution G is called the *Green function* [24, 28, 46].

Suppose that the function $G(x,\xi)$ has been constructed. Then the boundary value problem (1.1.1), (1.1.2) can be reduced to an equivalent homogeneous Fredholm integral equation of the second kind [24, 33],

$$u(x) = \lambda \int_0^1 r(\xi)G(x,\xi)u(\xi)d\xi, \quad 0 \le x \le 1. \tag{1.2.25}$$

Observe that this equation can be made symmetric by means of the substitution $v = \sqrt{r}u$.

The problem for equation (1.2.25) is stated as follows: *Find values of* $\lambda > 0$ *for which there exists a nontrivial solution* $u(x, \lambda)$.

Note that the function $G(x,\xi)$ does not depend on the coefficient $r(x)$ and is determined by the coefficients $p(x), q(x)$ in (1.2.24). In general, it is hardly possible to obtain an analytic expression of the function G. In some special cases, if the corresponding homogeneous equation can be integrated in analytic form, the solution of problem (1.2.24) can be obtained in quadratures or in explicit form.

For example, if p, q are constant, the function $G(x,\xi)$ has the form

$$\begin{aligned} G(x,\xi) &= \frac{\sinh(kx)\sinh[k(1-\xi)]}{\sqrt{pq}\,\sinh k} && \text{if} \quad x \le \xi, \\ G(x,\xi) &= \frac{\sinh(kx)\sinh[k(1-\xi)]}{\sqrt{pq}\,\sinh k} - \frac{\sinh[k(x-\xi)]}{\sqrt{pq}} && \text{if} \quad x \ge \xi, \end{aligned} \tag{1.2.26}$$

where $k = \sqrt{q/p}$. Direct verification shows that the derivative $G'(x,\xi)$ (in x) at the point $x = \xi$ has a jump of the first kind equal to $-1/p$. This property can also be obtained directly from equation (1.2.24),

$$G'(\xi+0,\xi) - G'(\xi-0,\xi) = -\frac{1}{p(\xi)},$$

where ξ a point of continuity of $p(x)$. By arguments similar to those used in Examples 3 and 4, one can construct the function $G(x,\xi)$ in the case of analytic integrability of the homogeneous equation (1.2.24). In the case of piecewise continuous coefficients p, r, q (see Example 2), the reduction procedure has to be modified accordingly.

Another case when the function $G(x,y)$ can be constructed analytically, in the form of a quadrature, is that of $q(x) \equiv 0$. Indeed, introducing the unit step-function $h(x)$, we obtain

$$G(x,\xi) = \int_0^1 \frac{h(y-\xi)}{p(y)}\,dy \int_0^x \frac{dz}{p(z)} \left[\int_0^1 \frac{dy}{p(y)}\right]^{-1} - \int_0^x \frac{h(z-\xi)}{p(z)}\,dz, \tag{1.2.27}$$

$$h(x) \equiv 0, \quad x < 0; \quad h(x) \equiv 1, \quad x \ge 0.$$

Differentiating the function G in x for $x < \xi$ and $x > \xi$, one finds that G' has a jump at the point $x = \xi$. In order to use this expression in equation (1.2.25), one has to resort to numerical integration methods.

Traditionally, eigenvalue problems in the form Fredholm integral equations are investigated by the algebraic method (the Fredholm theory) and methods of functional analysis (the theory of continuous symmetric operators) [21–24, 53]. However, these methods are not very efficient from the computational standpoint. If the kernel of the integral equation (1.2.25) can be approximated by some degenerate kernel, then an approximate solution of the eigenvalue problem can be obtained by solving a certain algebraic eigenvalue problem (in matrix form) [23, 50].

1.2.3. Reduction to a Volterra integral equation of the second kind. The Sturm–Liouville eigenvalue problem (1.1.1), (1.1.2) can be reduced to a nonhomogeneous integral equation with the parameter λ. For this purpose, in the original equation we pass to another independent variable, $x \to z$, by

$$z = \frac{1}{\ell}\int_0^x \left(\frac{r(\xi)}{p(\xi)}\right)^{1/2} d\xi, \quad \ell = \int_0^1 \left(\frac{r(x)}{p(x)}\right)^{1/2} dx, \quad 0 \le z \le 1, \tag{1.2.28}$$

and change the unknown function, $u(x) \to v(z)$, by

$$v = v(z) = (r(x)p(x))^{1/4}u(x), \quad x = x(z). \tag{1.2.29}$$

In (1.2.29), it is assumed that the function $z = z(x)$ defined by (1.2.28) can be inverted (explicitly or implicitly). For the new unknown function v we obtain a standard eigenvalue problem similar to (1.1.10) ($\Lambda = \ell^2\lambda$, $S = -\ell^2 s$),

$$v'' + (\Lambda^2 + S(z))v = 0, \qquad v(0) = v(1) = 0. \tag{1.2.30}$$

Equations of the type (1.2.30) often occur in theoretical physics. The parameter Λ^2 is interpreted as energy and the function S as a perturbed potential. It should be kept in mind that transformations (1.2.28), (1.2.29) make sense, if the functions $p(x), r(x)$ are twice continuously differentiable and conditions (1.1.3) are fulfilled.

In order to construct an equivalent Volterra integral equation of the second kind, the term Sv is moved to the right-hand side of equation (1.2.30) and regarded as given. Then, with the help of an impulse transition function $W(z)$, the solution $v(z)$ satisfying the condition $v(0) = 0$ can be written in the form

$$\begin{gathered} v(z) = c\sin\Lambda z + \int_0^z W(z-\xi)S(\xi)v(\xi)d\xi, \\ W(z) = \frac{1}{\Lambda}\sin\Lambda z, \qquad c \neq 0, \qquad v(1) = 0. \end{gathered} \tag{1.2.31}$$

By construction, the boundary condition $v(0) = 0$ is fulfilled automatically. In order to find the parameter Λ and the function $v(z)$ (to within an arbitrary constant coefficient c), one utilizes the boundary condition (1.2.31),

$$c\sin\Lambda + \frac{1}{\Lambda}\int_0^1 \sin\Lambda(1-x)S(x)v(x)\,dx = 0. \tag{1.2.32}$$

Without the loss of generality, one can take $c = 1$, i.e., make the replacement $v = c\Lambda$, and then find c from a suitable normalization condition. Note that $r(z) \equiv 1$ in equation (1.2.30). In particular, we can take $c = 1/\Lambda$, which agrees with the condition $v'(0) = 1$.

Relations (1.2.31), (1.2.32) are convenient (and often used) for approximate asymptotic solution of the problem for large Λ, i.e., for the construction of higher modes Λ_n, v_n with $n \gg 1$. In particular, the first approximation has the form [24, 46]

$$\begin{gathered} \Lambda_n = \pi n + O\Big(\frac{1}{n}\Big), \quad V_n(z) = \sin(\pi n z) + O\Big(\frac{1}{n}\Big); \quad \sqrt{\lambda_n} = \frac{\pi}{\ell}n + O\Big(\frac{1}{n}\Big), \\ u_n(x) = c(p(x)r(x))^{-1/4}\sin\left[\frac{\pi n}{\ell}\int_0^x \left(\frac{r(\xi)}{p(\xi)}\right)^{1/2} d\xi\right] + O\Big(\frac{1}{n}\Big). \end{gathered} \tag{1.2.33}$$

Here, ℓ is defined by (1.2.28). Formulas (1.2.33) are widely used in mechanics and physics, although they give a fairly rough approximation, since the solution does not contain the function $q(x)$. It should be observed that these formulas have asymptotic character. Subsequent approximations with respect to $1/n$ require asymptotic methods (the averaging method). This problem is considered in detail in Chapter 5. In the case of sufficiently smooth coefficients $p(x)$, $r(x)$, $q(x)$, it is possible to obtain finite expressions of λ_n, $u_n(x)$ in quadratures, and those expressions turn out fairly precise starting from $n \sim 1$. This fact was also noticed in [48]. Application of regular expansions without averaging is unjustified.

The above approach is connected with the intricate transformation (1.2.28), (1.2.29) of our equation to the form (1.2.30), but this cannot be done analytically in explicit form, in general. However, there are no limitations of this kind in the asymptotic approach based on the method of averaging and described in Chapter 5. Moreover, the latter approach can be used for all types of boundary conditions (see Subsection 1.1.1).

Further development of the methods of analytic investigation of eigenvalue problems is related to the development of regular and asymptotic methods of the perturbation theory. In this connection, wide utilization of numerical methods is presumed, because the above formulas cannot be realized in explicit analytical form, in general. Notions like "analytical solution" will not be misleading for a thoughtful and experienced specialist. Investigation of analytical solutions of meaningful real problems by numerical methods is often a very complicated task requiring much effort. This fact suggests that it is necessary to create numerical procedures for an effective highly precise solution and analysis of the Sturm–Liouville problem. Great scientists of the past were aware of the importance of numerical methods for solving fundamental problems, contributed to the development of these methods, and widely used them for studying natural phenomena and technical objects.

1.3. Solving the Sturm–Liouville Problem by the Method of Regular Perturbations

1.3.1. Statement of the perturbed problem. The methods of the theory of perturbations, both regular and asymptotic, are widely used in mathematical physics and vibration theory. These methods are based on the introduction of a small parameter [19, 22, 24, 25, 27, 33–35, 42, 46–49, 62, 68]. Next, we describe a simple algorithm of analytical calculations aimed at obtaining an approximate solution of an eigenvalue problem.

Consider the so-called perturbed problem with a numerical parameter ε

$$
\begin{gathered}
(pu')' + (\lambda r - q)u = 0, \qquad u(0) = u(1) = 0; \\
p = p_0(x) + \varepsilon p_1(x), \quad r = r_0(x) + \varepsilon r_1(x), \quad q = q_0(x) + \varepsilon q_1(x), \\
p_0, r_0 > 0, \qquad q_0 \geq 0, \qquad |\varepsilon| \leq \varepsilon_0.
\end{gathered}
\tag{1.3.1}
$$

For the sake of brevity, we do not indicate possible dependence of the functions p_1, r_1, q_1 on ε. As usual, the parameter ε is assumed small, $|\varepsilon| \ll 1$. It is assumed that we know the solution of the unperturbed problem (with $\varepsilon = 0$)

$$
\begin{gathered}
(p_0(x)u')' + (\lambda r_0(x) - q_0(x))u = 0, \quad u(0) = u(1) = 0, \\
\lambda_k = \lambda_{0k}, \quad u_k(x) = u_{0k}(x), \quad k = 1, 2 \ldots
\end{gathered}
\tag{1.3.2}
$$

1.3.2. Standard procedure of asymptotic expansions. For a fixed index k, consider the problem of finding approximations of λ_k, u_k having a given error with respect to ε. We seek λ_k, u_k in the form of the expansions (the index k is dropped for the sake of brevity)

$$
\begin{gathered}
\lambda = \lambda(\varepsilon) = \lambda_0 + \varepsilon\lambda_1 + \varepsilon^2\lambda_2 + \cdots + \varepsilon^N\lambda_n + \cdots, \\
u = u(x, \varepsilon) = u_0(x) + \varepsilon u_1(x) + \varepsilon^2 u_2(x) + \cdots + \varepsilon^N u_n + \cdots.
\end{gathered}
\tag{1.3.3}
$$

Here, λ_j, u_j are the quantities to be found.

The standard procedure consists of substituting (1.3.3) into (1.3.1), gathering the terms containing equal powers of ε, and equating to zero the coefficients of these powers. As a result, we obtain the following sequence of boundary value problems:

$$
\begin{gathered}
(p_0(x)u_j')' + (\lambda_0 r_0(x) - q_0(x))u_j = -\lambda_j r_0(x)u_0 + H_j, \qquad u_j(0) = u_j(1) = 0, \\
H_j = H_j\left(x, \lambda_0, \ldots, \lambda_{j-1}, u_0, u_0', u_0'', \ldots, u_{j-1}, u_{j-1}', u_{j-1}''\right), \quad j = 1, 2, \ldots
\end{gathered}
\tag{1.3.4}
$$

Without the loss of generality, it may be assumed that $u_0'(0) = 1$, $u_j'(0) = 0$.

For $j = 0$, we have the unperturbed eigenvalue problem (1.3.2), whose solution λ_0, $u_0(x)$ is known. The linear operators H_j are defined in terms of the perturbations $p_1(x)$, $r_1(x)$, $q_1(x)$ and the solutions $\lambda_0, u_0, \lambda_1, u_1, \ldots, \lambda_{j-1}, u_{j-1}$ found on previous steps. In particular,

$$H_1 = -(\lambda_0 r_1(x) - q_1(x))u_0(x) - (p_1(x)u_0'(x))'. \tag{1.3.5}$$

Similar but more lengthy expressions are obtained for $H_2, H_3, \ldots$ These can be constructed on a computer with the help of symbolic calculations. From (1.3.5) and the expressions of subsequent functions H_j, $j \geq 2$, it follows that $\lambda_j = 0$, $u_j(x) \equiv 0$, if $p_1(x) = r_1(x) = q_1(x) \equiv 0$.

1.3.3. Finding the expansion coefficients. The sequence of nonhomogeneous boundary value problems (1.3.4) consists of recurrent equations, which can be solved consecutively. For $j = 1$, the unknown quantities λ_1, $u_1(x)$ can be found by standard methods [22, 24, 27, 33, 34, 42, 46, 47]. According to the Fredholm alternative, this nonhomogeneous boundary value problem has a solution, provided that the right-hand side of equation (1.3.4) is orthogonal to the eigenfunction $u_0(x)$ of the homogeneous problem (since the corresponding operator is self-adjoint). We have

$$\begin{aligned} \lambda_j &= \int_0^1 H_j u_0(x)\,dx \left[\int_0^1 r_0(x)u_0^2(x)\,dx\right]^{-1}, \quad j = 1, 2, \ldots, \\ \lambda_1 &= \int_0^1 [p_1(x)u_0'^2(x) - (\lambda_0 r_1(x) - q_1(x))u_0^2(x)]\,dx. \end{aligned} \tag{1.3.6}$$

After the insertion of the unknown quantities λ_j defined by (1.3.6) into (1.3.4), the right-hand side will be completely determined. This allows us to construct the sought functions $u_j(x)$ in the form of quadratures by the method of variation of integration constants, since the second linearly independent solution of the homogeneous equation is easily constructed with the help of the Liouville formula. If a complete system of eigenfunctions $u_{0m}(x)$ of the generating problem is known, then the sought solutions $u_{kj}(x)$ corresponding to $m = k$ can be represented as a Fourier series based on that system,

$$u_{kj}(x) = \sum_{m \neq k}^{\infty} C_{mj} u_{0m}(x), \quad C_{mj} = \frac{1}{\lambda_{0k} - \lambda_{0m}} \int_0^1 H_j u_{0m}(x)\,dx \left[\int_0^1 r_0(x)u_{0m}^2(x)\,dx\right]^{-1}. \tag{1.3.7}$$

Formulas (1.3.6), (1.3.7) for $j = 2$ are rather cumbersome and are not used in practical calculations. In order to evaluate the effect of the perturbations $p_1(x)$, $r_1(x)$, $q_1(x)$, one usually considers only the first approximation

$$\lambda_{(1)} = \lambda_0 + \varepsilon\lambda_1, \quad u(x) = u_0(x) + \varepsilon u_1(x). \tag{1.3.8}$$

Note that the assumption that there is a small parameter ε in problem (1.3.1) is a kind of tribute to tradition. In order that the above procedure could be applied, it is important only that $|\varepsilon p_1| \ll p_0$, $|\varepsilon r_1| \ll r_0$. It is possible to do without the parameter ε, or take $\varepsilon = 1$, or introduce it just for the sake of convenience.

In spite of the seeming limitations of the perturbation theory, its methods are at the basis of outstanding discoveries in astronomy (planets Neptune and Pluto), quantum mechanics (computation of thin and superthin structures of atomic spectra), calculations of complex vibration systems in mechanics and radio electronics, quantum electronics (parametric vibrations), etc. From the standpoint of calculations, an essential drawback of the perturbation theory is the rapidly growing complexity of the expressions obtained when constructing higher-order approximations, and this entails growing roundoff errors. Moreover, no significant increase of the parameter ε can be attained by increasing the order of the expansion in powers of ε.

1.3.4. Justification questions. The above algorithm of the perturbation theory needs mathematical justification, in particular, one has to prove convergence of the approximations as $N \to \infty$, both for the modes with a fixed index k, and for those with $k \to \infty$, and also to determine the convergence radius $|\varepsilon| \le \varepsilon_0$. For applications, it is important to have an effective method for the evaluation of the constants c_λ, c_u in the Nth approximation ($n = 1, 2, \dots$)

$$\begin{aligned} \lambda &= \lambda_0 + \varepsilon\lambda_1 + \dots + \varepsilon^N \lambda_N + c_\lambda \varepsilon^{N+1}, \\ u &= u_0 + \varepsilon u_1 + \dots + \varepsilon^N u_N + c_u \varepsilon^{N+1}. \end{aligned} \tag{1.3.9}$$

Results of this kind are obtained in Chapter 3 in the special case of p_0, r_0, q_0 being constant. A remarkable feature of that case is that if the perturbation terms $p_1(x)$, $r_1(x)$, $q_1(x)$ are quasipolynomials, then the corresponding coefficients in the expansions (1.3.3), (1.3.9) can be constructed in explicit analytical form.

Note that for the construction of a solution to problem (1.3.1) a recurrent algorithm with successive approximations may be used. In order to implement that algorithm, problem (1.3.1) is transformed into an equivalent problem which takes into account the solution λ_0, u_0 of the unperturbed problem (1.3.2),

$$\begin{aligned} L_0[U] &= -\Lambda r_0 u_0 - L_1[u_0] - \varepsilon L_1[U] - \varepsilon \Lambda r_1 u_0 - \varepsilon \Lambda r_0 U - \varepsilon^2 \Lambda r_1 U, \\ U(0) &= U(1) = 0, \quad L_{0,1}[U] = (p_{0,1}U')' - (\lambda_0 r_{0,1} - q_{0,1})U. \end{aligned} \tag{1.3.10}$$

Here, for the sake of brevity, we have introduced linear differential operators $L_{0,1}$ constructed from the coefficients $p_{0,1}$, $r_{0,1}$, $q_{0,1}$ and the known $\lambda = \lambda_0$. Without loss of generality, one can take $U'(0) = 0$ while constructing a solution of the nonhomogeneous boundary value problem (1.3.10) and finding Λ. The recurrent algorithm consists in consecutively solving the boundary value problem (1.3.10): on the jth iteration step, the terms proportional to ε are taken for Λ_{j-1}, U_{j-1} obtained on the previous steps. For $j = 1$, we obtain the known solution λ_1, u_1. This procedure is also rather cumbersome and can hardly be implemented without a computer. Moreover, as the number of iterations increases, the roundoff errors tend to accumulate, which prevents highly precise fast calculations.

Other modifications of the perturbation method for solving the Sturm–Liouville problem without an explicit small parameter ε are described in subsequent chapters. The theoretical approaches and analytical methods presented here can be extended to problems with other boundary conditions, namely, those of the second and the third kinds, as well as the periodic conditions.

1.4. Numerical Methods for Solving the Sturm–Liouville Problem

As mentioned in Section 1.2, the distinction between analytical and numerical approaches to the study of eigenvalue problems is mostly a matter of convention. However, numerical methods are particularly effective for a fairly thorough investigation of difficult eigenvalue problems and allow us to trace the dependence of the solutions on the parameters of the system and to detect their special features. Such methods are expected to possess high efficiency in the sense of: 1) simplicity of implementation, 2) fast rate of convergence, 3) high precision and operation speed, 4) no saturation effect caused by roundoff errors, 5) stability of calculations with respect to computer bugs, etc. To a certain extent, these features characterize the Rayleigh–Ritz method, the shooting method, the Bubnov–Galiorkin method (projection methods), the finite-element (spline) method, difference methods, iteration methods, and some other methods. High efficiency in the said sense can be claimed of the original numerical-analytical method of accelerated convergence and its modifications developed by the authors. The main part of this book is dedicated to that method (Chapters 2, 4, 6, 7, 9).

1.4.1. The Rayleigh–Ritz method. The equivalence of the Sturm–Liouville problem and the corresponding isoperimetric variational problem was used for the creation of a very effective method for the approximation of eigenvalues and eigenfunctions. This method is based on a functional or a numerical (discrete) approximation of the sought solution. The idea of the method belongs to Lord Rayleigh [62], and its generalizations and developments that resulted in an effective algorithm are due to Ritz [22, 28, 33, 37, 46, 64]. The Rayleigh–Ritz method is usually applied for studying lower modes.

Our exposition of the Rayleigh–Ritz method is based on the variational statement of problem (1.1.12), (1.1.13). For the sake of clarity, we use both vector and component notation. Let $\psi_j(x)$, $j = 1, 2, \dots, n$, be a system of continuously differentiable linearly independent functions satisfying zero boundary conditions at $x = 0$, 1,

$$\begin{gathered}\psi(x) = (\psi_1(x), \psi_2(x), \dots, \psi_n(x))^T,\ \psi_j(0) = \psi_j(1) = 0,\ j = 1, 2, \dots, n,\\ (c, \psi(x)) = c^T\psi(x) = c_1\psi_1(x)c_2\psi_2(x) + \dots + c_n\psi_n(x) \neq 0,\ c \neq 0.\end{gathered} \tag{1.4.1}$$

Here, the subscript T indicates transpose of a matrix (vector), i.e., ψ is a column-vector, c is an arbitrary constant n-vector, which may be assumed to have unit length. The functions $\psi_j(x)$ are called *test functions*.

Let us construct the function minimizing the functional $J[u]$ (1.1.12) with the isoperimetric condition $\Phi[u] = 1$ on the n-dimensional space consisting of functions

$$u_n(x, c) = (c, \psi(x)) = \sum_{j=1}^{n} c_j\psi_j(x), \tag{1.4.2}$$

where ψ_j satisfy conditions (1.4.1). The functions u_n are called *n-coordinate approximations.*

Substituting the expressions (1.4.2) into the functionals J and Φ, we obtain a finite-dimensional algebraic problem for the quadratic forms $J_n(c)$ and $\Phi_n(c)$:

$$\begin{aligned} J[u_n] &= J_n(c) = (Ac, c) = \sum_{j,k=1}^{n} a_{jk}c_jc_k, \quad a_{jk} = a_{kj},\\ \Phi[u_n] &= \Phi_n(c) = (Bc, c) = \sum_{j,k=1}^{n} b_{jk}c_jc_k, \quad b_{jk} = b_{kj},\\ a_{jk} &= \int_0^1 [p(x)\psi_j'(x)\psi_k'(x) + q(x)\psi_j(x)\psi_k(x)]\,dx,\\ b_{jk} &= \int_0^1 r(x)\psi_j(x)\psi_n(x)\,dx, \quad j, k = 1, 2, \dots, n. \end{aligned} \tag{1.4.3}$$

Here, A, B are symmetric positive definite matrices (by construction). The original variational problem is replaced by the following problem:

$$J_n(c) \to \min_c, \quad \Phi_n(c) = 1 \quad (n \geq 2). \tag{1.4.4}$$

Problem (1.4.4) is solved by the method of Lagrange multipliers, which yields a necessary condition of the constrained minimum and the normalization condition (in vector and component forms)

$$\frac{1}{2}\frac{\partial}{\partial c}(J_n(c) - \lambda \Phi_n(c)) = (A - \lambda B)c = 0, \quad (Bc, c) = 1,$$
$$\frac{1}{2}\sum_{j=1}^{n}(a_{jk} - \lambda b_{jk})c_k = 0, \quad \sum_{j,k=1}^{n} b_{jk}c_j c_k = 1. \tag{1.4.5}$$

Note that the matrix B becomes diagonal ($b_{jk} = 0, j \neq k$), if the functions $\psi_j(x)$ are orthogonalized with weight $r(x)$; their orthonormalization with weight $r(x)$ results in B being the identity matrix.

The linear homogeneous algebraic system (1.4.5) admits a nontrivial solution c^*, if and only if the following characteristic equation holds:

$$\Delta(\lambda) = \det(A - \lambda B) = 0. \tag{1.4.6}$$

This is an algebraic (nontranscendental) equation of order n. According to the theory of symmetric positive definite matrices [22, 36, 50], equation (1.4.6) has exactly n positive roots λ_m^*, which may be enumerated in increasing order. Each λ_m^* corresponds to an eigenvector $c^{(m)}$. Thus, dropping the dependence on n, we have

$$0 < \lambda_1^* < \lambda_2^* < \cdots < \lambda_m^* < \lambda_{m+1}^* < \cdots < \lambda_n^*,$$
$$(A - \lambda_m^* B)c^{(m)} = 0, \quad m = 1, 2, \ldots, n, \tag{1.4.7}$$
$$\sum_{k=1}^{n}(a_{jk} - \lambda_m^* b_{jk})c_k^{(m)} = 0.$$

Each vector $c^{(m)}$ is defined to within an arbitrary constant coefficient. Substituting $c^{(m)}$ into (1.4.2), we obtain the approximations of "eigenfunctions"

$$u_{(n)}^{(m)} = u_{(n)}(x, c^{(m)}) = (c^{(m)}, \psi(x)) = \sum_{j=1}^{n} c_j^{(m)}\psi_j(x), \tag{1.4.8}$$

where each $u_{(n)}^{(m)}$ is defined to within a nonzero constant coefficient, just as the vector $c^{(m)}$. These constant coefficients are chosen to fulfill the normalization condition (1.4.5) for each λ_m^*. Direct calculations show that

$$J_n(c^{(m)}) = (Ac^{(m)}, c^{(m)}) = \sum_{j,k=1}^{n} a_{jk}c_j^{(m)}c_k^{(m)} = \lambda_m^*; \qquad m = 1, 2, \ldots, n. \tag{1.4.9}$$

The values λ_m^* determine the minimum of the functional $J[u]$ on the class of functions $u_{(n)}(x, c)$. This class is narrower than the class of all admissible functions, and therefore, λ_m^* cannot be smaller than the actual eigenvalues, and we have thus obtained upper bounds for the actual eigenvalues,

$$0 < \lambda_1 \leq \lambda_1^*, \quad \lambda_{m-1} < \lambda_m \leq \lambda_m^*, \quad m = 2, \ldots, n. \tag{1.4.10}$$

The special case $n = 1$ of the above procedure is called *the Rayleigh principle* and is generally used for the construction of un upper bound λ_1^* for the first eigenvalue λ_1. The calculations in this case are simplified, because there is no minimization of $J_1(c)$ and the desired expression takes the form $\lambda_1^* = J_1(1)/\Phi_1(1)$,

$$0 < \lambda_1 \leq \lambda_1^* = \int_0^1 [p(x)\psi_1'^2(x) + q(x)\psi_1^2(x)]\,dx \left[\int_0^1 r(x)\psi_1^2(x)\,dx\right]^{-1}. \tag{1.4.11}$$

The deviation of the upper bounds λ_m^* (1.3.6) and λ_1^* (1.4.11) from the actual eigenvalues depends on the choice of the test functions $\psi_j(x)$, which should naturally take into account global properties of the functions $p(x)$, $r(x)$, $q(x)$ for all $0 < x < 1$. General recommendations for choosing the test functions pertain to the shape of the curves $\psi_j(x)$ and the number of their intermediate null-points corresponding to the lower vibration modes under consideration. Thus, $\psi_1(x)$ is chosen to have no intermediate null-points (bell-shaped function), say $x(1-x)$ or $\sin \pi x$. The function $\psi_2(x)$ should have an intermediate root at some point $x = x_1$. In particular, one may take $x_1 = 1/2$ and the function $x(1-x)(x-1/2)$ or $\sin 2\pi x$, etc. Similarly, as $\psi_j(x)$ with $j-1$ intermediate null-points one may take Lagrange polynomials or trigonometric functions:

$$\begin{aligned} \psi_j(x) &= x\Big(\frac{1}{j} - x\Big)\Big(\frac{2}{j} - x\Big)\ldots\Big(\frac{j-1}{j} - x\Big)(1-x), \\ \psi_j(x) &= \sin(\pi j x). \end{aligned} \tag{1.4.12}$$

Functions $\psi_j(x)$ of the type (1.4.12) have equidistant null-points and symmetrical structure. If necessary, the shape and the null-points (nodes) $x = x_i$ of the curves $\psi_j(x)$ may be changed in a suitable manner. Utter neglect of the properties of eigenfunctions, when choosing the system of test functions $\psi_j(x)$, may lead to very rough bounds λ_m^* and distorted shapes $u_{(n)}^{(m)}(x)$. Computational experience shows that if the variation of the coefficients $p(x)$, $r(x)$, $q(x)$ is "not too fast", then the Rayleigh principle and the Rayleigh-Ritz method with test functions (4.1.12) give quite acceptable results, which explains why this approach is very popular and widely used in applied calculations connected with eigenvalue problems. However, these approaches should be supplemented with an effective method for the construction of lower bounds λ_{m*} for the eigenvalues λ_m, since only two-sided estimates $\lambda_{m*} \leq \lambda_m \leq \lambda_m^*$ allow us to judge about the precision of the obtained approximate values.

Assuming that the upper bound λ_m^* is sufficiently accurate, one may obtain a lower bound λ_{m*} by the shooting method or a fairly cumbersome procedure of the Wainstein–Aronszajn method [28].

Of great theoretical importance is the problem of convergence of the Rayleigh–Ritz method as $n \to \infty$. Krylov, who originally introduced the method of averaging (in collaboration with Bogolyubov), proved by analytical means that the approximations λ_m^*, $u_n^{(m)}(x)$ converge to the exact quantities and also obtained estimates of this convergence for various classes of test functions [37].

Let us formulate some results from [37] which are useful for applications. To simplify formulas, we assume that $p(x) \equiv 1$, $q(x) \equiv 0$ and take the system of test functions $\psi_j(x) = x^j(1-x)$, $j = 1, 2, \ldots, n$. It should be mentioned that this system is not very convenient for practical calculations. If the function $r(x)$ is continuous, then the relative error is estimated by

$$0 \leq \frac{\lambda_m^* - \lambda_m}{\lambda_m} < \frac{\lambda_m^* r^+}{(n+1)(n+2)}, \qquad m = 1, 2, \ldots, n. \tag{1.4.13}$$

Here, r^+ is the maximum of $r(x)$ in x $(0 \leq x \leq 1)$. From (1.4.13), it follows that for a fixed m and $n \to \infty$, the relative error tends to zero.

If the function $r(x)$ has a continuous derivative, then a more precise estimate holds for the relative error,

$$\begin{aligned} &0 \leq \frac{\lambda_m^* - \lambda_m}{\lambda_m} < \frac{\lambda_m^* N}{n(n+1)^2(n+2)}, \qquad m = 1, 2, \ldots, n, \\ &N = \left(\max_{0 \leq x \leq 1} \left|\frac{r'}{\sqrt{r}}\right| + \sqrt{\lambda_m^*}\left(\frac{(r^+)^5}{r^-}\right)^{1/4}\right)^2, \end{aligned} \tag{1.4.14}$$

where $r^{\pm}$ is the minimum of $r(x)$ in x. From (1.4.14), it follows that for a fixed m and $n \to \infty$, the convergence rate is very high. Moreover, for asymptotically large values of the index m, say $m \sim [\sqrt{n}]$, the Rayleigh-Ritz method for $n \gg 1$ gives highly precise bounds λ_m^* (both relative and absolute errors are very small).

For the system of trigonometric test functions $\psi_j(x) = \sin(\pi j x)$, $j = 1, 2, \dots n$, error estimates are similar to (1.4.13) and for a continuous function $r(x)$ have the form

$$0 \le \frac{\lambda_m^* - \lambda_m}{\lambda_m} < \frac{\lambda_m^* r^+}{\pi^2 (n+1)^2}, \qquad m = 1, 2, \dots, n. \tag{1.4.15}$$

The error bound in (1.4.15) is $\pi^2 \approx 10$ times smaller that in (1.4.13); for a fixed m, both the relative and the absolute errors tend to zero and are of the order n^{-2} as $n \to \infty$.

For a continuously differentiable $r(x)$, the following error estimates hold:

$$0 \le \frac{\lambda_m^* - \lambda_m}{\lambda_m} < \frac{\lambda_m^*}{\pi^4 (n+1)^4} \left(\lambda_m^* (r^+)^2 + \max_{0 \le x \le 1} \frac{r'^2}{r} \right). \tag{1.4.16}$$

For a fixed m and $n \to \infty$, high precision and high convergence rate are observed. However, for asymptotically large values of m, for instance $m \sim [\sqrt{n}]$, the error bound in (1.4.16) is greater than in (1.4.14).

Error estimates for the eigenfunction approximations $u_{(n)}^{(m)}(x)$ (1.4.8) constructed by the Rayleigh–Ritz method with trigonometric test functions $\psi_j(x)$ have the form

$$|u_{(n)}^{(m)}(x) - u_m(x)| \le \frac{\lambda_m^*}{\pi^2 (n+1)^{3/2}} \left(\frac{2r^+}{3} \right)^{1/2} \left(1 + M \left(\frac{3r^+}{r^-} \right)^{1/2} \right). \tag{1.4.17}$$

Here, M is a constant which can be effectively estimated in terms of λ_{m+1}^* and λ_{m-1}^* [37]. If the derivative $r'(x)$ is continuous, the estimates (1.4.17) can be substantially improved. Similar error estimates take place in the case of $p \ne 1$, $q \ne 0$. The above considerations show that for eigenvalues λ_m and eigenfunctions $u_m(x)$ corresponding to the lower modes $m = 1, 2, \dots$, the Rayleigh–Ritz method is fairly effective in the case of smooth coefficients p, r, q, provided that their variation is not too fast.

1.4.2. Some general facts and remarks pertaining to other numerical methods in the Sturm–Liouville problem. The above methods have been further developed and generalized as the so-called projection methods, which can be applied for studying much wider classes of eigenvalue problems. Those methods are based on the ideas of the method of moments — the Bubnov–Galiorkin method, in particular, the Galiorkin–Petrov method [15, 30, 33, 45, 64, 66]. These approaches are also connected with the approximation of solutions of linear and nonlinear operator equations in functional spaces. For the self-adjoint boundary value problems for eigenvalues and eigenfunctions considered above, the methods of Bubnov–Galiorkin and Rayleigh–Ritz are equivalent. The method of moments (projection methods) also presumes wide utilization of numerical integration procedures, minimization of functions depending on several variables, etc.

Difference methods [22, 23, 30, 38, 53] are not directly related to functional approximation of the unknown eigenfunctions. When solving the Sturm-Liouville problem problem by such methods, the interval $0 \le x \le 1$ is divided into sufficiently many segments and one seeks a finite-difference solution depending on the unknown parameter λ. The boundary conditions at $x = 0$, $x = 1$ and the criterion of the existence of a nontrivial solution (the determinant of the corresponding system should be equal to zero) yield an eigenvalue problem for high-dimensional matrices. Difference methods are not very efficient for eigenvalue problems of Sturm–Liouville type.

The widely used finite-element method [18, 23, 30, 53, 60] is, in essence, a modification of the Rayleigh–Ritz method and that of finite-differences. According to the finite element method, one considers a partition of the interval and a polynomial (spline) approximation of the unknown function on the segments of that partition, the degree of the polynomial being not too high. From publications known to the authors, it seems that the efficiency of the finite element method, with regard to the problems of Sturm–Liouville type, is not very high.

Chapter 2

The Method of Accelerated Convergence for the Sturm–Liouville Problem

In this chapter, we describe a new numerical-analytical method for solving the Sturm–Liouville problem. This method is based on a certain relation between eigenvalues and the length of the interval. We start with formal constructions and then obtain computational formulas and give examples with different types of boundary conditions (the first, second, and third kinds, as well as the conditions of periodicity).

2.1. Numerical-Analytical Upper and Lower Bounds for Eigenvalues

2.1.1. The problem of constructing two-sided estimates. Consider the Sturm-Liouville problem with the boundary conditions of the first kind

$$(p(x)u')' + [\lambda r(x) - q(x)]u = 0, \qquad u(0) = u(1) = 0. \tag{2.1.1}$$

It is assumed that the functions $p(x)$, $r(x)$, $q(x)$ are sufficiently smooth (this condition may be dropped), are defined on a larger segment $0 \le x \le 1+a$, $a > 0$, and satisfy the inequalities (1.1.3). These inequalities ensure that all eigenvalues are positive (this is important for many applied problems). Let us consider in detail the process of finding the first (the smallest) eigenvalue and the first eigenfunction.

Take any continuously differentiable function $\psi(x)$ satisfying the conditions $\psi(0) = \psi(1) = 0$ and resembling the first eigenfunction in the sense that on the segment $[0, 1]$ it has a single extremum and no intermediate null-points. The Rayleigh–Ritz principle yields the following estimate for the first eigenvalue:

$$\lambda_1 \le \lambda_1^* = \int_0^1 (p(x)\psi'^2 + q\psi^2)\,dx \left[\int_0^1 r(x)\psi^2\,dx\right]^{-1}.$$

We have $\lambda_1 = \lambda_1^*$, if $\psi = C\varphi_1(x)$, where C is an arbitrary constant and $\varphi_1(x)$ is the eigenfunction corresponding to the first eigenvalue. The estimate does not characterize closeness of the values λ_1^* and λ_1. In this connection, it seems important to implement the following plan:

1. Establish a criterion of closeness of λ_1^* to λ_1. As a rule, the estimates obtained by Krylov lead to fairly precise results only if a system of many test functions is used (n should be large; see Section 1.4), and it is implied that the integration is highly precise ("ideal").

2. Give a constructive method for obtaining a lower bound λ_{1*}:

$$0 < \lambda_{1*} \le \lambda_1 \le \lambda_1^*.$$

3. Develop a refinement procedure for eigenvalues and eigenfunctions.
4. Construct an algorithm for finding eigenvalues and eigenfunctions with arbitrary accuracy: both the absolute and the relative errors should be arbitrarily small.
5. Extend the method to subsequent λ_n, φ_n, $n \ge 2$.
6. Extend the results to other types of boundary conditions (the second, the third, and periodic).

2.1.2. Construction and analysis of comparison systems. To realize the above plan, we consider the Sturm–Liouville problem, together with the following Cauchy problem:

$$(p(x)u')' + [\lambda_1^* r(x) - q(x)]u = 0; \qquad u(0) = 0, \quad u'(0) = 1. \tag{2.1.2}$$

It is assumed that the solution $u = V_1(x, \lambda_1^*)$ of the Cauchy problem (2.1.2) has already been constructed by some analytical or numerical procedure. With modern computers, finding the solution of the Cauchy problem (2.1.2) is a simple task.

The solution $u = V_1(x, \lambda_1^*)$ has the following property:

$$V_1(\xi, \lambda_1^*) = 0, \qquad 0 < \xi \le 1,$$

where ξ is the first null-point of the function V_1. The inequality $\xi \le 1$ follows from Sturm's first oscillation theorem [24, 29, 33, 59] which claims that if the coefficient λ increases, the root of the solution of the Cauchy problem shifts to the left. The equality $\xi = 1$ is possible only for $\lambda_1^* = \lambda_1$. It would be natural to associate the bound λ_{1*} with the root ξ.

Consider another Cauchy problem

$$(p(x)u')' + [\lambda_1^* \xi^2 r(x) - q(x)]u = 0; \qquad u(0) = 0, \quad u'(0) = 1. \tag{2.1.3}$$

Theorem 2.1. *Let $u(x) = W_1(x, \lambda_1^* \xi^2)$ be a solution of the Cauchy problem* (2.1.3). *If $W_1(x, \lambda_1^* \xi^2) > 0$ for $0 < x \le 1$, then $\lambda_1^* \xi^2$ is a lower bound for the first eigenvalue. If $W_1(x, \lambda_1^* \xi^2) = 0$ for $x = \xi_1 < 1$, then $\lambda_1^* \xi^2$ is a refined lower bound, i.e., $\lambda_1 \le \lambda_1^* \xi^2 < \lambda_1^*$. Moreover, $\xi_1 > \xi$.*

Proof. Let λ_1 be the first eigenvalue corresponding to the eigenfunction $\varphi_1(x)$. For definiteness, we assume that $\varphi_1(x) > 0$ for $0 < x < 1$ (this is possible, since $\varphi_1(x)$ does not change sign). Since $\varphi_1(1) = 0$, we have $\varphi_1'(1) < 0$. Let us prove the first statement of the theorem.

We have two equations

$$(p(x)\varphi_1')' + [\lambda_1 r(x) - q(x)]\varphi_1 = 0,$$

$$(p(x)W_1')' + [\lambda_1^* \xi^2 r(x) - q(x)]W_1 = 0.$$

Let us multiply the first equation by W_1 and the second by φ_1, subtract one from the other, and integrate in x from 0 to 1. We obtain

$$\int_0^1 [(p\varphi_1')'W_1 - (pW_1')'\varphi_1]\,dx + (\lambda_1 - \lambda_1^* \xi^2) \int_0^1 rW_1\varphi_1\,dx = 0. \tag{2.1.4}$$

Integrating by parts in the first term and taking into account the boundary conditions, we get

$$p(1)\varphi_1'(1)W_1(1) + (\lambda_1 - \lambda_1^* \xi^2) \int_0^1 rW_1\varphi_1\,dx = 0. \tag{2.1.5}$$

Since

$$\varphi_1'(1) < 0, \qquad \int_0^1 rW_1\varphi_1\,dx > 0,$$

it follows from (2.1.5) that $\lambda_1^*\xi^2 < \lambda_1$, which means that we have obtained a lower bound for the first eigenvalue.

Let us prove the second statement of the theorem. Since $W_1(\xi_1) = 0$ and $W_1(x) > 0$ for $x < \xi_1$, it follows that $W_1'(\xi_1) < 0$. Let us replace the upper integration limit in (2.1.5) by ξ_1. Integrating by parts in the first term and taking into account the boundary conditions at the points $x = 0$, $x = \xi_1$, we obtain

$$-\varphi_1(\xi_1)p(\xi_1)W_1'(\xi_1) + (\lambda_1 - \lambda_1^*\xi^2)\int_0^{\xi_1} rW_1\varphi_1\,dx = 0. \tag{2.1.6}$$

Since $\varphi_1(\xi_1) > 0$ and $W_1'(\xi_1) < 0$, it follows from (2.1.6) that

$$\lambda_1 < \lambda_1^*\xi^2. \tag{2.1.7}$$

The theorem is proved.

Thus, we have obtained two useful inequalities:

1) the two-sided estimate

$$\lambda_1^*\xi^2 \le \lambda_1 \le \lambda_1^*; \tag{2.1.8}$$

2) the refined upper estimate

$$\lambda_1 \le \lambda_1^*\xi^2 < \lambda_1^*. \tag{2.1.9}$$

The inequalities (2.1.8) and (2.1.9) can be used for the construction of a convergent algorithm based on successive approximations, but its convergence would be slow.

2.2. Criterion of Closeness between the First Eigenvalue and its Upper (Lower) Bound. Introduction of a Small Parameter

Now we describe the main idea on which the method of accelerated convergence is based. Let λ_1^* be an upper bound for the first eigenvalue λ_1. After solving the Cauchy problem (2.1.2), we find the root ξ of the solution $V_1(x, \lambda_1^*)$. The authors have proposed [4, 5] the following criterion of closeness between λ_1^* and λ_1:

$$\varepsilon = 1 - \xi \ll 1. \tag{2.2.1}$$

If we have a lower estimate $\lambda_{1*} < \lambda_1$, then the closeness criterion is also valid with $\varepsilon = \xi - 1$. Unless indicated otherwise, we will always deal with upper bounds λ_1^*, because these are more easily obtained by the Rayleigh–Ritz method. Increasing the number of test functions, n, it is always possible to ensure the inequality (2.2.1). Criterion (2.2.1) can be introduced in a great variety of situations, and this makes the method expounded here fairly universal.

2.3. Theory of Perturbations

2.3.1. Construction of an equivalent perturbed problem. Suppose that the Cauchy problem (2.1.2) has been solved and we have found the point ξ. In problem (2.1.1), we introduce the new unknown variable $y = x\xi$, the parameter $\Lambda = \lambda/\xi^2$, and the function $U(y,\varepsilon) \equiv u(x)$. Then, the Sturm–Liouville problem (2.1.1) takes the form of a boundary value problem with the small parameter $\varepsilon = 1 - \xi$ (there is no small parameter in the original problem)

$$\begin{gathered}\left(p\left(\tfrac{y}{\xi}\right)U'\right)' + \left[\Lambda r\left(\tfrac{y}{\xi}\right) - \tfrac{1}{\xi^2}q\left(\tfrac{y}{\xi}\right)\right]U = 0,\\ U(0,\varepsilon) = U(\xi,\varepsilon) = 0; \qquad \lambda = \Lambda(\varepsilon)\xi^2.\end{gathered} \tag{2.3.1}$$

Since $\varepsilon = 1 - \xi$, the coefficients of this equation can be expanded in powers of ε (it suffices to consider one or two terms of the expansion),

$$\begin{aligned} p\left(\frac{y}{\xi}\right) &= p\left(\frac{y}{1-\varepsilon}\right) = p(y) + \varepsilon y p'(y) + \cdots,\\ r\left(\frac{y}{\xi}\right) &= r\left(\frac{y}{1-\varepsilon}\right) = r(y) + \varepsilon y r'(y) + \cdots,\\ \frac{1}{\xi^2}q\left(\frac{y}{\xi}\right) &= q(y) + 2\varepsilon q(y) + \varepsilon y q'(y) + \cdots. \end{aligned} \tag{2.3.2}$$

Using (2.3.2), we transform equation (2.3.1) to (the prime indicates differentiation in y)

$$\begin{gathered}(p(y)U')' + [\Lambda r(y) - q(y)]U = -\varepsilon[(yp'U')' + (\Lambda y r' - 2q - yq')U] + \varepsilon^2 H_2,\\ U(0,\varepsilon) = U(\xi,\varepsilon) = 0,\end{gathered} \tag{2.3.3}$$

where H_2 is a known function calculated through the derivatives of p, r, q, the parameter Λ, and the function U with the argument y.

The first eigenvalue Λ_1 and the function U_1 are to be determined as functions of the parameter $\varepsilon \ll 1$.

Note that for $\varepsilon = 0$, these quantities are known and have the form $\Lambda_1^{(0)} = \lambda_1^*$; $U_1^{(0)} = V_1(y, \lambda_1^*)$, where V_1 is the solution of the Cauchy problem (2.1.2).

2.3.2. Approximate solution of the perturbed problem.

We seek a solution of problem (2.3.3) in the form of expansions in powers of the parameter ε,

$$U = U_1^{(0)} + \varepsilon U_1^{(1)} + \varepsilon^2 U_1^{(2)} + \cdots, \qquad \Lambda_1 = \Lambda_1^{(0)} + \varepsilon \Lambda_1^{(1)} + \varepsilon^2 \Lambda_1^{(2)} + \cdots. \tag{2.3.4}$$

Substituting the series (2.3.4) into (2.3.3), we obtain a sequence of boundary value problems. In particular, for $\varepsilon = 0$ we have

$$\begin{gathered}\left(p(y)U_1^{(0)\prime}\right)' + (\Lambda_1^{(0)} r(y) - q(y))U_1^{(0)} = 0,\\ U_1^{(0)}(0) = U_1^{(0)}(\xi) = 0; \qquad U_1^{(0)} = V_1(y, \lambda_1^*), \qquad \Lambda_1^{(0)} = \lambda_1^*.\end{gathered} \tag{2.3.5}$$

As already mentioned, the solution of problem (2.3.5) is known and is given above.

The terms with the coefficient ε yield the problem

$$\begin{gathered}(p(y)U_1^{(1)\prime})' + (\lambda_1^* r(y) - q(y))U_1^{(1)} = -\Lambda_1^{(1)} r(y)V_1 - (yp'V_1')' - (\lambda_1^* y r' - 2q - yq')V_1,\\ U_1^{(1)}(0) = U_1^{(1)}(\xi) = 0.\end{gathered} \tag{2.3.6}$$

Boundary value problems similar to (2.3.6) are obtained, if subsequent terms of the expansions (2.3.4) with ε^2, $\varepsilon^3, \ldots$ are taken into account.

Since the function $U_1^{(0)}(y)$ is known, let us find the unknown quantities $\Lambda_1^{(1)}$ and $U_1^{(1)}(y)$. According to the Fredholm alternative, problem (2.3.6) admits a solution if the right-hand side of the equation is orthogonal to V_1, i.e.,

$$\Lambda_1^{(1)} \int_0^\xi r V_1^2 dy + \int_0^\xi [(yp'V_1')' + (\lambda_1^* yr' - 2q - yq')V_1)V_1]\, dy = 0. \tag{2.3.7}$$

Relation (2.3.7) allows us to find a correction $\Lambda_1^{(1)}$ to the eigenvalue $\Lambda_1^{(0)}$ in the form of quadrature of the known function,

$$\Lambda_1^{(1)} = -\frac{1}{\|V_1\|^2} \int_0^\xi [(yp'V_1')' + (\lambda_1^* yr' - 2q - yq')V_1]V_1\, dy, \quad \|V_1\|^2 = \int_0^\xi r(y)V_1^2(y, \lambda_1^*)\, dy. \tag{2.3.8}$$

2.3.3. Reduction of the correction term to differential form. The correction $\Lambda_1^{(1)}$ has an intricate structure, but it can be simplified to a great extent, so that the integral in the numerator can be found in explicit form. Indeed, we have

$$\int_0^\xi [(yp'V_1')' + (\lambda_1^* yr' - yq' - 2q)V_1]V_1\, dy \equiv I_1 + I_2 - I_3. \tag{2.3.9}$$

The integral (2.3.9) and the corresponding correction (2.3.8) will be calculated in detail, since our expression for the first eigenvalue correction $\Lambda_1^{(1)}$ cannot be found in other publications and integrals similar to (2.3.9) will frequently occur in the sequel.

By I_1, I_2, I_3 in (2.3.9) we have denoted the following quantities:

$$I_1 = \int_0^\xi (yp'V_1')'V_1\, dy, \qquad I_2 = \int_0^\xi (\lambda_1^* yr' - q'y)V_1^2\, dy, \qquad I_3 = \int_0^\xi 2qV_1^2\, dy.$$

Let us calculate these integrals in consecutive order, using the known properties of the function $V_1(y, \lambda_1^*)$ (2.3.5). Recall that $V_1(0, \lambda_1^*) = V_1(\xi, \lambda_1^*) = 0$, $V_1'(0, \lambda_1^*) = 1$. After elementary transformation of I_1, we get

$$I_1 = \int_0^\xi (p'yV_1')'V_1\, dy = p'yV_1'V_1\Big|_0^\xi - \int_0^\xi p'yV_1'^2\, dy = -p(\xi)\xi V_1'^2(\xi, \lambda_1^*) + \int_0^\xi p(yV_1'^2)'\, dy. \tag{2.3.10}$$

Similar transformations of I_2 yield

$$\begin{aligned} I_2 &= \int_0^\xi (\lambda_1^* yr' - yq')V_1^2\, dy = (\lambda_1^* r - q)yV_1^2\Big|_0^\xi - \int_0^\xi (\lambda_1^* r - q)(yV_1^2)'\, dy \\ &= -\int_0^\xi (\lambda_1^* r - q)(V_1^2 + 2yV_1V_1')\, dy = \int_0^\xi (pV_1)'(V_1 + 2yV_1')\, dy. \end{aligned} \tag{2.3.11}$$

Substituting these I_1, I_2, and I_3 into (2.3.9), we obtain

$$I_1 + I_2 - I_3 = p(\xi)\xi V'^2(\xi, \lambda_1^*) - 2\int_0^\xi (pV_1'^2 + qV_1^2)\, dy. \tag{2.3.12}$$

Since the equation for the function V_1 implies that

$$\int_0^\xi (pV_1'^2 + qV_1^2)\,dy = \|V_1\|^2\lambda_1^*,$$

the equality (2.3.12) takes the form

$$I_1 + I_2 - I_3 = p(\xi)\xi V_1'^2(\xi, \lambda_1^*) - 2\|V_1\|^2\lambda_1^*. \tag{2.3.13}$$

Substituting (2.3.13) into (2.3.8), we obtain the following expression for the correction $\Lambda_1^{(1)}$:

$$\Lambda_1^{(1)} = -\frac{1}{\|V_1\|^2}p(\xi)\xi V_1'^2(\xi, \lambda_1^*) + 2\lambda_1^*. \tag{2.3.14}$$

Using (2.3.14), we go back to the original problem (2.1.1) and obtain a refined value $\lambda_1 = \lambda_{1(1)} + O(\varepsilon^2)$, where

$$\lambda_{1(1)} = \xi^2(\lambda_1^* - \frac{\varepsilon}{\|V_1\|^2}p(\xi)\xi V_1'^2(\xi, \lambda_1^*) + 2\varepsilon\lambda_1^*) = \lambda_1^* - \frac{\varepsilon}{\|V_1\|^2}p(\xi)\xi V_1'^2(\xi, \lambda_1^*) + O(\varepsilon^2). \tag{2.3.15}$$

Note that the function $U_1^{(1)}$ has not been used for refining the value of λ_1.

Let us make some remarks.

1. In the case of boundary conditions of the second or the third kind, the corresponding corrections are calculated in a similar manner. The resulting expressions differ from (2.3.15) and are given below.

2. Since $\xi = 1 - \varepsilon$, $\varepsilon \ll 1$, we can take $\xi = 1$ in the final formula (2.3.15) without losing precision with respect to ε.

3. The approximate formula (2.3.15) implies that the following exact formula takes place as $\varepsilon \to 0$:

$$\left.\frac{\partial\lambda_1}{\partial\xi}\right|_{\xi=1} = -p(1)\frac{(U_1'(1))^2}{\|U_1\|^2} < 0. \tag{2.3.16}$$

Having found the refined eigenvalue $\lambda_{1(1)}$, we can use equation (2.3.6) for the calculation of $U_1^{(1)}$. Similar constructions can be used for finding all the next approximations $\lambda_1^{(2)}$, $U_1^{(2)}, \ldots$. However, the formulas obtained in this way are rather cumbersome and many analytical calculations have to be performed, their volume increasing rapidly. For this reason, it is proposed to utilize replacements like (2.3.1) on the basis of the refined values $\lambda_{1(1)}$, ξ_1, $\lambda_{1(2)}$, ξ_2, etc.

2.4. Description of the Method of Accelerated Convergence

The method or algorithm of accelerated convergence amounts to the implementation of the following operations:

1. Find an upper bound λ_1^* (by the Rayleigh–Ritz or any other method).

2. Insert this λ_1^* into the equation of the Sturm–Liouville problem and solve the Cauchy problem (2.1.2), i.e., find the function $V_1(x, \lambda_1^*)$.

3. Find, in some way, the root of the solution of the Cauchy problem, i.e., the point ξ such that

$$V_1(\xi, \lambda_1^*) = 0.$$

Numerically, the point ξ is found either by interpolation or by decreasing the step when integrating the Cauchy problem near the point x at which the function $V_1(x, \lambda_1^*)$ changes sign.

4. Determine the parameter ε and, using (2.3.15), calculate the refined eigenvalue

$$\lambda_{1(1)} = \lambda_1^* - \varepsilon p(1)\frac{V_1'^2(1, \lambda_1^*)}{\|V_1\|^2}. \tag{2.4.1}$$

5. Replacing λ_1^* in the Cauchy problem (2.1.2) by $\lambda_{1(1)}$, repeat the entire process, i.e., find the solution of the Cauchy problem, determine the new point ξ_1 and the new parameter $\varepsilon_1 = 1 - \xi_1$. The result is the refined value

$$\lambda_{1(2)} = \lambda_{1(1)} - \varepsilon_1 p(1)\frac{V_1'^2(1, \lambda_{1(1)})}{\|V_1\|^2}.$$

The process is continued in a recurrent manner, $\lambda_{1(2)} \to \xi_2 = 1 - \varepsilon_2 \to \lambda_{1(3)} \to \dots$

During the implementation of this algorithm, it may happen that one has to find a solution of the Cauchy problem (2.1.2) on a wider interval $0 \le x \le 1 + a$ with $a \sim \varepsilon^2$. This is why we have assumed that the smooth functions p, q, r are defined on a larger interval.

On each step, we find an approximate eigenvalue $\lambda_{1(k)}$ and the corresponding approximate eigenfunction $V_1(x, \lambda_{1(k)})$. The accuracy of the solution obtained is controlled in terms of two quantities: $|\xi_k - 1|$ and $|V_1(\xi_k, \lambda_{1(k)})|$ as $k \to \infty$. Computational experience shows that sufficient accuracy is ensured for k equal to 2 or 3. Sometimes, one has to consider additional iterations (with k from 5 to 7), if the variation of the functions p, r, q is very large and fast.

2.5. Some Applications of the Accelerated Convergence Method

2.5.1. Test model problems. Let us demonstrate the efficiency of the above algorithm for some numerical examples.

Example 1. Consider the Sturm–Liouville problem [33]

$$u'' + \lambda(1 + \sin \pi x)u = 0, \qquad u(0) = u(1) = 0. \tag{2.5.1}$$

By the Rayleigh–Ritz method with the test function $\psi_1 = \sin \pi x$ we obtain the estimate $\lambda_1 \le \lambda_1^* = 5.33827$.

Integration of the Cauchy problem yields the abscissa value $\xi = 0.99942$ and the parameter $\varepsilon = 5.8 \cdot 10^{-4} \ll 1$. Integration of the Cauchy problem (2.1.3) with the parameter $\lambda_1^* \xi^2 = 5.33205$ shows that the assumptions of the first part of Theorem 2.1 hold, and therefore, we have the two-sided estimate $5.33205 \le \lambda_1 \le 5.33827$. Tiresome calculations [28] yield the "exact" value $\lambda_1 = 0.54032 \cdot \pi^2 = 5.33274$. The refinement with the help of (2.4.1) yields the eigenvalue $\lambda_{1(1)} = 5.33284$, which determines λ_1 with the relative error 10^{-5}.

Example 2. The solution of the Sturm–Liouville problem

$$u'' + \lambda(1 + x^2)^{-2}u = 0, \qquad u(0) = u(1) = 0 \tag{2.5.2}$$

can be found in analytical form and expressed in terms of elementary functions. Thus, we obtain

$$\lambda_1 = 15, \qquad \varphi_1(x) = 2x(1 - x^2)(1 + x^2)^{-3/2}.$$

Taking $\psi_1(x) = \sin \pi x$ as a test function, we obtain the upper estimate $\lambda_1 < \lambda_1^* = 15.33728$. Integrating the Cauchy problem (2.1.2), we find the abscissa $\xi = 0.98383$ and the parameter $\varepsilon = 1.617 \cdot 10^{-2}$. By Theorem 2.1, $\lambda_1^* \xi^2 = 14.83533$ is a lower bound for λ_1. Thus, we have

$$\lambda_{1*} = 14.83533 \le \lambda_1 \le 15.33728 = \lambda_1^*.$$

The refined value $\lambda_{1(1)} = 14.99719$ determines the first eigenvalue with the relative error $2 \cdot 10^{-4}$. The next iteration yields $\lambda_{1(2)} = 15.000005$, which determines the eigenvalue with the relative error $4 \cdot 10^{-7}$.

Example 3. Consider the Sturm–Liouville problem

$$u'' + \lambda\big[1 + \sin(\tfrac{1}{2}\pi x)\big]u = 0, \qquad u(0) = u(1) = 0. \tag{2.5.3}$$

Using the test function $\psi_1 = \sin \pi x$, we obtain the following upper estimate $\lambda_1 < \lambda_1^* = 5.87805$. Integration of the Cauchy problem yields the point $\xi = 0.99805$, which shows that the second statement of Theorem 2.1 holds, since for $\lambda_1^*\xi^2$ the solution of the Cauchy problem vanishes at the point $\xi_1 = 0.99975 < 1$. Thus, we have a refined upper estimate,

$$\lambda_1 \le \lambda_1^*\xi^2 = 5.85515 < \lambda_1^* = 5.87805.$$

Moreover, using ξ_1, we obtain another refined estimate from above,

$$\lambda_1 \le \lambda_1^*\xi^2\xi_1^2 = 5.85222 = \lambda_1^{**}.$$

The refinement formula in which the bound $\lambda_1^{**} = 5.85222$ is used yields $\lambda_{1(1)} = 5.84500$ with the relative error 10^{-5}.

Example 4. To demonstrate the efficiency of the Rayleigh–Ritz method, consider the Sturm–Liouville problem

$$(\sqrt{1 + x}u')' + \lambda u = 0, \qquad u(0) = u(1) = 0, \tag{2.5.4}$$

and the system of test functions

$$\psi_1(x) = x(1 - x), \qquad \psi_2(x) = x^2(1 - x); \qquad \psi_3 = x^3(1 - x).$$

The upper bounds obtained in this case are fairly rough, even though there are three test functions. This is due to the improper choice of the test functions (see below).

By the Rayleigh–Ritz method with a single test function ψ_1 we obtain the upper bound $\lambda_1^* = 12.15$. With two test functions ψ_1, ψ_2, we get $\lambda_1^* = 12.12781$. Finally, with three test functions ψ_1, ψ_2, ψ_3, we obtain $\lambda_1^* = 12.122555$. Taking $\lambda_1^* = 12.15$ and solving the corresponding Cauchy problem, we obtain the point $\xi = 0.9885$. This shows that the assumptions of the first statement of Theorem 2.1 hold and $\lambda_{1*} = \lambda_1^*\xi^2$ is a lower bound, i.e., we have $\lambda_{1*} = \lambda_1^*\xi^2 = 11.86975$ and the two-sided estimate $11.86975 \le \lambda_1 \le 12.1226$ holds. The refinement by (2.3.15), after the first iteration, yields $\lambda_1 \approx \lambda_{1(2)} = 11.89546$ with the relative error $2.4 \cdot 10^{-4}$.

Using only the test function $\psi_1 = \sin \pi x$, which turns out more adequate, we obtain the upper bound $\lambda_1^* = 11.99558$. Solving the Cauchy problem (2.1.2) yields the point $\xi = 0.99563$. It is clear that for further refinement it would be better to take the bound $\lambda_1^* = 11.99558$, the abscissa $\xi = 0.99563$, and the parameter $\varepsilon = 0.00437$. According to (2.3.15), this brings us, after the first iteration, to the value $\lambda_1 \approx \lambda_{1(1)} = 11.8979$ with the relative error $3 \cdot 10^{-5}$.

This example shows that suitably chosen test functions may considerably reduce the amount of calculations and decrease the errors. It should be stressed once again that the above algorithm allows us to control (on every iteration step) the accuracy of our approximation of an eigenvalue and to establish an interval to which the eigenvalue belongs. Upon finding λ_1 with the desired accuracy, one has to solve the Cauchy problem in order to find u_1.

2.5.2. A method for the calculation of weighted norms. In conclusion, we describe a method for the calculation of the norm with weight $r(x)$,

$$\|u\|_r^2 = \int_0^\xi r u^2\, dx,$$

for the first eigenfunction. The common approach to this problem is the numerical integration according to any quadrature formula, since the values of u at the nodal points are known after solving the Cauchy problem. One can also write an auxiliary differential equation whose solution allows us to find $||u||_r^2$.

Here, we describe another approach which is realized on each iteration step i. Let λ_1 be the first eigenvalue corresponding to the eigenfunction $u(x)$, $0 \le x \le \xi$. It is assumed that $\lambda_1 = \lambda_{1(i)}$, $\xi = \xi_i$, $u_i = V$.

We introduce an auxiliary function h which is the solution of the following Cauchy problem:

$$(ph')' + (\lambda_1 r(x) - q(x))h = -r(x)u, \qquad h(0) = h'(0) = 0. \tag{2.5.5}$$

The unknown h is the so-called sensitivity of the eigenfunction to λ: $h = \partial u/\partial\lambda$.

The function u is a solution of the Sturm–Liouville problem on the interval $0 \le x \le \xi$, i.e., the solution of the Cauchy problem (2.1.2)

$$(pu')' + (\lambda_1 r(x) - q(x))u = 0, \qquad u(0) = u(\xi) = 0 \qquad (u'(0) = 1). \tag{2.5.6}$$

Multiplying the first equation by u and the second by h, taking the difference of the resulting relations and integrating, we obtain

$$\int_0^\xi [u(ph')' - h(pu')']\,dx = -\int_0^\xi r(x)u^2\,dx. \tag{2.5.7}$$

Integrating by parts in (2.5.7) and taking into account conditions (2.5.6), we obtain an explicit expression for the norm on the interval $0 \le x \le \xi$,

$$\int_0^\xi r(x)u^2\,dx = h(\xi)p(\xi)u'(\xi). \tag{2.5.8}$$

The expression (2.5.8) requires that two Cauchy problems (2.5.5), (2.5.6) be solved simultaneously, in order to find the functions h and u. Formula (2.5.8) is also valid for the exact value λ_1 and $\xi = 1$.

Since $1 - \xi = \varepsilon$, formula (2.5.8) can be written as

$$\int_0^\xi r(x)u^2\,dx \approx \int_0^1 r(x)u^2\,dx = h(1)p(1)u'(1) \tag{2.5.9}$$

without the loss of accuracy with respect to powers of ε ($\varepsilon = \varepsilon_i$). From (2.5.9), it follows that $h(1) < 0$, since $u(x) \ge 0$, $u'(1) < 0$.

2.6. The Method of Accelerated Convergence for Higher Eigenvalues

2.6.1. An example with the calculation of two eigenvalues. The above algorithm can be almost literally transferred to the problem of finding subsequent eigenvalues. In order to explain how the method of accelerated convergence can be used for finding higher eigenvalues λ_n and eigenfunctions $u_n(x) = u(x, \lambda_n)$, consider the following example.

Our aim is to find the eigenvalues λ_2 and λ_3 of the boundary value problem (see (2.5.1))

$$u'' + \lambda(1 + \sin \pi x)u = 0, \qquad u(0) = u(1) = 0, \tag{2.6.1}$$

provided that we have already found λ_1. Let us use the approximation obtained by the Rayleigh–Ritz method with four test functions,

$$u = C_1 \sin(\pi x) + C_2 \sin(2\pi x) + C_3 \sin(3\pi x) + C_4 \sin(4\pi x). \tag{2.6.2}$$

Substituting this expression into the corresponding functionals (1.4.3) and the condition (1.4.5), we obtain the characteristic equation

$$\det \|A - \lambda B\| = 0 \tag{2.6.3}$$

with the matrices A and B of the form

$$A = \begin{pmatrix} \pi^2 & 0 & 0 & 0 \\ 0 & 4\pi^2 & 0 & 0 \\ 0 & 0 & 9\pi^2 & 0 \\ 0 & 0 & 0 & 16\pi^2 \end{pmatrix}, \quad B = \begin{pmatrix} 1.8488264 & 0 & -0.1697652 & 0 \\ 0 & 1.679061 & 0 & -0.1940176 \\ -0.1697652 & 0 & 1.654809 & 0 \\ 0 & -0.1940176 & 0 & 1.6467247 \end{pmatrix}.$$

The first three roots of the characteristic equation (2.6.3) are

$$\lambda_1^* = 0.540322\pi^2; \quad \lambda_2^* = 2.37186\pi^2 = 23.40932; \quad \lambda_3 = 5.49612\pi^2 = 54.24453.$$

Substituting λ_2^* into the Cauchy problem (2.1.2), we find its solution $V(x, \lambda_2^*)$, the point $\xi = 0.999946$ nearest to $x = 1$, and the parameter $\varepsilon = 5.4 \cdot 10^{-5}$. Using the refinement formula (2.4.1), after the first iteration we obtain

$$\lambda_{2(1)} = 23.407218 = 2.371647\pi^2, \qquad V(1, \lambda_{2(1)}) = 5.4105 \cdot 10^{-5}.$$

The next iteration yields $\xi = 0.999999$, $\varepsilon = 10^{-6}$,

$$\lambda_{2(2)} = 23.407186, \qquad V(1, \lambda_{2(2)}) = 8.172 \cdot 10^{-7}.$$

This $\lambda_{2(2)}$ determines λ_2 with the relative error $4.5 \cdot 10^{-6}$.

Similarly, we calculate $\lambda_{3(2)} = 53.7720$ with the relative error 10^{-5}.

2.6.2. Some properties of the procedure of finding subsequent eigenvalues. When using the Rayleigh–Ritz method for finding higher eigenvalues corresponding to $n > 3$, one has to compute two square matrices with the total number of elements being equal to n^2 (due to symmetry), and also calculate the roots of the characteristic equation (2.6.3).

In some special cases, either the matrix A or B happens to be diagonal. However, as the index of the eigenvalue grows, the volume of calculations for the initial approximation of the eigenvalue grows rapidly. Therefore, for $n \geq 4$, it seems reasonable to use asymptotic formulas [29] for the initial approximation of λ_n. According to (1.2.33), we have the following estimates for λ_n in problem (2.6.1):

$$\tilde{\lambda}_n = \frac{\pi^2 n^2}{l^2}, \qquad l = \int_0^1 (1 + \sin \pi x)^{1/2}\, dx = 1.27324.$$

For $n = 4$, the bound is $\tilde{\lambda}_4 = 97.40902$, and for $n = 5$ we have $\tilde{\lambda}_5 = 152.2016$. The solution of the Cauchy problem shows that these two quantities are indeed upper bounds. Using the above algorithm, we find the refined values

$$\lambda_4 \approx \lambda_{4(4)} = 96.3500, \qquad \lambda_5 \approx \lambda_{5(4)} = 151.1162$$

with the relative error $5 \cdot 10^{-4}$.

It should be said that the relative error grows with the growth of n. This is mainly due to the fact that the requirements on the precision of calculations are not fully met, in particular, when integrating equations.

2.7. Problems with Boundary Conditions of the Second Kind

2.7.1. Construction of a comparison problem. The above method of accelerated convergence can be adapted to problems with boundary conditions of the second kind. Consider the Sturm–Liouville problem

$$(p(x)u')' + [\lambda r(x) - q(x)]u = 0, \qquad u'(0) = u'(1) = 0. \tag{2.7.1}$$

Our assumptions about the coefficients p, q, r are the same as above, in particular, the inequalities (1.1.3) hold.

The boundary value problem (2.7.1) is equivalent to the isoperimetric variational problem

$$J[u] = \int_0^1 [p(x)u'^2 + q(x)u^2]\,dx \to \min_u, \qquad \Phi[u] = \int_0^1 r(x)u^2\,dx = 1.$$

By the Rayleigh–Ritz method with test functions $\psi_1(x), \ldots, \psi_m(x)$ satisfying the boundary conditions $\psi_k'(0) = \psi_k'(1) = 0$, we obtain the upper bounds $\lambda_1^*, \ldots, \lambda_{m-1}^*$. Next, we solve the following Cauchy problem similar to (2.1.2):

$$(p(x)u')' + [\lambda_1^* r(x) - q(x)]u = 0, \qquad u(0) = 1, \qquad u'(0) = 0 \qquad (\lambda_1^* \geq \lambda_1). \tag{2.7.2}$$

Suppose that we have constructed the solution $u = V(x, \lambda_1^*)$ of this problem. The point ξ is found from the condition $V'(x, \lambda_1^*) = 0$.

2.7.2. Approximate solution of the problem. We introduce the parameter $\varepsilon = 1 - \xi > 0$. The closeness criterion $\varepsilon \ll 1$ has the same sense as above: $\lambda_1^* \to \lambda_1$ as $\varepsilon \to 0$.

Most of the calculations are similar to those of Sections 2.2–2.6. However, the formula refining λ_1 takes another form (see (2.4.1)),

$$\lambda_{1(1)} = \lambda_1^* - \frac{\varepsilon}{\|V\|^2}[\lambda_1^* r(\xi) - q(\xi)]V^2(\xi, \lambda_1^*). \tag{2.7.3}$$

Using the Cauchy problem and the refinement formula (2.7.3), one applies the algorithm of accelerated convergence as described in Section 2.4.

2.7.3. Test problem. As an example, consider a problem that admits an analytical solution,

$$u'' + \frac{\lambda}{(1+x^2)^2}u = 0, \qquad u'(0) = u'(1) = 0. \tag{2.7.4}$$

Passing to the variables $y = u\sqrt{1+x^2}$, $t = \arctan x$, one can solve problem (2.7.4) in terms of elementary functions and obtain $\lambda_n = \gamma_n^2 - 1$, where γ_n is a root of the transcendental equation

$$\tan\frac{\gamma\pi}{4} = \frac{1}{\gamma}. \tag{2.7.5}$$

The first root of equation (2.7.5) is $\gamma_0 = 1$. This root corresponds to the zero eigenvalue $\lambda_0 = 0$ and the eigenfunction $\varphi_0 \equiv 1$. The next root of equation (2.7.5) is found numerically (by the bisectional method or by the Newton method) and is equal to $\gamma_1 = 4.291488$. This corresponds to the eigenvalue $\lambda_1 = \gamma_1^2 - 1 = 17.416870$ and the eigenfunction

$$\varphi_1(x) = \sqrt{1+x^2}\cos(\gamma_n \arctan x).$$

Let us take the analytically obtained value $\lambda_1 = 17.416870$ as the exact one. In order to implement the method of accelerated convergence, we proceed as follows. We have $\lambda_0 = 0$ and $\varphi_0(x) \equiv 1$. Therefore, to obtain an upper estimate by the Rayleigh–Ritz method, we have to construct a test function which satisfies the boundary conditions and is orthogonal to $\varphi_0 \equiv 1$ with the weight $(1+x^2)^{-2}$. For this purpose, we take $\psi_1(x) = C_0 + C_1 \cos \pi x$ and find C_0, C_1 from the orthogonality condition

$$\int_0^1 \psi_1(x) \frac{dx}{(1+x^2)^2} = 0. \tag{2.7.6}$$

From (2.7.6) we find that for $C_1 = 1$ the test function has the form $\psi_1(x) = \cos \pi x - 0.27269$. The Rayleigh–Ritz method yields the upper bound $\lambda_1 \le \lambda_1^* = 18.06583$. In this case, all quadratures have to be calculated numerically. Integration of the Cauchy problem with the boundary conditions $u'(0) = 0$, $u(0) = 1$ yields the function $V'(x, \lambda_1^*) = 0$, the point $\xi = 0.96875$, and the small parameter $\varepsilon = 0.03125$. We also obtain a sufficiently precise lower bound $\lambda_{1*} = \lambda_1^* \xi^2 = 16.95440$, in view of (2.1.3) and analog of Theorem 2.1. Using the refinement formula (2.7.3), we find that $\lambda_1 \approx \lambda_{1(1)} = 17.48420$ with the relative error $4 \cdot 10^{-3}$. The next refinement $\lambda_{1(2)}$ practically coincides with the exact eigenvalue.

Consider a more complex test function

$$\psi_1(x) = C_1(\cos \pi x - 0.27269) + C_2(\cos 2\pi x + 0.00125).$$

By the Rayleigh–Ritz principle, we obtain the upper bound $\lambda_1 \le \lambda_1^* = 17.41833$, which is substantially closer to the exact value. Integration of the Cauchy problem yields the abscissa $\xi = 0.99993$ and the parameter $\varepsilon = 7 \cdot 10^{-5}$. We obtain the lower bound $\lambda_{1*} = \lambda_1^* \xi^2 = 17.41575$ in view of (2.1.3) and analog of Theorem 2.1. The average $1/2(\lambda_{1*} + \lambda_1^*) = 17.41704$ determines the exact λ_1 with the relative error 10^{-5}. Calculations by the refinement formula (2.7.3) yield the eigenvalue λ_1 with the relative error 10^{-8}. Subsequent eigenvalues $\lambda_2, \lambda_3, \ldots$ can be found in a similar way.

In order to find the corresponding eigenfunctions, it suffices to integrate the Cauchy problem (2.7.2) after inserting the eigenvalues calculated before.

2.8. Problems with Boundary Conditions of the Third Kind

2.8.1. Statement of the third boundary value problem. Consider the Sturm–Liouville problem with the third boundary condition [6]

$$\begin{gathered}(p(x)u')' + [\lambda r(x) - q(x)]u = 0, \\ \alpha_0 p(0)u'(0) - \beta_0 u(0) = 0; \qquad \alpha_0, \beta_0 \ge 0, \qquad \alpha_0 + \beta_0 = 1; \\ \alpha_1 p(1)u'(1) + \beta_1 u(1) = 0; \qquad \alpha_1, \beta_1 \ge 0, \qquad \alpha_1 + \beta_1 = 1.\end{gathered} \tag{2.8.1}$$

The nonnegative coefficients $\alpha_{0,1}$ and $\beta_{0,1}$ characterize the relative effect of elastic forces at the end-points $x = 0$, $x = 1$.

The corresponding variational statement of problem (2.8.1) has the form

$$\begin{gathered}\min_u J[u] = \lambda; \qquad J[u] = \int_0^1 [p(x)u'^2 + q(x)u^2]\,dx - p(x)u'u\Big|_0^1; \\ \Phi[u] = \|u\|^2 = \int_0^1 r(x)u^2\,dx = 1; \qquad [\alpha_{0,1}p(x)u' \mp \beta_{0,1}u]_{x=0,1} = 0.\end{gathered} \tag{2.8.2}$$

The absolute minimum of the functional in (2.8.2) under the isoperimetric condition $\Phi[u] = 1$ and the boundary conditions of the third kind is equal to the first eigenvalue λ_1, and the function $u(x, \lambda_1) = \varphi_1(x)$ on which this minimum is attained is the first eigenfunction. Subsequent eigenvalues λ_n, $n \geq 2$, are defined recurrently on narrower functional classes whose elements are orthogonal with weight $r(x)$ to the preceding eigenfunctions $u_1, \ldots, u_{n-1}$,

$$\Psi_k[u] = \int_0^1 r(x) u_k(x) u \, dx = 0. \tag{2.8.3}$$

Solving the variational problem (2.8.2), (2.8.3) amounts to finding a system of eigenvalues $\{\lambda_n\}$ and orthonormal eigenfunctions $\{u_n(x)\}$, which are at the same time solutions of the Sturm–Liouville problem (2.8.1). In this case, we also have the Rayleigh–Ritz principle:

$$0 < \lambda_1 \leq \lambda_1^* = \frac{J[\psi]}{\Phi[\psi]}, \tag{2.8.4}$$

where ψ is any test function satisfying boundary conditions (2.8.1) or (2.8.2). The accuracy of the bound (2.8.4) depends on the choice of the test function. The upper bound coincides with the eigenvalue, i.e., $\lambda_1^* = \lambda_1$, only if $\psi = c\varphi_1(x)$.

As mentioned above, the Rayleigh–Ritz method is theoretically capable of yielding upper bounds λ_1^*, $\lambda_2^* \ldots \lambda_n^*$ arbitrarily close to the desired eigenvalues. This approach, however, has its own difficulties and in order to have suitable accuracy, one has to obtain adequate lower bounds (see Section 2.1, the plan with 6 items).

2.8.2. Construction of a comparison system. Since no due attention is given in mathematical literature to the case of boundary conditions of the third kind, our exposition will contain all possible details, with inevitable repetitions of what has been said in previous sections.

Let us describe the procedure of finding the first eigenvalue λ_1 and the first eigenfunction φ_1. It is assumed that an upper bound λ_1^* has already been constructed. Inserting this into equation (2.8.1), we solve one of the following Cauchy problems:

$$(p(x)v')' + [\lambda_1^* r(x) - q(x)]v = 0;$$

$$1) \quad v(0) = \alpha_0 a, \qquad v'(0) = \frac{\beta_0 a}{p(0)} \qquad (a = 1); \tag{2.8.5}$$

$$2) \quad v(0) = \alpha_0 p(0) b, \qquad v'(0) = \beta_0 b \qquad (b = 1).$$

The solution $v = V(x, \lambda_1^*)$ of each Cauchy problem may be obtained either numerically or analytically. With this solution we construct the auxiliary function

$$E(x, \lambda_1^*) = \alpha_1 p(x) V'(x, \lambda_1^*) + \beta_1 V(x, \lambda_1^*). \tag{2.8.6}$$

In some cases, it is useful to know the differential equation which holds for the function $E(x, \lambda_1^*)$. Differentiating (2.8.6) in x, we obtain

$$E' = -\alpha_1(\lambda_1^* r(x) - q(x))V + \beta_1 V'. \tag{2.8.7}$$

The initial value $E(0, \lambda_1^*)$ is obtained from the corresponding condition in (2.8.5). Let ξ be the root of the equation $E(x, \lambda_1^*) = 0$ nearest to $x = 1$. Let us introduce the small parameter $\varepsilon = 1 - \xi$, $\varepsilon \ll 1$. The above considerations show that the solution of the Cauchy problem (2.8.5) is a solution of the Sturm–Liouville problem on the interval $0 \leq x \leq \xi$,

$$\begin{aligned}&(pv')' + (\lambda_1^* r - q)v = 0,\\ &\alpha_0 p(0)v'(0) - \beta_0 v(0) = 0,\\ &\alpha_1 p(\xi)v'(\xi) + \beta_1 v(\xi) = 0.\end{aligned} \tag{2.8.8}$$

According to (2.8.5), we have to solve the Cauchy problem either with the initial conditions 1) or 2). These conditions are equivalent and can be used both for $\alpha_0 = 0$ ($\beta_0 > 0$) and $\beta_0 = 0$ ($\alpha_0 > 0$). In case 1, the value $v(0)$ is prescribed and $v'(0)$ should be calculated, while in case 2 the value $v(0)$ should be calculated and $v'(0)$ is prescribed. Accordingly, the function v will contain either the parameter a or b, which can be taken equal to 1. Note that $E(1, \lambda_1^*) = 0$ only if $\lambda_1^* = \lambda_1$, and we have $\varepsilon \to 0$ if $\lambda_1^* \to \lambda_1$.

It is assumed above that the functions $p(x)$, $r(x)$, $q(x)$ have been smoothly extended to the interval $0 \le x \le 1 + \delta$, where $\delta > 0$, $\delta \sim \varepsilon^2$.

To ensure the smallness of the parameter $\varepsilon = 1 - \xi$, one can utilize the Rayleigh–Ritz method or the more cumbersome shooting method. Computational experience shows that the accelerated convergence method is applicable for $\varepsilon \sim 10^{-1} \div 10^{-2}$.

2.8.3. Solution of the perturbed problem. Suppose that the parameter ε is sufficiently small. Let us change the variables in the boundary value problem (2.8.1), using the transformations $y = x\xi$, $U(y, \varepsilon) = u$, $\Lambda = \xi^{-2}\lambda$, just as in Section 2.3.

The desired solution of the original boundary value problem (2.8.1) is constructed in the form of series, just as in Section 2.5,

$$\Lambda = \lambda_1^* + \varepsilon\Lambda^{(1)} + \cdots, \qquad U(y, \varepsilon) = U_0(y) + \varepsilon U_1 + \cdots. \tag{2.8.9}$$

The function $U_0(y)$ is taken equal to the solution $V(y, \lambda_1^*)$ of the Cauchy problem (2.8.5) for $x = y$. The unknown quantities $\Lambda^{(1)}$, U_1 are found by solving the following nonhomogeneous boundary value problem (similar to (2.3.6)):

$$\begin{aligned}&(p(y)U_1')' + [\lambda_1^* r(y) - q(y)]U_1 = -\Lambda^{(1)} r(y)U_0(y)\\ &\qquad -(yp'(y)U_0')' - \lambda_1^* y r'(y)U_0(y) + 2q(y)U_0 + yq'(y)U_0(y),\\ &\alpha_0 p(0)U_1'(0) - \beta_0 U_1(0) = \beta_0 U_0(0) \quad (U_0(y) \equiv V(y, \lambda_1^*)),\\ &\alpha_1 p(\xi)U_1'(\xi) + \beta_1 U_1(\xi) = -\alpha_1 p'(\xi)U_0'(\xi) - \beta_1 U_0(\xi).\end{aligned} \tag{2.8.10}$$

Similar relations are obtained for U_2, $\Lambda^{(2)}$, ... To determine the unknown constant $\Lambda^{(1)}$ we use the Fredholm alternative and the fact that the corresponding homogeneous problem is self-conjugate. Let us multiply equation (2.8.10) by U_0 and integrate the result in y over the interval $0 \le y \le \xi$. Integration by parts eliminates the terms with U_1. Thus, we obtain a linear equation for $\Lambda^{(1)}$. From this equation we find the desired value expressed in terms of the known quantities,

$$\begin{aligned}\Lambda_1 = {}&\|U_0\|^{-2}\int_0^{\xi} yp'(y)U_0'^2(y)\,dy + \|U_0\|^{-2}\int_0^{\xi}[-\lambda_1^* y r'(y) + 2q(y) + yq']U_0^2(y)\,dy\\ &+\|U_0\|^{-2}\left[\frac{\beta_1}{\alpha_1}U_0^2(\xi) + \frac{\beta_0}{\alpha_0}U_0^2(0)\right].\end{aligned} \tag{2.8.11}$$

When calculating $\Lambda^{(1)}$, U_1 in the first approximation with respect to ε (with the error $O(\varepsilon^2)$), we can take $\xi = 1$ without the loss of precision.

Having calculated the coefficient $\Lambda^{(1)}$ from (2.8.11), one can find the function U_1 which is a solution of problem (2.8.10). This is usually done by one of the known numerical methods.

Then, one recurrently determines $\Lambda^{(2)}$, U_2, etc. As mentioned earlier, this approach is not very productive. For this reason, we use the algorithm of accelerated convergence, as in the cases of boundary conditions of the first or the second kind.

2.8.4. Differential relation between eigenvalues and the interval length. The lengthy expression (2.8.11) with quadratures of functions, which are usually found by numerical integration of the Cauchy problem (2.8.5), can be substantially simplified. Integrating by parts in (2.8.11) and eliminating $r'(y)$, $q'(y)$, after elementary transformations we obtain

$$\lambda_1 = \lambda_1^* + \varepsilon\lambda_1^{(1)} + O(\varepsilon^2);$$

$$\lambda_1^{(1)} = -\frac{1}{\|U_0\|^2}\big(\xi p(\xi)U_0'^2(\xi) + [\lambda_1^* r(\xi) - q(\xi)]U_0^2(\xi)\big), \quad \|U_0\|^2 = \int_0^\xi r(y)U_0^2(y)\,dy. \tag{2.8.12}$$

The functions $U_0(\xi)$ and $U_0'(\xi)$ in (2.8.12) can be expressed one through the other with the help the boundary condition at the right end-point $y = \xi$.

The norm of the function $U_0(x)$ is found by quadrature formulas (for instance, the Simpson formula) or by the integration of an auxiliary Cauchy problem for U and its derivative $h = \partial U/\partial\lambda$,

$$(p(x)h')' + [\lambda_1^* r(x) - q(x)]h = -r(x)U_0; \qquad h(0) = h'(0) = 0, \tag{2.8.13}$$

$$\|U_0\|^2 = [U_0'(\xi,\lambda_1^*)h(\xi,\lambda_1^*) - U_0(\xi,\lambda_1^*)h'(\xi,\lambda_1^*)]p(\xi). \tag{2.8.14}$$

Formula (2.8.12) becomes much simpler in the case of boundary conditions of the first or the second kind and takes the form (2.4.1), (2.7.3).

2.8.5. The method of accelerated convergence. In the present situation, the algorithm of accelerated convergence amounts to the following operations:

1. Find an upper bound λ_1^* (by the Rayleigh–Ritz or any other method). Sometimes, it is possible to obtain a lower bound λ_{1*}, and then all calculations are carried out in a similar manner, but it should be kept in mind that $1 - \xi = \varepsilon < 0$.

2. Substitute this λ_1^* (or λ_{1*}) into the equation of the Sturm–Liouville problem and solve the Cauchy problem (2.8.5) (with one set of boundary conditions); find the function $V(x,\lambda_1^*)$.

3. Find in some way a root of the equation

$$E(x,\lambda_1^*) = \alpha_1 p(x)V'(x,\lambda_1^*) + \beta_1 V(x,\lambda_1^*) = 0, \qquad \xi = \xi(\lambda_1^*) = \arg_x E(x,\lambda_1^*). \tag{2.8.15}$$

The point ξ is found numerically, either by interpolation or by decreasing the integration step when solving the Cauchy problem.

4. Introduce the parameter $\varepsilon = 1 - \xi$ and, using (2.8.12), calculate the refined eigenvalue

$$\lambda_{1(1)} = \lambda_1^* + \varepsilon\mu(\lambda_1^*), \quad \mu \equiv \lambda_1^{(1)} = -\frac{1}{\|U_0\|^2}\left(\xi p(\xi)U_0'^2(\xi) + [\lambda_1^* r(\xi) - q(\xi)]U_0^2(\xi)\right),$$
$$U_0(x) \equiv V(x,\lambda_1^*), \qquad \xi = \xi(\lambda_1^*). \tag{2.8.16}$$

5. Insert $\lambda_{1(1)}$ into the Cauchy problem instead of λ_1^* and repeat the entire process. Thus, one obtains the refined eigenvalue $\lambda_{1(2)}$, etc.

On every iteration step, one finds an approximate eigenvalue $\lambda_{1(k)}$ and the corresponding eigenfunction $V(x,\lambda_{1(k)})$. The accuracy of the approximation is controlled with the help of two quantities: the abscissa $\xi_k \to 1$ and the error $E(\xi_{(k)},\lambda_{1(k)}) \to 0$ as $k \to \infty$. Computational experience shows that sufficient accuracy can be attained for $k = 2 \div 3$. Sometimes, additional iterations are needed, up to $k \approx 5$.

6. In order to find subsequent eigenvalues, one has to know upper bounds λ_2^*, λ_3^*, etc., (or lower bounds λ_{2*}, λ_{3*}). The algorithm proceeds in exactly the same manner.

2.8.6. Example. Consider a model example with the coefficient $r(x)$ having large variation,

$$u'' + \lambda(1 - 0.9\sin \pi x)^{-1} u = 0, \qquad u(0) = 0, \quad u'(1) + 2u(1) = 0. \tag{2.8.17}$$

The coefficient $r(x)$ in this equation varies ten times: for $0 \le x \le 1$, we have $r(0) = r(1) = 1$, $r(1/2) = 10$. The equivalent variational problem has the form

$$\begin{aligned} J[u] &= \int_0^1 u'^2\,dx + 2u^2(1) \to \min_u, \\ \Phi[u] &= \int_0^1 \frac{u^2\,dx}{1 - 0.9\sin \pi x} = 1, \end{aligned} \tag{2.8.18}$$

where u satisfies boundary conditions (2.8.17).

To obtain the initial bound λ_1^*, we use the Rayleigh principle with the system of test functions consisting of eigenfunctions of the Sturm–Liouville problem with constant coeficients

$$\psi'' + \lambda\psi = 0, \qquad \psi(0) = 0, \qquad \psi'(1) + 2\psi(1) = 0. \tag{2.8.19}$$

The eigenvalues and the eigenfunctions of problem (2.8.19) are given by the formulas

$$\lambda_n = \eta_n^2; \qquad \eta_n + 2\tan\eta_n = 0; \qquad \psi_n = \sin \eta_n x.$$

The smallest root of the equation $\eta + 2\tan\eta = 0$ is $\eta_1 = 2.28893$. Let us use $\psi_1 = \sin\eta_1 x$ as a test function in the Rayleigh principle. Calculation of the corresponding integral yields the upper bound $\lambda_1^* = 1.12451$. Then, integrating the Cauchy problem with $v(0) = 0$, $v'(0) = 1$, we obtain the desired root $\xi = 0.93321$ of the function

$$E(x, \lambda_1^*) = V'(x, \lambda_1^*) + 2V(x, \lambda_1^*).$$

The resulting value of the parameter $\varepsilon = 6.679 \cdot 10^{-2}$ turns out sufficiently small for the method of accelerated convergence to be applicable. Formula (2.8.16) yields the refined eigenvalue $\lambda_{1(1)} = 1.07907$, which gives us an upper bound and the corresponding root $\xi_1 = 0.99170$ and the parameter $\varepsilon_1 = 8.3 \cdot 10^{-3}$. Subsequent iterations yield the following quantities: on the second step $\lambda_{1(2)} = 1.0739955$, $\xi_2 = 0.9991044$, $\varepsilon_2 = 8.956 \cdot 10^{-4}$; on the third step $\lambda_{1(3)} = 1.0734555$, $\xi_3 = 0.9999726$, $\varepsilon_3 = 2.74 \cdot 10^{-5}$; and finally $\lambda_{1(4)} = 1.073440$. Verification shows that $\xi_3^2\lambda_{13}$ is a lower bound. Thus we have obtained the two-sided estimate $\lambda_{1*} = \xi_3^2\lambda_{1(3)} = 1.07338 \le \lambda_1 \le 1.073440 = \lambda_{1(4)}$. As a highly precise approximation of λ_1 we can take the mean value $(\lambda_{1*} + \lambda_{1(4)})/2$, which gives us λ_1 with the relative error of order $2 \cdot 10^{-5}$.

Now, let us take a two-coordinate approximation and apply the Rayleigh-Ritz method with $\psi = C_1 \sin\eta_1 x + C_2 \sin\eta_2 x$, where $\eta_1 = 2.28893$, $\eta_2 = 5.08699$ are the roots of the equation $\eta + 2\tan\eta = 0$. The Rayleigh-Ritz principle yields the matrices

$$A = \begin{pmatrix} 3.18666 & 0 \\ 0 & 13.80483 \end{pmatrix}, \qquad B = \begin{pmatrix} 2.83381 & 0.91900 \\ 0.91900 & 1.81257 \end{pmatrix}.$$

Next, we find the smallest root of the characteristic equation $\det\|A - \lambda B\| = 0$. This root $\lambda_1^* = 1.09432$ provides an upper bound for the first eigenvalue of problem (2.8.17). It is easy to see that this value is closer to λ_1 than the value obtained above with a single test function. In this case, we obtain $\xi = 0.97028$, $\varepsilon = 2.972 \cdot 10^{-2}$. As expected, the two-coordinate approximation yields a result which is (about two times) more precise with respect to ε. After the first iteration we obtain $\lambda_{1(1)} = 1.0754351$, $\varepsilon_1 = 3.0094 \cdot 10^{-3}$. The second and the third iterations yield $\lambda_{1(2)} = 1.0736612$, $\lambda_{1(3)} = 1.073413$. The result of the third iteration has nearly the same accuracy as that of the fourth iteration with the one-coordinate approximation.

2.9. Problems with Periodic Boundary Conditions

2.9.1. Statement of the periodic boundary value problem. In this case, the Sturm–Liouville problem is formulated as follows: *Find the values of the parameter λ for which there exists a nontrivial solution of the differential equation*

$$(p(x)u')' + [\lambda r(x) - q(x)]u = 0, \tag{2.9.1}$$

with the periodicity conditions [13, 24, 42, 68]

$$u(0) = u(1), \qquad u'(0) = u'(1). \tag{2.9.2}$$

Here, $p(x)$, $r(x)$, $q(x)$ are given sufficiently smooth functions of argument x, $0 \le x \le 1$, which may be extended for all $|x| < \infty$ as periodic functions with period 1. Moreover, it is assumed that the functions $p(x)$, $r(x)$ are strictly positive, $p(x) \ge p_0 > 0$, $r(x) \ge r_0 > 0$. To ensure sufficient smoothness of periodic solutions, it is assumed that $p(0) = p(1 \pm 0)$, $r(0) = r(1 \pm 0)$, $q(0) = q(1 \pm 0)$.

Note that problem (2.9.1), (2.9.2) describes parametric vibrations. Such boundary value problems are considered for the equations of Mathieu or Hill and are studied in a huge number of publications (see, for instance, [13, 16, 19, 42, 43, 61, 68]). Computational experience shows that it cannot be claimed that for all p, r, q a solution of problem (2.9.1), (2.9.2) is "unique". By the substitution $z = p(x)u'$, equation (2.9.1) is reduced to normal form which does not contain the derivative of $p(x)$, and we obtain the following boundary value problem with periodic boundary conditions

$$u' = \frac{z}{p(x)}, \quad z' = -[\lambda r(x) - q(x)]u; \qquad u(0) = u(1), \quad z(0) = z(1). \tag{2.9.3}$$

Boundary value problems (2.9.1), (2.9.2) and (2.9.3) are equivalent, provided that the above periodicity conditions hold for the coefficients p, q, r.

2.9.2. Main properties of the periodic problem. Observe that the boundary value problem (2.9.1), (2.9.2) is self-conjugate and the eigenfunctions corresponding to different λ_n are orthogonal with weight $r(x)$ on the interval $0 \le x \le 1$. This can be verified directly.

In contrast to the boundary value problems considered above, with each eigenvalue corresponding to a single eigenfunction (to within a constant coefficient), in the case of periodic boundary conditions (2.9.2), the same eigenvalue may correspond to two linearly independent eigenfunctions. A simple example illustrating this property of eigenvalue degeneration (multiplicity) is the problem

$$u'' + \lambda u = 0; \qquad u(0) = u(1), \quad u'(0) = u'(1);$$

$$\lambda_n = 4\pi^2 n^2, \quad \varphi_n^{(1)} = \cos(2\pi n x), \quad \varphi_n^{(2)} = \sin(2\pi n x).$$

In such cases, one speaks of degenerate (or multiple) eigenvalues. Below, we consider some examples of degenerate eigenvalues of problems with variable coefficients p, q, r; see Section 12.1, and also [61].

An analytical or exact solution of boundary value problems with periodic boundary conditions can be obtained only in rare cases of explicit integrability of the equation. Highly precise numerical approximation of the solutions of such problems encounters great computational difficulties. The theory and the methods of the solution of periodic boundary value problems have been developed to a great extent for the equations describing vibration systems (Floquet–Lyapunov theory, Lyapunov–Poincaré methods, Krylov–Bogolyubov theory). Regular methods of highly precise approximate solution of periodic boundary value problems (2.9.1), (2.9.2) or (2.9.3), in the general case, have not been developed to a desirable extent. Below, we describe the method (algorithm) of accelerated convergence for solving periodic boundary value problems (2.9.1), (2.9.2) or (2.9.3).

The periodic boundary value problem (2.9.1), (2.9.2) admits an equivalent variational formulation: *Find a function realizing the minimum of the functional*

$$\begin{gathered} J[u] = \int_0^1 [p(x)u'^2 + q(x)u^2]\,dx \to \min_u, \\ \Phi[u] = \|u\|^2 = \int_0^1 r(x)u^2\,dx = 1; \quad u(0) = u(1), \quad u'(0) = u'(1), \end{gathered} \tag{2.9.4}$$

with the additional orthogonality conditions (2.8.3).

2.9.3. Construction of upper bounds. A solution of the variational problem (2.9.4) is also a solution of the Sturm–Liouville problem. The variational interpretation of the self-conjugate problem (2.9.1) allows us to use the Rayleigh–Ritz method for finding approximations of the first several eigenvalues and obtain their upper bounds. In particular, the Rayleigh–Ritz method can be effectively used for the construction of an upper bound λ_1^* for the first eigenvalue λ_1. These bounds are used as initial approximations in more precise calculations. For definiteness, consider the problem of finding the first eigenvalue and the first eigenfunction.

The Rayleigh–Ritz principle yields the upper bound

$$\lambda_1 \le \lambda_1^* = \frac{J[\psi]}{\Phi[\psi]}, \qquad \psi(0) = \psi(1), \qquad \psi'(0) = \psi'(1). \tag{2.9.5}$$

Here, $\psi(x)$ is a continuously differentiable test function, which is chosen on the basis of intuitive physical considerations regarding the first eigenfunction $u_1(x)$. The inequality (2.9.5) turns into equality for $\psi(x) = Cu_1(x)$, but in general, we have $\lambda_1^* > \lambda_1$. Computational experience shows that a fairly rough approximation $\psi(x)$ yields a fairly precise bound λ_1^*. As mentioned above, the Rayleigh–Ritz method justified by Krylov yields arbitrarily precise bounds for the eigenvalues, provided that sufficiently many linearly independent periodic test functions are used. Together with highly precise upper bounds, for practical applications it is important to have lower bounds, which are often used for evaluating operational quality of physical and mechanical systems described by the corresponding boundary value problems. As a rule, the construction of lower bounds encounters great difficulties. In this case, too, it is desirable to implement the plan formulated in Section 2.1.

2.9.4. Construction of the comparison system. Let us turn to the realization of the above plan for the boundary value problem (2.9.1), (2.9.2).

Let $\lambda^{(0)}$ be an approximation for the first eigenvalue. In particular, we can take $\lambda^{(0)} = \lambda_1^*$. Using some numerical or analytical method, we construct two linearly independent solutions of equation (2.9.1) or system (2.9.3). These solutions can be obtained from the following two Cauchy problems:

$$\begin{gathered}(p(x)v')' + (\lambda^{(0)}r(x) - q)v = 0,\\ 1)\quad v_1(0) = 0, \qquad v_1'(0) = 1; \qquad 2)\quad v_2(0) = 1, \qquad v_2'(0) = 0.\end{gathered} \tag{2.9.6}$$

Denote by $v_1(x) = V_1(x, \lambda^{(0)})$, $v_2(x) = V_2(x, \lambda^{(0)})$ these solutions and suppose that they have been found in numerical or analytical form. The general solution of equation (2.9.6) can be represented as the linear combination

$$v_0(x) = C_1 V_1(x, \lambda^{(0)}) + C_2 V_2(x, \lambda^{(0)}), \qquad v_0' = \frac{dv_0}{dx}, \tag{2.9.7}$$

where C_1, C_2 are arbitrary constants. Let us require that the functions v_0, v_0' satisfy the conditions of periodicity at some point $x = \xi$ (cf. (2.9.2)),

$$v_0(0) = v_0(\xi), \qquad v_0'(0) = v_0'(\xi). \tag{2.9.8}$$

The abscissa ξ and the constants C_1, C_2 are to be determined. Substituting the expressions (2.9.7) into (2.9.8), we obtain a homogeneous system of linear equations for the unknown constants C_1 and C_2. Since we are seeking a nontrivial solution of this system, its determinant should be equal to zero. This gives us the characteristic equation,

$$\Delta(\xi, \lambda^{(0)}) = \det\begin{pmatrix} V_1(\xi, \lambda^{(0)}) & V_2(\xi, \lambda^{(0)}) - 1 \\ V_1'(\xi, \lambda^{(0)}) - 1 & V_2'(\xi, \lambda^{(0)}) \end{pmatrix} = 0, \tag{2.9.9}$$

from which the point ξ is found. It is clear that for $\lambda^{(0)} = \lambda_1$ (exact eigenvalue), we have $\xi = 1$. Using the Liouville theorem, we can write equation (2.9.9) in concise form,

$$\begin{gathered}\Delta(\xi, \lambda^{(0)}) = V_1'(\xi, \lambda^{(0)}) + V_1'(\xi, \lambda^{(0)}) - 2 = 0,\\ \xi = \xi(\lambda^{(0)}) = \arg_\xi \Delta(\xi, \lambda^{(0)}).\end{gathered} \tag{2.9.10}$$

For $\lambda^{(0)} = \lambda_1$, equation (2.9.10) has the root $\xi = 1$, and therefore, (for $\lambda^{(0)}$ sufficiently close to the exact λ_1) there exists a root ξ arbitrarily close to $\xi = 1$.

2.9.5. Introduction of a small parameter. As a measure of closeness (criterion of closeness) between $\lambda^{(0)}$ and λ_1 we introduce a parameter ε, which is assumed sufficiently small,

$$\begin{gathered}\varepsilon = 1 - \xi, \quad |\varepsilon| \ll 1; \quad \xi(\lambda^0) = \min \arg_\xi \Delta(\xi, \lambda^{(0)}) > 0;\\ |\varepsilon| \le d|\lambda_1 - \lambda^{(0)}|, \quad |\lambda_1 - L^{(0)}| \le D|\varepsilon|, \quad d, D, = \text{const}.\end{gathered} \tag{2.9.11}$$

The root ξ is assumed simple, i.e., $d\xi/d\lambda \neq 0$ for $\lambda = \lambda^{(0)}$, and $\partial\Delta/\partial\xi \neq 0$ for $\xi = \xi(\lambda^{(0)})$. It is easy to show that the points λ for which $d\xi/d\lambda = 0$, $\partial\Delta/\partial\xi = 0$ form a set of zero measure.

Having found ξ, we can represent the solutions $v(x)$ (2.9.7) of the Cauchy problems (2.9.6), (2.9.8) in two equivalent forms

$$\begin{aligned} v_0(x) &= A[V_2'(\xi, \lambda^{(0)})V_1(x, \lambda^{(0)}) + (1 - V_1'(\xi, \lambda^{(0)}))V_2(x, \lambda^{(0)})],\\ v_0(x) &= B[(1 - V_2(\xi, \lambda^{(0)}))V_1(x, \lambda^{(0)}) + V_1(\xi, \lambda^{(0)})V_2(x, \lambda^{(0)})].\end{aligned} \tag{2.9.12}$$

Here, A and B are arbitrary constants which may be taken equal to unity $(A = B = 1)$ or chosen from the condition of normalization with weight $r(x)$ on the interval $0 \le x \le \xi$,

$$\Phi[v_0] = \int_0^{\xi} r(x) v_0^2(x)\, dx = 1.$$

Both functions (2.9.12) satisfy the periodicity condition on the interval $0 \le x \le \xi$. We assume that the functions V_1 and V_2 in (2.9.12) have nonzero coefficients. Thus, the expressions (2.9.12) define a solution of the periodic boundary value problem with period ξ and the first eigenvalue $\lambda^{(0)}$. It is assumed that the coefficients $p = p(x)$, $r = r(x)$, $q = q(x)$ have been periodically extended with period $\xi = \xi(\lambda^{(0)})$.

Note that in actual calculations the value of the parameter ε can be made arbitrarily small by a proper choice of the test function ψ or by the Rayleigh–Ritz process. In practice, the method of accelerated convergence is applicable for $\varepsilon = 0.1 \div 0.01$. Such values of ε can be guaranteed without too many restrictions on $\psi(x)$. In order to construct the algorithm of accelerated convergence, we introduce new variables and renormalize the eigenvalue (cf. (2.3.1)),

$$y = x\xi, \quad 0 \le y \le \xi, \quad U = U(y, \varepsilon) \equiv u(x), \quad \Lambda = \xi^{-2}\lambda. \tag{2.9.13}$$

In these variables, the boundary value problem (2.9.1), (2.9.2) takes the form

$$\begin{gathered}\left(p\left(\frac{y}{\xi}\right)U'\right)' + \left[\Lambda r\left(\frac{y}{\xi}\right) - \frac{1}{\xi^2} q\left(\frac{y}{\xi}\right)\right]U = 0,\\ U(0,\varepsilon) = U(\xi,\varepsilon), \qquad U'(0,\varepsilon) = U'(\xi,\varepsilon).\end{gathered} \tag{2.9.14}$$

2.9.6. Approximate solution of the perturbed problem. As the generating solution we take the solution of the problem with $\varepsilon = 0$. This solution is known: $U_0(y) = v_0(y)$ (2.9.7), $\Lambda_0 = \lambda^{(0)}$ (in particular, $\Lambda = \lambda_1^*$; see (2.9.5)).

Let us seek the perturbed solutions U, Λ in the form of expansions in powers of ε, similar to (2.3.4),

$$U(y,\varepsilon) = U_0(y) + \varepsilon U_1 + \varepsilon^2 U_2 + \cdots, \qquad \Lambda(\varepsilon) = \Lambda^{(0)} + \varepsilon \Lambda_1 + \varepsilon^2 \Lambda_2 + \cdots. \tag{2.9.15}$$

Substituting the series (2.9.15) into (2.9.14), we obtain the following nonhomogeneous problem for the unknown $U_1(y)$, Λ_1 (the function $v_0(y)$ is defined by (2.9.7)):

$$\begin{aligned}(p(y)U_1')' + [\lambda^{(0)} r(y) - q(y)]U_1 &= -\Lambda_1 r(y) U_0(y) - [\lambda^{(0)} y r'(y)'\\ &\quad -2q(y) - yq'(y)]U_0(y) - (yp'(y)U_0'(y))',\\ U_1(0) = U_1(\xi), \qquad U_1'(0) &= U_1'(\xi); \qquad U_0(y) \equiv v_0(y).\end{aligned} \tag{2.9.16}$$

Since the unperturbed problem is self-conjugate, the Fredholm alternative implies that the right-hand side of equation (2.9.16) should be orthogonal to the function $U_0(y)$. The orthogonality condition allows us to find Λ_1 in the form of quadrature, similarly to (2.3.8),

$$\begin{gathered}\Lambda_1 = \frac{1}{\|U_0\|^2}\left\{\int_0^{\xi} [-\lambda^{(0)} y r'(y) + 2q + yq']U_0^2\, dy - \int_0^{\xi} (yp'(y)U_0')' U_0\, dy\right\},\\ \|U_0\|^2 = \int_0^{\xi} r(y) U_0^2\, dy.\end{gathered} \tag{2.9.17}$$

Relation (2.9.17) is a necessary and sufficient condition for the existence of a solution V_1 of the boundary value problem (2.9.16). The function V_1 may be constructed by various methods, in particular, by the method of variation of arbitrary constants. In a similar way, one can find $\varLambda_2$, V_2, etc. However, calculations of further coefficients of the series (2.9.15) are extremely lengthy and ineffective.

For the original eigenvalue λ_1, using (2.9.13), we obtain the expression $\lambda_1 = \lambda^{(1)} + O(\varepsilon^2)$, where

$$\lambda^{(1)} = \xi^2(\lambda^{(0)} + \varepsilon\varLambda_1) = \lambda^{(0)} + \varepsilon(\varLambda_1 - 2\lambda^{(0)}), \tag{2.9.18}$$

the coefficient $\varLambda_1$ is defined by (2.9.17) on the basis of the generating solution $\lambda^{(0)}$, v_0.

As mentioned above, the expression (2.9.17) of $\varLambda_1$ in (2.9.18) can be much simplified by the integration by parts. Finally, for $\lambda^{(1)}$ we obtain the concise expression

$$\begin{gathered}\lambda^{(1)} = \lambda^0 + \varepsilon\mu(\xi,\lambda^0), \qquad \xi = \xi(\lambda^0), \qquad \varepsilon = 1-\xi, \qquad \lambda^{(0)} = \lambda^0, \\ \mu(\xi,\lambda^0) = -\xi\|v_0\|^{-2}[(\lambda^0 r(\xi) - q(\xi))v_0^2(\xi) + p(\xi)v_0'^2(\xi) + p'(\xi)v_0(\xi)v_0'(\xi)],\end{gathered} \tag{2.9.19}$$

where the squared norm with weight $r(x)$ on the interval $0 \le x \le \xi$ is defined in (2.9.17). If the function p is differentiable and p', r, q satisfy the Lipschitz condition, then without the loss of accuracy, we can take $\xi = 1$ in (2.9.19). By the method of regular perturbations, we obtain the following estimates:

$$|\lambda - \lambda^{(1)}| \le C\varepsilon^2, \qquad 0 \le |\varepsilon| \le \varepsilon_0, \tag{2.9.20}$$

which characterize the error and the convergence interval. The constants C and ε_0 can be estimated in terms of the coefficients of system (2.9.1) and their derivatives. In what follows, it is assumed that the abscissa ξ calculated by (2.9.10) satisfies the second inequality in (2.9.20).

Next, we briefly describe an algorithm for the consecutive refinement of the approximations of eigenvalues and eigenfunctions. This algorithm has accelerated convergence and is similar to the Newton method of tangential lines. Let us take the refined value $\lambda^{(1)}$ given by (2.9.19) and insert it into (2.9.6) instead of the initial approximation λ^0. Then, using a numerical or analytical method, we construct the functions $v_{1,2}(x,\lambda^{(1)})$ which represent eigenfunctions on some interval $0 \le x \le \xi_{(1)}$, $\xi_{(1)} > 0$. Then, from equation (2.9.10), we determine the abscissa $\xi_{(1)} = \xi(\lambda^{(1)})$ and the parameter $\varepsilon_{(1)} = 1 - \xi_{(1)}$, according to (2.9.11), and we also construct the eigenfunction $v_{(1)}(x) = v(x,\lambda^{(1)})$ by (2.9.12). Thus, we obtain the estimates

$$\begin{gathered}|\varepsilon_{(1)}| \le d|\lambda - \lambda^{(1)}| \le dC\varepsilon^2, \qquad |v(x,\lambda) - v_{(1)}(x)| \le K|\varepsilon_{(1)}|, \\ |v'(x,\lambda) - v'_{(1)}(x)| \le K|\varepsilon_{(1)}|, \qquad 0 \le x \le \eta_{(1)}, \qquad \eta_{(1)} = \max(1,\xi_{(1)}).\end{gathered} \tag{2.9.21}$$

2.9.7. The method of accelerated convergence. For definiteness, assume that both the exact eigenfunction $v(x,\lambda)$ and its approximation $v(x,\lambda^{(1)})$ are normalized; it is also possible to take $A = 1$ or $B = 1$ in (2.9.12). Using the known quantities $\xi_{(1)}$, $\varepsilon_{(1)}$ and $v_{(1)}(x)$, $v'_{(1)}(x)$, with the help of a formula similar to (2.9.19), we define the next approximation of λ: $\lambda^{(2)} = \lambda^{(1)} + \varepsilon_{(1)}\mu(\xi_{(1)},\lambda^{(1)})$. Here, we can take $\xi_{(1)} = 1$ instead of $\xi_{(1)} = \xi(\lambda^{(1)})$, without losing accuracy with respect to powers of $\varepsilon_{(1)}$, $\varepsilon_{(1)} = O(\varepsilon^2)$. This process can be continued to infinity $(k = 0, 1, 2, \ldots)$,

$$\begin{gathered}(p(x)v')' + [\lambda^{(k)}r(x) - q(x)]v = 0, \qquad v_{(k)}(x) = v(x,\lambda^{(k)}), \\ \varDelta(\xi,\lambda^{(k)}) = v_1'(\xi,\lambda^{(k)}) + v_2(\xi,\lambda^{(k)}) - 2 = 0, \\ \xi_{(k)} = \xi(\lambda^{(k)}) = \min\arg_\xi \varDelta(\xi,\lambda^{(k)}) > 0, \qquad \varepsilon_{(k)} = 1 - \xi_{(k)}, \\ \lambda^{(k+1)} = \lambda^{(k)} + \varepsilon_{(k)}\mu(\xi_{(k)},\lambda^{(k)}) = \lambda^{(k)} + \varepsilon_{(k)}\mu(1,\lambda^{(k)}) + O(\varepsilon_{(k)}^2).\end{gathered} \tag{2.9.22}$$

The functions $v_{1,2}(x, \lambda^{(k)})$, $v_{(k)}(x)$ are defined by (2.9.6), (2.9.12) for $\lambda^0 = \lambda^{(k)}$, $\xi = \xi(\lambda^{(k)})$. Thus, the algorithm of refining the eigenvalue λ and the eigenfunction $u(x)$ reduces to a sequence of operations (usually numerical) of the same type: 1) integration of two Cauchy problems (2.9.6); 2) determination of the root of the equation (2.9.10); 3) refinement of the eigenvalue λ by a formula like (2.9.19); see (2.9.22).

Note that the calculation of the squared norm $\|v_{(k)}\|^2$ can be replaced by the procedure of auxiliary integration of the following equation for the unknown function $w = \partial v/\partial\lambda$:

$$\begin{aligned}
&(p(x)w')' + [\lambda^{(k)} r(x) - q(x)]w = -r(x)v(x, \lambda^{(k)}), \qquad w(0) = w'(0) = 0,\\
&\|v_{(k)}\|^2 \equiv \int_0^{\xi_{(k)}} r(x)v^2(x, \lambda^{(k)})\, dx \\
&\qquad = [v'(\xi_{(k)}, \lambda^{(k)})w(\xi_{(k)}, \lambda^{(k)}) - v(\xi_{(k)}, \lambda^{(k)})w'(\xi_{(k)}, \lambda^{(k)})]p(\xi_{(k)}).
\end{aligned} \tag{2.9.23}$$

As mentioned above, without the loss of precision with respect to powers of $\varepsilon_{(k)}$, on the kth step we can take $\xi_{(k)} = 1$, $k = 0, 1, \ldots$. This simplifies calculations, because in highly precise calculations one has to match step lengths with the length of the interval, $\xi_{(k)}$.

Let us evaluate the error of the recurrent procedure (2.9.22). This error is determined by the value of the parameter $\varepsilon_{(k)}$, i.e.,

$$\begin{gathered}
|\varepsilon_{(k)}| \le dC\varepsilon_{(k-1)}^2 \le (dC)^3\varepsilon_{(k-2)}^4 \le \cdots \le (dC)^{n_k-1}\varepsilon_{(1)}^{n_k} = (dC)^{-1}(\varepsilon dC)^{n_k},\\
n_k = 2^k, \qquad k = 0, 1, 2, \ldots
\end{gathered} \tag{2.9.24}$$

The procedure described above yields fast convergence of the approximations to the exact solution of the periodic boundary value problem, provided that $\delta = |\varepsilon| dC < 1$. In practice, even for p, r, q of a relatively complex structure, one has $\delta = 0.1 \div 0.01$, which allows us to determine eigenvalues and eigenfunctions with great precision after two or three iterations. Four or five iterations practically exhaust the precision capacities of modern software. The approach proposed here can be easily implemented and does not require costly software.

Next, we discuss the problem of finding the initial approximation λ^0. This bound can be refined by the Rayleigh–Ritz method with a small number of test functions. In practice, one often has to deal with functions $p = p(x, \alpha)$, $r = r(x, \alpha)$, $q = q(x, \alpha)$ depending on a parameter α, $\alpha \in A$, and the problem is to construct a solution of the periodic problem depending on α: $\lambda = \lambda(\alpha)$, $u = u(x, \alpha)$. In such a case, it is possible to use an effective algorithm of continuation of the solution with respect to the parameter α, under the assumption that for some $\alpha_0 \in A$, a fairly precise solution is either known or can somehow be constructed, for instance, by the Rayleigh–Ritz method, or the shooting method, or the perturbation method describe above. The highly precise solution $v(x, \alpha_0)$ obtained by the above algorithm is used in what follows as the initial approximation (test function) for the construction of the desired bound $\lambda^0(\alpha)$ for other values of $\alpha \in A$ close to $\alpha_0 \in A$. This means that the highly precise bound $\lambda^0(\alpha)$ is regarded as the initial approximation $\lambda^0(\alpha)$ for $\alpha \in A$ close enough to α_0. The extension parameter can be introduced artificially. For example, let $p = P$, $r = R$, $q = Q$ in equations (2.9.11), where

$$\begin{gathered}
P(x, \alpha) = p_0(x) + \alpha(p(x) - p_0(x)), \quad R(x, \alpha) = r_0(x) + \alpha(r(x) - r_0(x)),\\
Q(x, \alpha) = q_0(x) + \alpha(q(x) - q_0(x)), \quad p_0(x), r_0(x) > 0, \quad 0 \le x, \alpha \le 1.
\end{gathered} \tag{2.9.25}$$

Here, $p_0(x)$, $r_0(x)$, $q_0(x)$ are some approximations of the functions $p(x)$, $r(x)$, $q(x)$ for which a solution can be constructed in a relatively simple manner (in particular, p_0, r_0, q_0 = const; see examples). Taking $\alpha = \alpha_j = j/N$ in (2.9.25), where $j = 0, 1, 2, \ldots, N$ and N is sufficiently large, and using the method of continuation in the parameter α_j with step $\Delta\alpha = 1/N$, we obtain a highly precise solution of the periodic boundary value problem. The implementation of the above approaches requires numerical experiments which can be easily realized on modern personal computers.

In practice, it may happen that for some $\alpha = \alpha_0$ there is simple degeneration, i.e., $\partial\Delta/\partial\xi = 0$ for $\xi = \xi(\lambda^0)$, but $\partial^2\Delta/\partial\xi^2 \neq 0$. For instance, this is the case for a system with constant coefficients (equations of the Mathieu or Hill type). Then, it turns out that the periodicity condition holds for both functions

$$v_{1,2}(0,\lambda^0) = v_{1,2}(\xi,\lambda^0), \qquad v'_{1,2}(0,\lambda^0) = v'_{1,2}(\xi,\lambda^0) \tag{2.9.26}$$

at the same point $\xi(\lambda^0)$. With the variation of the parameter α, the eigenvalue $\lambda(\alpha)$ will split into two eigenvalues corresponding different functions $v_1(x,\alpha)$ and $v_2(x,\alpha)$. In this case, more accurate and careful calculations are needed as α diverges from α_0.

In particular, for p, r, q = const, there are two eigenfunctions $v_1 = C_1 v_1$ and $v = C_2 v_2$, where $v_1 = \nu_0^{-1}\sin\nu_0 x$, $v_2 = \cos\nu_0 x$. By (2.9.10), we have $\Delta(\xi\nu_0) = -\sin^2(\xi\nu_0/2) = 0$. The abscissa $\xi = \xi(\nu_0) = 2\pi/\nu_0$ is the same for both functions. It is assumed that the value ν_0 is close to 2π, i.e., $v_0^2 = (\lambda^0 r - q)/p \approx (2\pi)^2$, which corresponds to $\lambda^0 \approx ((2\pi)^2 p + q)/r$. Expressions (2.9.12) should be transformed as follows:

$$\begin{aligned} v(x,\lambda^0) &= C_1[v_1(x,\lambda^0) + (1 - v_1'(\xi,\lambda^0))(v_2'(\xi,\lambda^0))^{-1} v_2(x,\lambda^0)] \\ &= C_2[(1 - v_2(\xi,\lambda^0))v_1^{-1}(\xi,\lambda^0))v_1^{-1}(\xi,\lambda^0)v_1(x,\lambda^0) + v_2(x,\lambda_0)]. \end{aligned} \tag{2.9.27}$$

If the conditions (2.9.26) hold, then, using (2.9.27), we obtain in the limit as $\xi \to \xi(\lambda^0)$:

$$v(x,\lambda^0) = C_1 v_1(x,\lambda^0), \qquad v(x,\lambda^0) = C_2 v_2(x,\lambda^0), \tag{2.9.28}$$

since $(1 - v_1')$ and $(1 - v_2)$ tend to zero faster than v_2' and v_1 as $\xi \to \xi(\lambda^0)$. For different branches of $\lambda(\alpha)$ one utilizes different expressions of the functions $v(x,\lambda,\alpha)$, according to (2.9.28).

In order to demonstrate the efficiency of the method of perturbations and the algorithm of accelerated convergence, let us consider examples with calculations for some model problems.

2.9.8. Examples. Here, we describe calculations for some model periodic eigenvalue problems containing parameters and examine how the eigenvalues and eigenfunctions depend on these parameters.

Example 1. We start with the problem

$$u'' + \lambda(\theta + x)^{-2} u = 0; \qquad u(0) = u(1), \quad u'(0) = u'(1),$$

where $\theta > 0$ is an arbitrary numerical parameter, $0 \le x \le 1$. This Euler-type equation admits an exact analytical solution

$$u = (\theta + x)^{1/2}(a\sin\varphi + b\cos\varphi), \qquad \varphi = \varphi(\nu,\theta,x) = \nu\ln(\theta + x), \tag{2.9.29}$$

where a, b are integration constants, $\nu = (\lambda - 1/4)^{1/2} \ge 0$, and $\lambda \ge 1/4$ is the unknown to be determined. The characteristic equation (2.9.10) for $\nu = \nu_n(\theta)$ corresponding to (2.9.29) can be reduced to

$$\begin{aligned} &\cos\Delta\varphi = 2\alpha(\alpha^2+1)^{-1} - 1/2\nu^{-1}(\alpha^2 - 1)(\alpha^2+1)^{-1}\sin\Delta\varphi, \\ &\Delta\varphi \equiv \varphi(\nu,\theta,1) - \varphi(\nu,\theta,0) = 2\nu\ln\alpha, \quad \alpha = (1+\theta^{-1})^{1/2} > 1. \end{aligned} \tag{2.9.30}$$

The roots $\nu_n(\theta)$ of equation (2.9.30) are simple and can be found numerically on a computer. By analytic means, one can establish their asymptotic behavior for $\theta \to 0$ ($\alpha \to \infty$) and $\theta \to \infty$ ($\alpha \to 1$):

$$\nu_n \sim \frac{\pi n}{2|\ln \theta|} \to 0, \qquad \theta \to 0; \qquad \nu_n \sim 2\pi n\theta \to \infty, \qquad \theta \to \infty. \tag{2.9.31}$$

The expressions (2.9.31) coincide with the leading terms of the corresponding expansions. With the help of the Weierstrass theorems, subsequent terms of their analytic expansions can be constructed. We see that the behavior of the roots $\nu_n(\theta)$ is fairly simple for all n, θ and does not exhibit singularities.

Example 2. Consider a more edifying problem that can be solved in terms of elementary functions

$$u'' + \lambda[1 + (x-\theta)^2]^{-2}u = 0; \qquad u(0) = u(1), \quad u'(0) = u'(1). \tag{2.9.32}$$

Extending $r(x, \theta) = [1 + (x-\theta)^2]^{-2}$ as a 1-periodic function of x, we obtain a function with jumps of the first kind equal to

$$\Delta r(\theta) = r(1,\theta) - r(0,\theta) = (\theta - 1/2)(4\theta^2 - 4\theta + 3)r(1,\theta)r(0,\theta),$$
$$\Delta r(1/2) = 0, \qquad \Delta r(\theta) \sim |\theta|^{-5}, \qquad r(x,\theta) \sim \theta^{-4}, \qquad |\theta| \to \infty. \tag{2.9.33}$$

From (2.9.33), it follows that the problem is symmetric with respect to $\theta = 1/2$, i.e., it suffices to examine its solution for $-\infty < \theta \le 1/2$ or $1/2 \le \theta < \infty$; see below.

In order to construct the general solution of this equation, we change the argument $x \to y$, the parameter $\lambda \to \nu$, and the function $u(x,\lambda) \to V(y,\nu)$:

$$x - \theta = \tan y, \quad y_0 \le y \le y_1, \quad y_0 = -\arctan\theta, \quad y_1 = \arctan(1-\theta),$$
$$\nu^2 = \lambda + 1, \quad u(x,\lambda,\theta) = \frac{V(y,\nu,\theta)}{\cos y}. \tag{2.9.34}$$

In terms of these new variables and parameters, our periodic boundary problem becomes

$$V'' + \nu^2 V = 0, \qquad y_0 \le y \le y_1, \qquad \frac{V(y_0)}{\cos y_0} = \frac{V(y_1)}{\cos y_1},$$
$$V'(y_0)\cos y_0 + V(y_0)\sin y_0 = V'(y_1)\cos y_1 + V(y_1)\sin y_1. \tag{2.9.35}$$

The equation for V in (2.9.35) can be solved in terms of trigonometric functions: $V = a\sin\nu y + b\cos\nu y$, where a, b are integration constants chosen to ensure the periodicity conditions. As a result, we obtain the corresponding characteristic equation for the eigenvalues $\nu = \nu_n(\theta)$ of problem (2.9.35),

$$\Delta(\nu,\theta) = \nu[(\cos^2 y_1 + \cos^2 y_0)\cos\nu(y_1 - y_0) - 2\cos y_1 \cos y_0]$$
$$+(\cos y_1 \sin y_1 - \cos y_0 \sin y_0)\sin\nu(y_1 - y_0) = 0. \tag{2.9.36}$$

Here, the quantities $y_{0,1} = y_{0,1}(\theta)$ are defined by (2.9.34). The transcendental equation (2.9.36) has countably many roots $\nu_n = \nu_n(\theta)$, $n = 0, 1, 2, \ldots$ for any θ, $|\theta| < \infty$. These roots can be found numerically with suitable software; see Fig. 2.1. Note that there is a root $\nu \equiv 0$ ($\lambda \equiv -1$), which yields the trivial solution $V(y,\nu,\theta) \equiv 0$. It is easy to see that $\nu_0(\theta) \equiv 1$ ($\lambda_0(\theta) \equiv 0$) is also a root of the equation, and this obviously corresponds to the function $V_0 = b\cos y$ ($u_0 \equiv b = \text{const}$). Figure 2.1 shows a numerical solution of equation (2.9.36): for different values of θ we have constructed the first seven eigenvalues $\nu_n(\theta)$, $n = 0, 1, \ldots, 6$, of problem (2.9.35).

Consider some values of θ for which equation (2.9.36) can be substantially simplified. Let $\theta = 1/2$. Then the solution of the problem splits: there are two sets of roots $\{\nu_k^*\}$, $\{\nu_m^{**}\}$ corresponding to eigenvalues $\{\lambda_k^*\}$, $\{\lambda_m^{**}\}$ (see (2.9.34)) and eigenfunctions $\{u_k^*(x)\}$, $\{u_m^{**}(x)\}$:

$$\nu_k^* = \pi k/y^0, \quad y^0 = -\arctan 1/2 \approx -0.4636, \quad \lambda_k^* = \nu_k^{*2} - 1 > 0, \quad k = 1, 2, \dots,$$
$$u_k^*(x) = a_k(1 + (x - 1/2)^2)^{1/2} \sin(\nu_k^* \arctan(x - 1/2));$$
$$\nu_m^{**} = \operatorname{Arg}(\tan \nu y^0 - 1/2\nu^{-1}), \quad m = 0, 1, 2, \dots, \quad \nu_0^{**} = 1, \qquad \lambda_0^{**} = 0, \tag{2.9.37}$$
$$\lambda_m^{**} = \nu_m^{**2} - 1 > 0, \quad m \geq 1; \quad \nu_m^{**} \to \pi m/y^0, \quad m \to \infty,$$
$$u_m^{**}(x) = b_m(1 + (x - 1/2)^2)^{1/2} \cos(\nu_m^{**} \arctan(x - 1/2)).$$

The roots $z_m(\gamma)$ of the equation $\tan z = \gamma z^{-1}$ for different values of γ can be constructed numerically.

Taking $\theta = 0$ (or $\theta = 1$), we obtain the following characteristic equation and approximate values of the roots for $n \gg 1$:

$$\cos\left(\tfrac{1}{4}\pi\nu\right) = \tfrac{2\sqrt{2}}{3} - \tfrac{1}{3}\nu^{-1} \sin\left(\tfrac{1}{4}\pi\nu\right), \quad \nu = \nu_n, \quad \lambda_n = \nu_n^2 - 1 \quad (n = 0, 1, \dots)$$
$$\nu_0 = 1; \qquad \nu_n = \pm 4\pi^{-1} \arccos\left(\tfrac{2\sqrt{2}}{3}\right) + 8n + (6\pi n)^{-1} + O(n^{-2}), \quad n \gg 1. \tag{2.9.38}$$

The eigenfunctions $u_n(x)$ for $\theta = 0, 1$ can be represented in the form

$$u_n(x) = C_n\sqrt{1 + x^2}\,[(1 - \sqrt{2}\cos(\tfrac{1}{4}\pi\nu_n)) \sin(\nu_n \arctan x) + \sqrt{2}\sin(\tfrac{1}{4}\pi\nu_n) \cos(\nu_n \arctan x)]. \tag{2.9.39}$$

Let us go back to equation (2.9.38). Its roots ν_0, ν_1, ..., ν_6 as functions of θ are shown in Fig. 2.1 for $-1.5 \leq \theta \leq 0.5$, and also for $0.5 \leq \theta \leq 2.5$, since our problem is symmetric with respect to $\theta = 1/2$. Note that the function $r(x, \theta)$ for $\theta = -1.5$ or $\theta = 2.5$ varies about 53 times. It follows from (2.9.37) that for $\theta = 1/2$, the values ν_{2k-1} and ν_{2k} become infinitely close to one another as $k \to \infty$, i.e., degeneration is observed (the eigenvalues become multiple; see Fig. 2.1). However, for $\theta \neq 1/2$, these values diverge by a finite distance, in particular, for $\theta = 0$, $\theta = 1$ and $k \to \infty$, we have $\nu_{2k} - \nu_{2k-1} \approx (8/\pi) \arccos(2\sqrt{2}/3) \approx 0.87$. For $|\theta| \to \infty$, it is easy to obtain the following estimate for ν_n: $\nu_n(\theta) \approx 2\pi n \theta^2$, i.e., $\lambda_n(\theta) \approx (2\pi n \theta^2)^2$, $n = 1, 2, \dots$

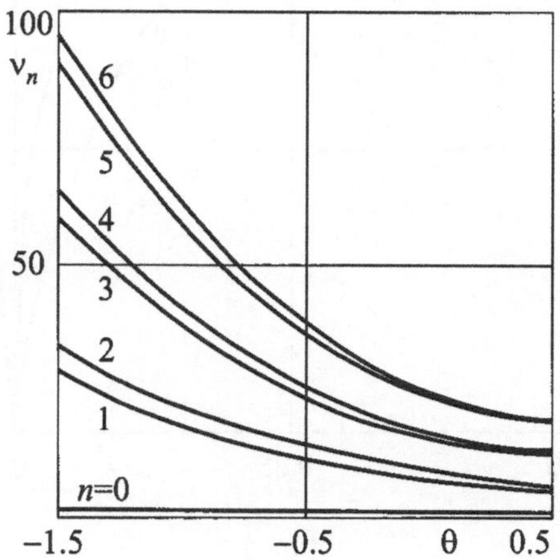

Fig. 2.1

Example 3. In order to demonstrate the efficiency of the accelerated convergence algorithm, we perform calculations for the periodic boundary value problem for a modified Mathieu equation (equation of Hill's type; see Section 12.1):

$$u'' + [\lambda - q(\varphi, \beta)]u = 0, \qquad u(-1) = u(1), \quad u'(-1) = u'(1),$$

$$q(\varphi, \beta) = \beta \cos\varphi(1 + \beta\cos\varphi)^{-1} = \beta\cos\varphi - \beta^2\cos^2\varphi + \beta^3\cos^3\varphi - \cdots, \tag{2.9.40}$$

$$\varphi = 2\pi x, \qquad -1 \le x \le 1.$$

Here, β is a scalar parameter and we assume that $|\beta| < 1$, since for $|\beta| \ge 1$ the function q in (2.9.40) becomes unbounded. For $|\beta| \ll 1$, equation (2.9.40) is close to the Mathieu equation (to within terms of the order $O(\beta^2)$).

In order to find the eigenvalues $\lambda = \lambda_n(\beta)$ and the corresponding 2-periodic eigenfunctions $u = u_n(x, \beta)$, we apply the above algorithm of accelerated convergence (see formulas (2.9.22) from Subsection 2.9.7) and the procedure of continuation in the parameter $\beta \ge 0$ (2.9.25), starting from $\beta = 0$. For $\beta = 0$, a solution of our boundary value problem can be easily constructed. In this case, the value $\lambda_0(0) = 0$ is nondegenerate and corresponds to the eigenfunction $u_0(x, 0) = b = \text{const}$. Subsequent eigenvalues $\lambda_n(0) = (\pi n)^2$, $n \ge 1$, have double multiplicity and each corresponds to two eigenfunctions $u_n^{(1)}(x, 0) = a_n(\pi n)^{-1}\sin\pi nx$ and $u_n^{(2)}(x, 0) = b_n\cos\pi nx$. For $\beta \ne 0$, the eigenvalue $\lambda_n(\beta)$ splits into two branches $\lambda_n = (\lambda_n^{(1)}, \lambda_n^{(2)})$, the first corresponding to odd eigenfunction $u_n^{(1)}(x, \beta)$ and the second to an even eigenfunction $u_n^{(2)}(x, \beta)$.

Figure 2.2 is a graphical representation of our calculation results for the lowest eigenvalues $\lambda_0(\beta)$ (the lower abscissa axis) and $\lambda_1(\beta) = (\lambda_1^{(1)}(\beta), \lambda_1^{(2)}(\beta))$ (the upper abscissa axis) for $0 \le \beta < 1$. It has been shown that for $\beta \ll 1$ these curves are close to the corresponding curves on the Ince–Strutt diagram [61]. With the increase of β, the divergence grows and the behavior of $\lambda_{0,1}(\beta)$ differs considerably from the values obtained for the Mathieu equation. Numerical and analytical investigations show that $\lambda_0, \lambda_1^{(1)} \to -\infty$ as $\beta \uparrow 1$, and this dependence is monotone; see Fig. 2.2.

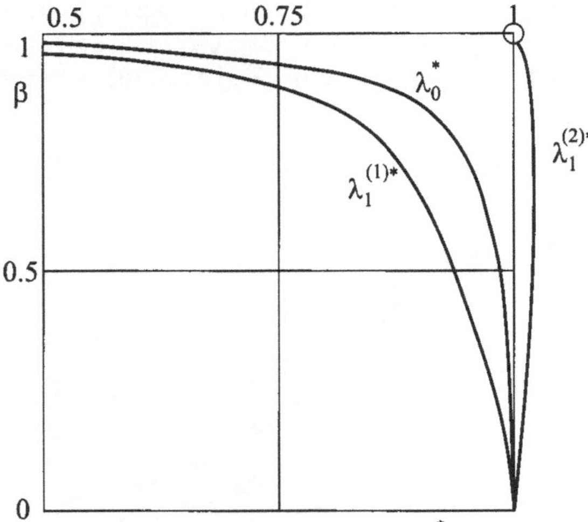

Fig. 2.2

The eigenvalues λ_0, $\lambda_1^{(1)}$ were calculated for $0 \le \beta \le 0.999$, with large variations of the function q, from $|q|_{\min} = 0$ to $|q|_{\max} = \beta(1 - \beta)^{-1} \approx 10^3$.

The investigation of the function $\lambda_1^{(2)}(\beta)$ is more difficult. In contrast to $\lambda_0(\beta)$, this function, just as $\lambda_1^{(1)}$, has no points of contact (is nontangential) with the corresponding vertical line for

$\beta = 0$. The function $\lambda_1^{(2)}(\beta)$ grows slowly for $0 \le \beta \le 0.7$, attains its maximum $\lambda_{1\,\max}^{(2)} \approx 10.040$, and monotonically decreases for $0.7 \le \beta \le 1$. However, $\lambda_1^{(2)}$ decreases much slower than λ_0, $\lambda_1^{(1)}$. Thus, the value $\lambda_1^{(2)} \approx \pi^2 (\lambda_1^{(2)^*} \approx 1)$ is attained only for $0.999 < \beta < 0.9995$. For $\beta_i = 1 - 10^{-i}$ $(i = 1, 2, 3)$ we have obtained the values of $\lambda_1^{(2)}(\beta_i)$ with the error of order 10^{-8}. These values with three significant digits (with no roundoff) are the following: 10.009; 9.918; 9.871.

It should be mentioned that for $\beta \to 1$ $(\beta < 1)$, the problem admits a limit solution in weak sense, $\lambda_1^{(2)} = \pi^2$ (Fig. 2.2, the dot in the right top corner). The corresponding function $u_1^{(2)}(x, \beta)$ converges to $\cos \pi x$ in the norm of the Sobolev space $W_2^{(1)}$ [25]. This can be shown by integrating equation (2.9.40) (the integral is understood in the sense of principal value). The value $\lambda_1^{(2)} = \pi^2$ is an upper bound for the eigenvalue of the corresponding variational problem (2.9.4) for $\beta \to 1 - 0$. This can be proved on the basis of the Rayleigh principle with the test function $\Psi(x) = \cos \pi x$. The integral $J[\Psi]$ in (2.9.4) is calculated by the methods the theory of residues, and one passes to the limit as $\beta \to 1$.

Note that our calculations of eigenvalues and eigenfunctions based on the recurrent procedure of the accelerated convergence method were performed with the error $\varepsilon^{(k)} \sim 10^{-8}$ (the values of the abscissa $\xi^{(k)}$ were obtained with the accuracy $10^{-10} \div 10^{-11}$) by the procedure of continuation in the parameter β. This usually required three or four iterations. The algorithm ensured high efficiency of calculations: precision, stability, and fast speed. Further progress in the numerical-analytical investigation of the modified Mathieu equation (2.9.40) may be important for understanding the behavior of eigenfrequencies in systems with singular points, when it is necessary to determine the boundaries of stability and instability regions for different resonance zones.

2.10. Proof of Convergence of Successive Approximations. Existence Theorem

The reader who is mainly interested in solving applied problems may omit this section, for it is not very important for the basic understanding of the above algorithms of the method of accelerated convergence. For mathematicians, it may be interesting to get acquainted with a new nontraditional proof of the existence and uniqueness of a solution of the Sturm–Liouville problem. This proof is based on the method of successive approximations in terms of powers of a small parameter ε. For simplicity, we consider the case of boundary conditions of the first kind (see Section 2.3).

2.10.1. Transformation of the perturbed boundary value problem. We have described an algorithm for the construction of solutions by the method of perturbations based on expanding the sought solutions in power series (2.3.4). Here we describe a nonstandard version of the method of successive approximations. This will allow us to construct an exact solution theoretically, and also to justify the method of perturbations. The above nonstandard procedure with the introduction of a small parameter ε (see Section 2.2) and the application of the perturbation method for the construction of an approximate solution of the boundary value problem (2.3.1) or (2.3.3) can be continued by taking into account higher powers ε^j, $j \ge 1$. However, the approach described in Section 2.3 is unsuitable for computer-aided construction of higher-order approximations. Here, we use another consideration and transform the original perturbed Sturm–Liouville problem (2.3.1) to

$$u'' + [\mu r(y) - q(y)]u + h(\mu, y, \varepsilon)u = 0,$$

$$u = u(y, \mu, \varepsilon), \quad u = u(0, \mu, \varepsilon) = u(\xi, \mu, \varepsilon) = 0, \quad y \in [0, \xi], \tag{2.10.1}$$

$$\mu = \lambda d^2, \quad d = (1 - \varepsilon)^{-1}, \quad h \equiv \mu[r(dy) - r(y)] + q(y) - d^2 q(dy)$$

(to simplify formulas we have taken $p(x) \equiv 1$, which causes no loss of generality; see Subsection 1.1.1). Assuming that the functions $r(x)$, $q(x)$ are smooth on the interval $x \in [0, 1]$, we find that $h = O(\varepsilon)$ for $y \in [0, \xi]$, $\mu \sim \lambda_1^* \sim 1$. Since system (2.10.1) is linear in u and in μ (bilinear with respect to u, μ jointly), it suffices to assume in what follows that the functions r, q are Lipschitz continuous in x. Taking $\varepsilon = 0$, we obtain an unperturbed problem of the type (2.3.5), whose solution (i.e., the first eigenvalue $\mu_0 = \lambda_1^*$ and the eigenfunction $u_0(y, \mu_0) = v_1(y, \lambda_1^*)$) is assumed known.

Direct solution of the boundary value problem (2.3.1) by the methods of the perturbation theory is rather cumbersome and requires much effort, because one has to fulfill the boundary condition for $y = \xi$. To overcome this difficulty, we use the relation

$$\mu \int_0^\xi u^2 r(y)\, dy = \int_0^\xi \{u'^2 + [q(y) - h(\mu, y, \varepsilon)]u^2\}\, dy, \tag{2.10.2}$$

which follows from (2.10.1), and drop the said boundary condition (2.3.1), replacing it by an arbitrary nonzero condition for the derivative u' at $y = 0$, say, $u'(0, \mu, \varepsilon) = 1$. This replacement is admissible, since $u'(0, \mu, \varepsilon) \neq 0$ and the functions u are defined to within a nonzero coefficient.

Thus, instead of the perturbed boundary value problem (2.3.1), we consider the Cauchy problem

$$\begin{gathered} u'' + [\mu r(y) - q(y)]u + h(\mu, y, \varepsilon)u = 0, \\ u(0, \mu, \varepsilon) = 0, \qquad u'(0, \mu, \varepsilon) = 1, \end{gathered} \tag{2.10.3}$$

which, together with (2.10.2), simplifies our task of finding the first eigenvalue $\mu = \mu_1(\varepsilon)$ and the corresponding eigenfunction $u = u_1(y, \varepsilon)$. Let us consider the standard procedure of successive approximations for the problem (2.10.2), (2.10.3). To that end, we change the variables and the parameters as follows:

$$\mu = \mu_0 + \nu, \quad \mu_0 = \lambda_1^*; \quad u = u_0 + z, \quad u_0(y, \mu_0) = v_1(y, \lambda_1^*). \tag{2.10.4}$$

Substituting these quantities into (2.10.2), we obtain a quasilinear functional equation of the form

$$\begin{gathered} \nu = \nu_1^* + N(\nu, I[z], I[z'], \varepsilon), \qquad \nu_1^* = \nu_1^*(\varepsilon) = -\|u_0\|^{-2} I[h_0 u_0^2], \\ I[\varphi] \equiv \int_0^\xi \varphi(y)\, dy, \quad \|u_0\|^2 = I[u_0^2 r], \quad N = I[\Psi(\nu, z, z', y, \varepsilon)], \\ \Psi \equiv z'^2 + (q - \mu_0 r)z^2 - (h_0 + \nu r)(2u_0 z + z^2) - \nu h_\nu(u_0 + z), \\ h_0 \equiv h(\mu_0, y, \varepsilon), \quad h_\nu \equiv r(dy) - r(y), \quad h_\nu \sim \varepsilon. \end{gathered} \tag{2.10.5}$$

For the sake of brevity, we will not indicate the dependence of u_0 on μ_0, since $\mu_0 = \lambda_1^*$ is known by assumption as the first eigenvalue of the unperturbed problem.

In a similar way we transform the Cauchy problem (2.10.3) for the perturbed function z introduced by (2.10.4). We get

$$\begin{gathered} z'' + [\mu_0 r(y) - q(y)]z = -(\nu_1^* r + h_0)u_0 - F(\nu, z, I[z], I[z'], y, \varepsilon), \\ F \equiv (\nu r + h)z - \nu(h_\nu + N u_0 r); \qquad z(0, \nu, \varepsilon) = 0, \qquad z'(0, \nu, \varepsilon) = 0. \end{gathered} \tag{2.10.6}$$

For $\varepsilon = 0$, problem (2.10.5), (2.10.6) admits the trivial solution. The quantities N, F are of the second order of smallness with respect to the small quantities ε, ν, ν_1, z, z'.

2.10.2. Proof of convergence of successive approximations. We seek a solution $\nu = \nu(\varepsilon)$, $z = z(y, \varepsilon)$ in the form of the following successive approximations:

$$\begin{aligned}
&\nu_{j+1} = \nu_1^* + N(\nu_j, I[z_j], I[z_j'], \varepsilon), \qquad j \geq 1; \qquad \nu_1 = \nu_1^*(\varepsilon) \sim \varepsilon, \\
&z_{j+1} = z_1^*(y, \varepsilon) + L(\nu_j, I[z_j], I[z_j'], y, \varepsilon), \qquad z_{j+1}' = \frac{dz_{j+1}}{dy}, \\
&z_1^*(y, \varepsilon) \equiv -\int_0^y W(y, s)[\nu_1^* r(s) + h(\mu_0, s, \varepsilon)]u_0(s)ds; \qquad z_1^*, z_1'^* \sim \varepsilon, \\
&L_j \equiv -\int_0^y W(y, s)F(\nu_j, z_j(s, \varepsilon), I[z_j], I[z_j'], s, \varepsilon)ds; \qquad L_j, L_j' \sim \varepsilon^2.
\end{aligned} \tag{2.10.7}$$

Here, $W(y, s)$ is an impulse transition function of the unperturbed equation (2.10.6). This function is known, because we can construct the general solution z_0 of the homogeneous equation in terms of the Lyapunov function,

$$\begin{aligned}
&z_0 = C_1 u_0(y) + C_2 w_0(y), \qquad w_0 \equiv u_0(y) \int_0^y \frac{ds}{u_0^2(s)}, \\
&W(y, s) = -u_0(y)w_0(s) + w_0(y)u_0(s), \qquad W(s, s) = 0, W_y'(s, s) = 1.
\end{aligned} \tag{2.10.8}$$

The function $W(y, s)$ defined by (2.10.8) is uniformly bounded for all y, s, $(\xi \geq y \geq s \geq 0)$.

Theorem 2.2. *Successive approximations ν_j, z_j, z_j', defined by* (2.10.7) *are uniformly convergent for sufficiently small ε, $\varepsilon \in [0, \varepsilon_0]$, as $j \to \infty$. Their limit is the unique exact solution $\nu^*(\varepsilon)$, $z^*(y, \varepsilon)$, $z^{*\prime}(y, \varepsilon)$ of problem* (2.10.5), (2.10.6), *which is identically equal to zero for $\varepsilon = 0$.*

This statement will be proved in several steps.

First, it can be shown by induction that for sufficiently small $\varepsilon > 0$, successive approximations (2.10.7) are uniformly bounded independently of $j = 1, 2, ..., k, k+1, ...$, and the following estimates hold:

$$|\nu_j| \leq \varepsilon C_\nu, \qquad |z_j| \leq \varepsilon C_z, \qquad |z_j'| \leq \varepsilon C_{z'}, \qquad j \geq 1. \tag{2.10.9}$$

The constants C_ν, C_z, $C_{z'}$ can be constructively estimated in terms of the bounds of the functions W, r, q, h, h_ν.

On the second step, using the inequalities (2.10.9), we show that the operators in (2.10.7) are contractions. By the Banach theorem about contracting operators, there exists a fixed point (see [17, 66]). Thus, we have shown that there exist the limits $\nu^*(\varepsilon)$, $z^*(y, \varepsilon)$, $z'^*(y, \varepsilon)$ for sufficiently small $\varepsilon > 0$. These limits are bounded, in view of (2.10.9), and identically vanish for $\varepsilon = 0$.

The parameter ε_0, which determines the region of convergence in ε, can be evaluated in a constructive manner as follows. For the differences

$$\Delta\nu_{j+1} = |\nu_{j+1} - \nu_j|, \qquad \Delta z_{j+1} = \max_{0 \leq y \leq \xi} |z_{j+1} - z_j|, \qquad \Delta z_{j+1}' = \max_{0 \leq y \leq \xi} |z_{j+1}' - z_j'|,$$

using (2.10.7) and (2.10.9), we obtain the estimates

$$\Delta g_{j+1} \leq \varepsilon(\alpha_g \Delta\nu_j + \beta_g \Delta z_j + \gamma_g \Delta z_j'), \qquad g = \nu, z, z'. \tag{2.10.10}$$

The constants (α, β, γ) in (2.10.10) can be effectively expressed in terms of the right-hand sides in (2.10.7), similarly to C_g in (2.10.9). Summing up the inequalities (2.10.10), we obtain the estimate

$$\begin{aligned}
&\Delta_{j+1} \leq \varepsilon C \Delta_j; \qquad \Delta_j \equiv \Delta\nu_j + \Delta z_j + \Delta z_j', \qquad C = \max(A, B, \Gamma), \\
&A = \alpha_\nu + \alpha_z + \alpha_{z'}, \qquad B = \beta_\nu + \beta_z + \beta_{z'}, \qquad \Gamma = \gamma_\nu + \gamma_z + \gamma_{z'}.
\end{aligned} \tag{2.10.11}$$

From (2.10.11), we obtain the basic condition of the theorem about contracting operators,

$$\varepsilon_0 C = \theta < 1, \qquad \varepsilon_0 = \theta C^{-1}, \qquad \varepsilon \in [0, \varepsilon_0]. \tag{2.10.12}$$

The convergence $\nu_j \to \nu^*$, $z_j \to z_j^*$, $z_j' \to z'^*$ under the condition (2.10.12) follows from the absolute and uniform convergence of partial sums $\nu_{j+1} = \nu_1 + (\nu_2 - \nu_1) + \cdots + (\nu_{j+1} - \nu_j)$ as $j \to \infty$, as well as similar partial sums for z_{j+1}, z_{j+1}'. Since the estimates (2.10.11) hold for the terms of these partial sums, the corresponding series are convergent, because their terms are majorized by terms of a geometric progression with ratio $\theta < 1$.

On the third step, we prove that the limit functions $\nu^*(\varepsilon)$, $z^*(y,\varepsilon)$, $z'^*(y,\varepsilon)$ satisfy equations (2.10.5), (2.10.6). To that end, using the transition function $W(y,s)$, we represent the Cauchy problem (2.10.6) in the form of an integral equation with the operator L (see (2.10.7)). Then these functions are inserted into N, F, L, and we obtain the corresponding values of ν, z, z'. Using (2.10.7), we calculate the quantities $\max \Delta g_{j+1}$. These satisfy inequalities of the type (2.10.10), with Δg_i in the right-hand side replaced by the differences Δg_j^*. As shown above, these differences tend to zero as $j \to \infty$, and therefore,

$$\nu = \nu^*(\varepsilon), \quad z = z^*(y,\varepsilon), \quad z' = z'^*(y,\varepsilon), \quad y \in [0,\xi], \quad \varepsilon \in [0,\varepsilon_0] \tag{2.10.13}$$

is a solution of problem (2.10.5), (2.10.6). In view of (2.10.4), the functions

$$\begin{gathered}\mu = \mu_0 + \nu^*(\varepsilon), \quad u = u_0(y) + z^*(y,\varepsilon), \quad u' = \frac{du}{dy}, \\ \lambda = d^{-2}[\mu_0 + \nu^*(\varepsilon)], \quad u = u_0(d^{-1}x) + z^*(d^{-1}x,\varepsilon), \quad u_x' = d^{-1}\frac{du}{dy}\end{gathered} \tag{2.10.14}$$

yield solutions of the Sturm–Liouville problems (2.10.1) and (2.1.1), respectively.

The fourth step consists of establishing the uniqueness of the solution (2.10.13), as well as the solution (2.10.14), for sufficiently small $\varepsilon > 0$ (see (2.10.12)). Assume the contrary, i.e., that problem (2.10.5), (2.10.6) admits another solution ν_*, z_*, z_*' satisfying the inequalities (2.10.9). Then, for the differences $\Delta\nu = \nu^* - \nu_*$, $\Delta z = |z^* - z_*|_{\max}$, $\Delta z' = |z'^* - z_*'|_{\max}$ we obtain inequalities of the type (2.10.10) with the same quantities $\Delta\nu$, Δz, $\Delta z'$ in the right-hand sides. It follows that $\Delta\nu = 0$, $\Delta z = 0$, $\Delta z' = 0$, i.e., $\nu^* \equiv \nu_*$, $z* \equiv z_*$, $z'^* \equiv z_*'$. Thus, the uniqueness of the solutions (2.10.13) (2.10.14) is proved.

Note that the desired solution $\nu(\varepsilon)$, $z(y,\varepsilon)$, $z'(y,\varepsilon)$ can be constructed by the procedure of Section 2.3 based on expansions in powers of the small parameter ε. This construction utilizes the fact that the expressions (2.10.2), (2.10.3) analytically depend on μ, u, u' or that the relations (2.10.5), (2.10.6) analytically depend on ν, z, z'. No analyticity of r, q in y, ε is required. The existence and the uniqueness of a solution, the convergence of the series, and an estimate for the convergence radius are proved by the method of the Cauchy majorizing functions.

There seem to be no difficulties of principal nature in extending the approach of Section 2.10 to the construction of subsequent eigenvalues λ_n and eigenfunctions $u_n(x)$, $n \geq 2$, by the Rayleigh–Ritz method.

Thus, the proposed nonstandard technique of the introduction of a small parameter ε allows us to refine approximations of eigenvalues and eigenfunctions. The refinement in this case is based on taking into account powers of ε^j. Formulas (2.3.15) can be used for developing a numerical-analytical procedure that ensures accelerated convergence (see Section 2.4).

2.11. Proof of Quadratic Convergence

The reader interested mainly in applied problems may omit this section with no loss to the understanding of our further exposition [6].

Consider again the third boundary value problem (2.8.1). Integrating the Cauchy problem (2.9.5) and finding the root $x = \xi_1$ of the equation $E(x, \lambda_1^*) = 0$ nearest to $x = 1$, we introduce the parameter $\varepsilon = 1 - \xi$. On the next step, we construct a refined eigenvalue λ_1, the root $x = \xi_1$ of the equation $E(x, \lambda_{1(1)}) = 0$ nearest to $x = 1$ and calculate the parameter ε_1. Since the root ξ_1 is simple, the following estimates hold for the parameter ε_1 and the solutions U_1, U_1':

$$|\varepsilon_1| \le d|\lambda_1 - \lambda_{1(1)}| \le dC\varepsilon^2, \quad |U - U_{(1)}| \le K|\varepsilon_1|, \quad |U' - U'_{(1)}| \le K|\varepsilon_1|. \tag{2.11.1}$$

Next, we use formulas of the type (2.8.16) for refining the value of λ,

$$\lambda_{1(2)} = \lambda_{1(1)} + \varepsilon_1 \mu(\lambda_{1(1)}); \qquad |\lambda_1 - \lambda_{1(2)}| \le C\varepsilon_1^2 \le d^2C^3\varepsilon^4. \tag{2.11.2}$$

Then the process of refining ξ, ε, U, U' and λ_1 should be repeated. As a result, we obtain the following recurrent relations:

$$\begin{gathered} \lambda_{1(k+1)} = \lambda_{1(k)} + \varepsilon_k \mu(\lambda_{1(k)}), \qquad |\lambda_1 - \lambda_{1(k+1)}| \le C\varepsilon_k^2, \\ \varepsilon_k = 1 - \xi_k, \qquad \xi_k = \min \arg_x E(x, \lambda_{1(k)}) > 0; \\ |\varepsilon_k| \le d|\lambda_1 - \lambda_{1(k)}| \le dC\varepsilon_{k-1}^2; \qquad |U - U_{(k)}| \le K|\varepsilon_k|; \\ |U' - U'_{(k)}| \le K|\varepsilon_k|. \end{gathered} \tag{2.11.3}$$

From (2.11.3) it follows that the iteration convergence rate is rather high and has quadratic order with respect to ε (as in the Newton method):

$$|\varepsilon_k| \le |dC|^{-1}(\varepsilon dC)^{n_k}, \qquad n_k = 2^k, \quad k = 0, 1, 2, \dots \tag{2.11.4}$$

The inequality (2.11.4) shows that few iterations (usually two or three) are needed for obtaining highly precise approximations of λ_1 and U_1. The recurrent refinement procedure (2.11.3) can be used for finding subsequent eigenvalues λ_n and eigenfunctions $U_n(x)$, $n \ge 2$. This procedure deserves attention, because it is easy to implement, spares resources, gives highly precise results, and can be realized on a PC.

2.12. The Method of Hyperaccelerated Convergence

2.12.1. Third-order refinement procedure. Consider the Sturm–Liouville problem (2.8.1)

$$\begin{gathered} (p(x)u')' + [\lambda r(x) - q(x)]u = 0; \\ \alpha_0 p(0)u'(0) - \beta_0 u(0) = 0; \qquad \alpha_0, \beta_0 \ge 0; \qquad \alpha_0 + \beta_0 = 1; \\ \alpha_1 p(1)u'(1) + \beta_1 u(1) = 0; \qquad \alpha_1, \beta_1 \ge 0; \qquad \alpha_1 + \beta_1 = 1. \end{gathered} \tag{2.12.1}$$

Arguing as in Section 2.8, we can continue the calculation of the coefficients of series (2.8.9) up to the terms with ε^2. Thus, we come to an algorithm of hyperaccelerated convergence. Omitting simple but rather lengthy calculations, we will only give the formulas and describe the operations that constitute the algorithm (method) of hyperaccelerated convergence. The procedure consists of the following steps.

1. Solve the Cauchy problem (2.8.5).
2. Find the root $\xi(\lambda_1^*)$ of the function $E(x, \lambda_1^*)$ (2.8.6) nearest to $x = 1$, and introduce the parameter $\varepsilon = 1 - \xi$.
3. For the same value $\lambda = \lambda_1^*$, solve (numerically) another Cauchy problem,

$$(p(x)w')' + [\lambda_1^* r(x) - q(x)]w = -r(x)v,$$
$$w(0) = \frac{\alpha_0 p(0)}{p(0)+\beta_0}; \qquad w'(0) = \frac{\beta_0}{p(0)+\beta_0}. \tag{2.12.2}$$

4. Compute the corrections

$$\mu_1 = -[\lambda_1^* r(\xi) - q(\xi)]\frac{u^2(\xi)}{\|u\|^2} - p(\xi)\frac{u'^2(\xi)}{\|u\|^2}, \qquad \|u\|^2 = \int_0^\xi r(x)u^2(x)\,dx, \tag{2.12.3}$$

$$\mu_2 = -2\mu_1^2\frac{u^2(\xi)}{\|u\|^2} + [\lambda_1^* r'(\xi) - q'(\xi)]\frac{u'^2(\xi)}{\|u\|^2} + 2\mu_1 r(\xi)\frac{u^2(\xi)}{\|u\|^2} - p'(\xi)\frac{u'^2(\xi)}{\|u\|^2}$$
$$+[2(\lambda_1^* r(\xi) - q(\xi))u(\xi)w(\xi) + p(\xi)u'(\xi)w'(\xi)]\frac{1}{\|u\|^2}. \tag{2.12.4}$$

One can take $\xi = 1$ in (2.12.4).

After calculating the corrections (2.12.3) and (2.12.4), one finds a refined approximation of the first eigenvalue with the error $O(\varepsilon^3)$,

$$\lambda_{1(1)} = \lambda_1^* + \varepsilon\mu_1 + \tfrac{1}{2}\varepsilon^2\mu_2, \qquad \varepsilon = 1 - \xi, \qquad \xi = \xi(\lambda_1^*). \tag{2.12.5}$$

The refined $\lambda_{1(1)}$ is inserted into (2.8.5), (2.12.2) instead of λ_1^*, and the whole process is repeated.

From (2.12.5), it follows that the convergence rate is ε^{n_k}, $n_k = 3^k$. This convergence is much faster than that in the previous sections, but the calculations are more lengthy. For this reason, the algorithm of hyperaccelerated convergence is more difficult to realize than the method of accelerated convergence of Newton type.

2.12.2. An application of the method of hyperaccelerated convergence. As an example, consider the problem of finding the first eigenvalue of the Sturm–Liouville problem for an equation that admits an explicit solution,

$$u'' + \lambda(1 + x^2)^{-2}u = 0, \qquad u(0) = u(1) = 0. \tag{2.12.6}$$

Let us take $\lambda_1^* = 16$ as the initial approximation of the exact eigenvalue $\lambda_1 = 15$. After the first iteration, the method of hyperaccelerated convergence gives us the value $\lambda_{1(1)} = 15.01941$, which determines the first eigenvalue with the relative error $\Delta\lambda/\lambda = 1.2 \cdot 10^{-3}$. If the algorithm of accelerated convergence is used with the same initial approximation $\lambda_1^* = 16$, we obtain the value $\lambda_{1(1)} = 15.0656$, which determines the first eigenvalue with the error $\Delta\lambda/\lambda = 4.25 \cdot 10^{-3}$. The method of hyperaccelerated convergence with one more iteration yields a practically precise result.

Computational experience shows that the method of accelerated convergence is more convenient for implementation. However, the algorithm of hyperaccelerated convergence may also be useful in certain situations.

2.13. Taking into Account Explicit Dependence of Boundary Conditions on Eigenvalues

Here, we describe a modification of the accelerated convergence method which can be used in the case of boundary conditions containing the spectral parameter.

Consider the Sturm–Liouville problem

$$\begin{aligned} &(p(x)u')' + (\lambda r(x) - q(x))u = 0, \\ E_0 &= (k_0 - m_0\lambda)u(0) - p(0)u'(0) = 0, \qquad m_0 \geq 0, \\ E_1 &= (k_1 - m_1\lambda)u(1) + p(1)u'(1) = 0, \qquad m_1 \geq 0. \end{aligned} \tag{2.13.1}$$

For mechanical vibration systems, the coefficients m_0, m_1 have the meaning of inertial concentrated loads.

For $m_{0,1} = 0$, boundary conditions (2.13.1) turn into the usual conditions of the third kind. The formulas of the accelerated convergence algorithm in the case of boundary conditions (2.13.1) can be obtained in the same manner as above. The accelerated convergence algorithm consists of the following operations.

1. For the initial approximation (bound) λ_1^* of the first eigenvalue λ_1, solve the Cauchy problem

$$(p(x)U')' + [\lambda_1^* r(x) - q(x)]U = 0, \qquad U_0(0) = A, \qquad U_0'(0) = B, \tag{2.13.2}$$

where A and B are related by the first boundary condition in (2.13.1),

$$(k_0 - m_0\lambda_1^*)A - p(0)B = 0. \tag{2.13.3}$$

Either A or B may be taken equal to unity.

2. Calculate the function

$$E_1 = (k_1 - m_1\lambda_1^*)U(x) + p(\xi)U'(x). \tag{2.13.4}$$

Find the root $x = \xi$ of the equation $E(x, \lambda_1^*) = 0$ closest to $x = 1$ and calculate the small parameter $\varepsilon = 1 - \xi$.

3. Calculate the refined eigenvalue

$$\lambda_{1(1)} = \lambda_1^* - \varepsilon \frac{p(\xi)U'^2(\xi) + (\lambda_1^* r(\xi) - q(\xi))U^2(\xi)}{\|U\|^2 + m_0 U^2(0) + m_1 U^2(\xi)}. \tag{2.13.5}$$

4. Insert $\lambda_{1(1)}$ into the Cauchy problem (2.13.2), find its solution, and repeat steps 1 to 3 of the algorithm.

5. In order to find subsequent eigenvalues λ_2, λ_3, ... and eigenfunctions u_2, u_3, ..., one has to construct the initial bounds (approximation) λ_2^*, λ_3^*. The refinement procedures are similar. In this case, one obtains quadratic convergence.

2.14. Exercises

Here we give a number of examples which may help the reader to acquire some skill in applying the method of accelerated convergence described in Chapter 2. For numerical calculations, one may use a PC with standard software for solving the Cauchy problem (usually, it suffices to use the fourth-order Runge–Kutta method).

Exercise 1. Find the first eigenvalue λ_1 with the relative error $\leq 3 \cdot 10^{-5}$ for the Sturm–Liouville problem

$$u'' + \lambda(4 - x^2)^{-2}u = 0, \qquad u(0) = u(1) = 0. \tag{2.14.1}$$

The exact eigenvalues of this problem can be expressed analytically:

$$\lambda_n = 4 + \frac{16\pi^2}{\ln^2 3} n^2, \qquad n = 1, 2, \ldots$$

In particular, we can take $\lambda_1 = 134.837074$ as the first eigenvalue. The problem is to calculate λ_1 with the said relative error, using the initial approximation $\lambda_1^* = 135.5$. After integrating the Cauchy problem, construct (on computer display) the graph of the eigenfunction $u_1(x)$ corresponding to the eigenvalue λ_1.

Exercise 2. Find the first eigenvalue λ_1 with the relative error 10^{-5} for the Sturm–Liouville problem with the third boundary condition,

$$u'' + \lambda(1 + x^2)^{-2} u = 0; \qquad u(0) = 0, \quad u(1) + u'(1) = 0. \tag{2.14.2}$$

The exact first eigenvalue is $\lambda_1 = 8.00000$. Using the test function $\psi_1(x) = \sin \eta_1 x$ with $\eta_1 = 2.28893$, obtain an upper bound λ_1^*. Then, using the accelerated convergence methods, calculate λ_1 with the relative error specified above. After integrating the Cauchy problem, construct (on computer display) a graphical representation of the eigenfunction $u_1(x)$ corresponding to the eigenvalue λ_1. By direct calculations, check that $|u_1'(1)| \le 10^{-5} \div 10^{-6}$.

Exercise 3. Find the eigenvalues λ_1, λ_2, λ_3 with the relative error $10^{-4} \div 10^{-5}$ for the Sturm–Liouville problem

$$u'' + \lambda(1 + x)^{-2} u = 0, \qquad u(0) = u(1) = 0. \tag{2.14.3}$$

Use $\psi_1 = \sin(\pi x)$, $\psi_2 = \sin(2\pi x)$, $\psi_3 = \sin(3\pi x)$ as test functions. By the accelerated convergence method, find λ_1, λ_2, λ_3 with the said relative error. After integrating the Cauchy problem, construct (on computer display) graphs of the eigenfunctions $u_1(x)$, $u_2(x)$, $u_3(x)$. In order to test the accuracy of these approximations, use the following analytical (exact) expressions of eigenvalues and eigenfunctions:

$$\lambda_n = \frac{1}{4} + \left(\frac{\pi n}{\ln 2}\right)^2, \qquad u_n(x) = \sqrt{1 + x} \sin\left(\frac{\pi n}{\ln 2} \ln(1 + x)\right).$$

Chapter 3

Approximate Analytical Solution of Perturbed Eigenvalue Problems

Regular perturbation methods are very useful for evaluating the effect of various kinds of perturbations. However, the application of these methods is strongly limited by our capacity to construct and analyze solutions of the unperturbed problem, the so-called generating problem, and require numerical analysis (see Sections 1.3, 2.3, 2.10). One of the simplest classes of perturbed problems is that of systems with slowly varying parameters which are close to constants. For such problems it is possible to obtain fairly complete theoretical results and relatively concise formulas, often with explicit analytical expressions.

3.1. Statement and Analysis of the Perturbed Sturm–Liouville Problem

3.1.1. Properties of the perturbed spectrum. Consider the eigenvalue problem for the following perturbed second-order equation (it is convenient to introduce the parameter λ^2) [14]

$$((1+\varepsilon\sigma(x))X')' + \lambda^2(1+\varepsilon\delta(x))X = 0, \tag{3.1.1}$$

$$X(0) = X(1) = 0, \quad x \in [0,1], \quad \varepsilon \in [0,\varepsilon_0], \quad \lambda^2 > 0.$$

Here, ε is a real parameter, σ, δ are given functions of class $C^{(1)}$, the prime indicates the derivative in the scalar argument x. It is known [22, 27, 33, 46] that for small enough $\varepsilon_0 > 0$ such that $1+\varepsilon\sigma \geq \zeta > 0$, $1+\varepsilon\delta \geq \zeta > 0$, there exists a system of eigenvalues $\{\lambda_n(\varepsilon)\}$ and eigenfunctions $\{X_n(x,\varepsilon)\}$, $n = 1, 2, \ldots$ which form an orthonormal system in L^2 with the weight $\mu(x,\varepsilon) = 1+\varepsilon\delta(x)$, and these functions form a basis in L^2.

It is required to construct approximations for these eigenvalues and eigenfunctions. The approximations should have given accuracy with respect to ε, and the corresponding error estimates should be uniform with respect to n [14].

Problem (3.1.1) for $\varepsilon = 0$ can be solved explicitly,

$$\lambda_n^{(0)} = \pi n, \quad X_n^{(0)}(x) = \sqrt{2}\sin\lambda_n^{(0)}x, \quad n = 1, 2, \ldots \tag{3.1.2}$$

Formal application of the perturbation method [27, 46] to problem (3.1.1) with $\varepsilon > 0$ leads to considerable complications connected with the appearance of "secular" terms of the form $O(\varepsilon^p n^r)$, where p is the highest power exponent in the expansion. Indeed, in the approximation of the first order in ε, we obtain the expressions

$$
\begin{aligned}
\lambda_n^{(1)} &= \lambda_n^{(0)} + \varepsilon\lambda_{n,1}, \quad X_n^{(1)} = X_n^{(0)} + \varepsilon X_{n,1}, \\
\lambda_{n,1} &= \lambda_n^{(0)} \int_0^1 [-\delta(x)\sin^2(\pi n x) + \sigma(x)\cos^2 \pi x]\, dx \sim n, \\
X_{n,1}(x) &= a_{n,1}X_n^{(0)}(x) + \lambda_n^{(0)}\varphi_{n,1}(x) \sim n, \\
a_{n,1} &= -\frac{1}{2}\int_0^1 (\delta X_n^{(0)2} + \lambda_n^{(0)}X_n^{(0)}\varphi_{n,1})\, dx, \\
\varphi_{n,1}(x) &= \frac{1}{\lambda_n^{(0)2}}\int_0^x [\sigma(s)X_n^{(0)\prime}(s) + 2\lambda_n^{(0)}\lambda_{n,1}X_n^{(0)}(s) \\
&\quad + \lambda_n^{(0)^2}\delta(s)X_n^{(0)}(s)]\sin[\lambda_n^{(0)}(x-s)]\, ds,
\end{aligned}
\tag{3.1.3}
$$

where $\lambda_n^{(0)}$, $X_n^{(0)}(x)$ are defined by (3.1.2).

In similarity to (3.1.3), higher-order expansions yield approximations with the estimates $\lambda_{n,p} \sim n^p$, $X_{n,p} \sim n^p$. Therefore, for fixed ε and p, the sought quantities $\lambda_n(\varepsilon)$, $X_n(x,\varepsilon)$ with $n > [\varepsilon^{-1}]$ may greatly differ from the actual values, and the expansions will be divergent. Thus, for a given sufficiently small $\varepsilon > 0$, the formal expansions are convergent and yield good approximations of the sought quantities λ_n, X_n only for sufficiently small n, $n\varepsilon_0 \ll 1$, and this makes them unfit for being taken as an approximate basis.

These facts make the application of the perturbation method for system (3.1.1) very difficult and hardly justifiable, in contrast to the well-studied case of problems like

$$X'' + [\lambda^2 + \varepsilon V(x)]X = 0, \quad X(a) = X(b) = 0. \tag{3.1.4}$$

Here, the function $\varepsilon V(x)$ may be regarded as a perturbing potential. Perturbed problems like (3.1.4) often occur in theoretical physics [19, 27, 46, 51, 55, 58], for instance, in quantum mechanics. For such problems, regular expansions of eigenvalues and eigenfunctions in powers of ε produce no secular terms.

3.1.2. The problem of secular terms and regularization of the problem. In order to identify the causes of secular terms and construct asymptotic expansions entailing no secular terms, it would be useful to consider a simple model problem of the type (3.1.1) for the Euler equation

$$X'' + \lambda^2(1+\varepsilon x)^{-2}X = 0, \quad X(0) = X(1) = 0, \qquad 0 \le |\varepsilon| < 1,$$

which admits a closed analytical solution of the form

$$
\begin{aligned}
\lambda_n^2(\varepsilon) &= (\pi n)^2 \frac{\varepsilon^2}{\ln^2(1+\varepsilon)} + \frac{\varepsilon}{4} = (\pi n)^2 \left[1 + \sum_{j=1}^{\infty} \frac{(-\varepsilon)^j}{j}\right]^{-2} + \frac{\varepsilon^2}{4}, \\
X_n(x,\varepsilon) &= a_n(x,\varepsilon)\sin\psi_n(x,\varepsilon), \quad (X_n, X_m)_\mu = \delta_{nm}, \quad \mu(x,\varepsilon) = (1+\varepsilon x)^{-2}, \\
a_n(x,\varepsilon) &= \left[\frac{2\varepsilon(1+\varepsilon x)}{\ln(1+\varepsilon)}\right]^{1/2}, \quad \psi_n(x,\varepsilon) = \pi n \frac{\ln(1+\varepsilon x)}{\ln(1+\varepsilon)}.
\end{aligned}
\tag{3.1.5}
$$

From (3.1.5), it follows that $\lambda_n(\varepsilon) = \pi n(1+O(\varepsilon))$ and $X_n - X_n^{(p)} = O((\varepsilon n)^{p+1})$, where $p \ge 0$ is the order of the expansion of the functions $X_n(x,\varepsilon)$ in powers of ε. The functions $X_n(x,\varepsilon)$ are assumed orthonormal with weight $\mu = (1+\varepsilon x)^{-2}$. The mechanism of the appearance of secular terms is obvious and is connected with frequency perturbations, i.e., the phase derivative $\psi_n' = \lambda_n^{(0)}(1+O(\varepsilon))$.

A similar situation arises in the theory of nonlinear vibrations [19, 42, 48] and is also due to frequency perturbations. For large time intervals, approximate analysis of perturbed vibrations is based on consecutive corrections of vibration frequencies, i.e., the corresponding expansions of periodic solutions have asymptotic character.

To avoid secular terms, it is proposed to correct the phases ψ_n, since $\psi_n' = \lambda_n(\varepsilon)(1+O(\varepsilon))$. This correction can be achieved by suitably changing the argument, $x \to y$,

$$y = \frac{x+\varepsilon\xi(x,\varepsilon)}{1+\varepsilon\xi(1,\varepsilon)}, \quad \xi(x,\varepsilon) = \int_0^x g(s,\varepsilon)ds, \tag{3.1.6}$$

$$g(x,\varepsilon) = (\delta(x)-\sigma(x))[1+\varepsilon\sigma(x)+(1+\varepsilon\sigma(x))^{1/2}(1+\varepsilon\delta(x))^{1/2}]^{-1},$$

$$x = y+\varepsilon\eta(y,\varepsilon), \quad y \in [0,1], \quad X(x,\varepsilon) \equiv Y(y,\varepsilon).$$

Indeed, after this transformation we obtain the following boundary value problem of the form (3.1.1) for the unknown function $Y(y,\varepsilon)$:

$$Y'' + \varepsilon h(y,\varepsilon)Y' + \nu^2 Y = 0, \quad Y(0) = Y(1) = 0, \tag{3.1.7}$$

where

$$\nu^2 = \lambda^2(1+\varepsilon\xi(1,\varepsilon))^2, \quad h(y,\varepsilon) = [g'(1+\varepsilon\sigma)+\sigma'(1+\varepsilon g)](1+\varepsilon\delta)^{-1}|_{x=y+\varepsilon\eta}(1+\varepsilon\xi(1,\varepsilon))^{-1}.$$

The function $h(y,\varepsilon)$ is obtained with the help of the transformation (3.1.6). Calculations show that formal application of perturbation methods to problem (3.1.7) yields regular expansions of the eigenvalues $\nu_n(\varepsilon)$ and eigenfunctions $Y_n(y,\varepsilon)$. The domain of convergence in ε and the error estimates do not depend on $n = 1, 2, \ldots$

It is easy to show that the functions Y_n are mutually orthogonal with the weight

$$\chi(y,\varepsilon) = \exp\left[\varepsilon \int_0^y h(s,\varepsilon)\,ds\right] = (1+\varepsilon\delta(y+\varepsilon\eta))(1+\varepsilon\eta').$$

Remark. Analysis of expressions (3.1.3) and higher-order expansion coefficients $\lambda_{n,p}$, $X_{n,p}$, $a_{n,p}$, $\varphi_{n,p}$ with $p > 1$ shows that if the functions $\delta(x), \sigma(x) \in C^{(1)}$ have zero mean values ($\delta_0 = \sigma_0 = 0$), then the said coefficients are of the order of unity for all $n \geq 1$. Moreover, $\varphi_{n,p} = O(1/\lambda_n^{(0)})$. The expansions of the sought solution $\lambda_n(\varepsilon)$, $X_n(x,\varepsilon)$ have regular character. For δ_0, $\sigma_0 \neq 0$, the system can be reduced to the said form by a linear transformation of the parameter λ in (3.1.1),

$$\left(\left(1+\varepsilon\frac{\sigma-\sigma_0}{1+\varepsilon\sigma_0}\right)X'\right)' + \Lambda^2\left(1+\varepsilon\frac{\delta-\delta_0}{1+\varepsilon\delta_0}\right)X = 0, \quad \Lambda^2 = \lambda^2\frac{1+\varepsilon\delta_0}{1+\varepsilon\sigma_0}.$$

Thus, for $\delta, \sigma \in C^{(1)}$, it can be assumed, without the loss of generality, that $\delta_0 = \sigma_0 = 0$, and we can consider regular expansions of the unknown λ_n, X_n in powers of ε.

3.1.3. Separation of variables. For our further investigation, it would be much more convenient to write equation (3.1.1) in the "amplitude–phase" variables a, ψ by letting

$$X = a\sin\psi, \quad X' = a\omega\cos\psi, \quad \omega = \lambda(1+\varepsilon\delta)^{1/2}(1+\varepsilon\sigma)^{-1/2}. \tag{3.1.8}$$

In the variables a, ψ, the boundary value problem (3.1.1) becomes

$$a' = 2\varepsilon f(x,\varepsilon)a\cos^2\psi, \quad \psi' = \lambda[1+\varepsilon d(x,\varepsilon)] + \varepsilon f(x,\varepsilon)\sin 2\psi;$$

$$a(0) = a^0 \sim 1, \quad \psi(0) = 0, \quad \psi(1) = \pi n, \tag{3.1.9}$$

$$f \equiv -\tfrac{1}{2}\delta'(1+\varepsilon\delta)^{-1} - \tfrac{1}{2}\sigma(1+\varepsilon\sigma)^{-1}, \quad \varepsilon d \equiv (1+\varepsilon\delta)^{1/2}(1+\varepsilon\sigma)^{-1/2} - 1.$$

Passing from the argument x to y and from the function X to Y by (3.1.6), and introducing the phase–amplitude variables b, φ by the transformation $Y = b\sin\varphi$, $Y' = b\nu\cos\varphi$ similar to (3.1.8), we can write our boundary value problem as

$$b' = \varepsilon h(y,\varepsilon) b\cos^2\varphi, \quad \varphi' = \nu + \tfrac{1}{2}\varepsilon h(y,\varepsilon)\sin 2\varphi;$$
$$b(0) = b^0 \sim 1, \quad \varphi(0) = 0, \quad \varphi(1) = \pi n. \tag{3.1.10}$$

Eigenvalue problems written in the form (3.1.9) or (3.1.10) are more convenient from the standpoint of asymptotic methods with a small parameter, because the equation for the plase ψ or φ can be solved independently of the amplitude a or b, respectively. From the boundary conditions for ψ (or φ), we find the eigenvalues $\lambda_n(\varepsilon)$ (or $\nu(\varepsilon)$). When the phases $\psi_n(x,\varepsilon)$, $\varphi_n(y,\varepsilon)$ have been found, the equations for the amplitudes a_n, b_n can be explicitly solved in quadratures. The corresponding integration constants a_n^0, b_n^0 can be easily calculated from the normalization condition for the functions $X_n(x,\varepsilon)$ with the weight $\mu(x,\varepsilon)$ and the functions $Y_n(y,\varepsilon)$ with the weight $\chi(y,\varepsilon)$. The model problem (3.1.5) in the form (3.1.9) or (3.1.10) can also be easily solved.

3.2. Approximate Solution of the Boundary Value Problem

3.2.1. Construction of eigenfrequencies and phases of partial vibrations. For the sake of definiteness, we consider problem (3.1.10). Problem (3.1.9) can be studied in a similar manner. The sought quantities $a_n(x,\varepsilon)$, $\psi_n(x,\varepsilon)$, $\lambda_n(\varepsilon)$ can be obtained from $b_n(y,\varepsilon), \varphi_n(y,\varepsilon)$, $\nu_n(\varepsilon)$ with the help of (3.1.6), (3.1.7).

The sought function $\varphi = \varphi(y,\nu,\varepsilon)$, which vanishes at $y = 0$, is constructed in the form $\varphi = \nu y + \alpha$, where the unknown $\alpha = \alpha(y,\nu,\varepsilon)$ is a solution of the integral equation

$$\alpha(y,\nu,\varepsilon) = \frac{\varepsilon}{2}\int_0^y h(s,\varepsilon)\sin 2(\nu s + \alpha(s,\nu,\varepsilon))ds. \tag{3.2.1}$$

The function α can be obtained by the Picard method as the limit of the sequence [29, 42, 48, 59] $(p = 0, 1, \ldots)$

$$\alpha_{p+1} = \frac{\varepsilon}{2}\int_0^y h(s,\varepsilon)\sin 2(\nu s + \alpha_p)ds, \quad \alpha_0 \equiv 0. \tag{3.2.2}$$

The integral operator in (3.2.1) satisfies the assumptions of the Banach theorem about contracting operators [25, 42] for all $y \in [0,1]$, $\nu \in R^1$, provided that the parameter ε is sufficiently small,

$$\max_{\varepsilon,y} \varepsilon|h(y,\varepsilon)| \le \kappa < 1, \quad \varepsilon \in [0,\varepsilon_0], \quad y \in [0,1], \quad \kappa = \text{const}. \tag{3.2.3}$$

Using the Banach theorem, we see that (3.2.2) is a Cauchy sequence, i.e., there exists a unique limit function $\alpha^*(y,\nu,\varepsilon)$ such that

$$|\alpha^*| \le c\varepsilon, \quad |\alpha'^*| \le c\varepsilon, \quad \left|\frac{\partial\alpha^*}{\partial\nu}\right| \le c\varepsilon, \quad c = \text{const}, \quad \varepsilon \in [0,\varepsilon_0], \quad y \in [0,1], \quad \nu \in R^1. \tag{3.2.4}$$

Since the integrand in (3.2.1) is a smooth function, the element α^* is a continuously differentiable solution of the Cauchy problem

$$\alpha' = \tfrac{1}{2}\varepsilon h(y,\varepsilon)\sin 2(\nu y + \alpha), \quad \alpha(0) = 0. \tag{3.2.5}$$

It is easy to show by induction that conditions (3.2.4) hold for the successive approximations $\alpha_p(y,\nu,\varepsilon)$ defined by (3.2.2), and the constant c can be constructively estimated in terms of the function $h(y,\varepsilon)$.

For sufficiently small $\varepsilon > 0$, the successive approximations α_p defined by (3.2.2) have power convergence with respect to ε, together with their derivatives in y and ν,

$$|\alpha^* - \alpha_p| \leq c\varepsilon^{p+1}, \quad |\alpha'^* - \alpha'_p| \leq c\varepsilon^{p+1}, \quad \left|\frac{\partial \alpha^*}{\partial \nu} - \frac{\partial \alpha_p}{\partial \nu}\right| \leq c\varepsilon^{p+1}, \tag{3.2.6}$$

where $c = \text{const}$, $p = 0, 1, 2, \dots$ The first estimate in (3.2.6) is obtained in the standard way from the inequalities

$$\Delta_{p+1} \leq \varepsilon B \Delta_p, \quad \Delta_{p+1} \leq D\varepsilon^{p+1}, \quad \Delta_p \equiv \max_y |\alpha_p - \alpha_{p-1}|, \tag{3.2.7}$$

where $B, D = \text{const}$, $y \in [0,1]$, $\varepsilon \in [0, \varepsilon_0]$, $p = 1, 2, \dots$ These inequalities follow directly from (3.2.2). Now, according to (3.2.7), we have

$$\max_{y \equiv [0,1]} |\alpha^* - \alpha_p| \leq \sum_{k=p}^{\infty} \Delta_{k+1} \leq D \sum_{k=p}^{\infty} \varepsilon^{k+1} \leq c\varepsilon^{p+1}. \tag{3.2.8}$$

Similarly, using (3.2.2), we obtain the inequalities

$$\Delta'_{p+1} \leq \varepsilon B \Delta_p \leq D\varepsilon^{p+1}, \quad \varepsilon \in [0, \varepsilon_0], \quad p = 1, 2, \dots, \tag{3.2.9}$$

where Δ'_p is the maximum modulus of the difference of the derivatives in y. Using this estimate, one obtains the second inequality in (3.2.6), just as in the case of (3.2.8). The last estimate in (3.2.6) can be obtained with the help of the inequalities (the quantities $\Delta_{\nu,p}$ are defined similarly to Δ_p and Δ'_p)

$$\begin{aligned} \Delta_{\nu,p+1} \leq \varepsilon \max_{y \in [0,1]} \int_0^y \Bigg| h\left(s + \frac{\partial \alpha_p}{\partial \nu}\right) \cos 2(\nu s + \alpha_p) \\ - h\left(s + \frac{\partial \alpha_{p-1}}{\partial \nu}\right) \cos 2(\nu s + \alpha_{p-1}) \Bigg| ds \leq \varepsilon B \Delta_{\nu,p} + \varepsilon D \Delta_p. \end{aligned} \tag{3.2.10}$$

From (3.2.10), we infer by induction that $\Delta_{\nu,p+1} \leq E\varepsilon^{p+1}$, $E = \text{const}$, and from an estimate of the type (3.2.8), we obtain the third estimate in (3.2.6).

Thus, the recurrent procedure (3.2.2) allows us to construct the desired phase $\varphi^* = \varphi(y, \nu, \varepsilon)$, $\varphi(0, \nu, \varepsilon) = 0$, with any given accuracy with respect to ε,

$$\varphi^* = \nu y + \alpha_p(y, \nu, \varepsilon) + \Delta\varphi_p, \quad |\Delta\varphi_p| \leq C\varepsilon^{p+1}, \quad C = \text{const}. \tag{3.2.11}$$

Substituting $\varphi^* = \nu y + \alpha^*$ given by (3.2.11) into the first equation (3.1.10), we obtain a simple explicit quadrature expression of the unknown amplitude $b(y, \nu, \varepsilon)$ ($b \geq b_* > 0$) (see below).

An approximate solution of the Cauchy problem (3.1.9) is constructed in the form $\psi(x, \lambda, \varepsilon) = \psi_0(x, \lambda, \varepsilon) + \beta$, where ψ_0 also depends on ε, and the unknown function β is calculated similarly to α. The phase ψ can also be found with the help of the transformations (3.1.6), (3.1.7), and the amplitude a can be found by quadrature according to (3.1.9).

The desired eigenvalues $\nu_n(\varepsilon)$ of the boundary value problem are obtained from the corresponding condition in (3.1.10) for

$$\varphi^*(1, \nu, \varepsilon) = \nu + \alpha^*(1, \nu, \varepsilon) = \pi n, \quad n = 1, 2, \dots \tag{3.2.12}$$

For small enough $\varepsilon > 0$, the quantities ν_n are constructed as the limits of the successive approximations ($k = 0, 1, \dots$)

$$\nu^{(k+1)}(\varepsilon) = \pi n - \alpha^*(1, \nu^{(k)}(\varepsilon), \varepsilon), \quad \nu_n^{(0)} = \pi n. \tag{3.2.13}$$

From the estimates (3.2.4) for α^* and α_p, we immediately obtain the following result.

Theorem 3.1. *For sufficiently small $\varepsilon > 0$, the successive approximations* (3.2.13) *converge to the unique solution of equation* (3.2.12). *This convergence is uniform in n and*

$$\begin{gathered}\lim_{k\to\infty} \nu_n^{(k)} = \nu_n^*(\varepsilon), \quad n = 1,2,\ldots, \quad \varepsilon \in [0,\varepsilon_0], \quad \varepsilon_0 c < 1, \\ |\nu_n^*(\varepsilon) - \nu_n^{(k)}(\varepsilon)| \le K\varepsilon^{k+1}, \quad K = \text{const}, \quad \nu_n^{(k)} = \pi n + \gamma_n^{(k)}, \\ \nu_n^*(\varepsilon) = \pi n + \gamma_n^*(\varepsilon), \quad |\gamma_n^{(k)}|, |\gamma_n^*| \le \Gamma\varepsilon, \quad \Gamma = \text{const}.\end{gathered} \tag{3.2.14}$$

Taking the $(p+1)$-th approximation $\alpha_p(y,\nu,\varepsilon)$ in (3.2.12), (3.2.13) instead of α^*, we obtain the following uniform estimate

$$|\nu_n^*(\varepsilon) - \nu_{n,p}^{(k)}(\varepsilon)| \le K\varepsilon^{k+1}, \quad k = 0,1,\ldots,p, \quad K = \text{const}. \tag{3.2.15}$$

To construct the system of eigenvalues $\{\nu_n(\varepsilon)\}$, we can also use the method of tangential lines (the Newton method [48]), which yields quadratic convergence with respect to ε,

$$\begin{gathered}\nu_n^{(k+1)} = \left[\pi n - \alpha^*\left(1,\nu_n^{(k)},\varepsilon\right) + \frac{\partial}{\partial\nu}\alpha^*\left(1,\nu_n^{(k)},\varepsilon\right)\right]\left[1 + \frac{\partial}{\partial\nu}\alpha^*\left(1,\nu_n^{(k)},\varepsilon\right)\right]^{-1}, \\ \nu_n^{(0)} = \pi n, \quad \left|\nu_n^* - \nu_n^{(k+1)}\right| \le K(\varepsilon)^{2^k}, \quad K = \text{const}, \quad k = 0,1,\ldots\end{gathered} \tag{3.2.16}$$

3.2.2. Finding eigenfunctions and the construction of an orthonormal basis. After the eigenvalues $\nu_n(\varepsilon)$ have been found from (3.2.11), (3.1.10), or (3.2.16), we define the desired quantities $\varphi_n, b_n, b_n^0, Y_n$ as follows $(n,m = 1,2,\ldots)$:

$$\begin{gathered}\varphi_n(y,\varepsilon) = \nu_n^*(\varepsilon)y + \alpha^*(y,\nu_n^*(\varepsilon),\varepsilon) = \varphi_n^{(p)} + \Delta\varphi_n^{(p)}, \\ b_n(y,\varepsilon) = b_n^0(\varepsilon)\exp\left[-\varepsilon\int_0^y h(s,\varepsilon)\cos^2\varphi_n ds\right] = b_n^{(p)} + \Delta b_n^{(p)}, \\ b_n^0(\varepsilon) = \left\{\int_0^1 \cos^2\varphi_n \exp\left[-\varepsilon\int_0^y h\cos 2\varphi_n ds\right]dy\right\}^{1/2} = b^{0(p)} + \Delta b_n^{0(p)}, \\ Y_n(y,\varepsilon) = b_n(y,\varepsilon)\sin\varphi_n(y,\varepsilon), \quad (Y_n, Y_m)_\chi = \delta_{nm}, \\ |\Delta\varphi_n^{(p)}| \le C\varepsilon^{p+1}, \quad |\Delta b_n^{(p)}| \le C\varepsilon^{p+1}, \quad |\Delta b_n^{0(p)}| \le C\varepsilon^{p+1}, \quad |Y_n| \le C = \text{const}.\end{gathered} \tag{3.2.17}$$

Here, $\{Y_n\}$ is a complete system of eigenfunctions orthonormal with the weight $\chi(y,\varepsilon)$. Substituting the approximate expressions $\alpha_p, \nu_n^{(p)}$ into (3.2.17), we find the quantities φ_n, b_n and Y_n with the error of order ε^{p+1}. The functions $\{Y_n^{(p)}\}$ are orthonormal with the same error. The following result holds.

Theorem 3.2. *For sufficiently small $\varepsilon > 0$, the following estimates hold:*

$$Y_n = Y_n^{(p)} + \Delta Y_n^{(p)}, \quad Y_n^{(p)} = b_n^{(p)}\sin\varphi_n^{(p)}, \quad |\Delta Y_n^{(p)}| \le C\varepsilon^{p+1}. \tag{3.2.18}$$

For the approximate basis $\{Y_n^{(p)}\}$, the approximate orthonormality conditions hold (δ_{nm} is the Kronecker symbol)

$$(Y_n^{(p)}, Y_m^{(p)})_\chi \equiv \int_0^1 Y_n^{(p)}Y_m^{(p)}\chi dy = \delta_{nm} + D_{nm}, \quad |D_{nm}| \le D\varepsilon^{p+1}. \tag{3.2.19}$$

Relations (3.2.18), (3.2.19) are obtained immediately from the uniform error estimates for φ_n, b_n established above.

3.2.3. Remarks.

1. The above result can also be obtained by expanding the sought quantities in powers of the small parameter ε. This approach can be justified with the help of the Cauchy majorizing functions or by other analytical methods of the perturbation theory [15, 25, 27, 33, 34, 42, 46, 48].

2. The above results can be immediately adapted to the construction of an approximate solution $\psi_n(x,\varepsilon), a_n(x,\varepsilon), \lambda_n(\varepsilon), X_n(x,\varepsilon)$ of problem (3.1.9) and the equivalent problem (3.1.1). As mentioned above, this solution is found with the help of formulas (3.1.6), (3.1.7).

3. In a similar way, one can construct systems of eigenvalues and eigenfunctions of problems with other boundary conditions, for instance,

$$\psi(0) = -\frac{\pi}{2}, \quad \psi(1) = \left(n - \frac{1}{2}\right)\pi \quad (X'(0) = X'(1) = 0);$$

$$\psi(0) = 0, \quad \psi(1) = \left(n - \frac{1}{2}\right)\pi \quad (X(0) = X'(1) = 0); \tag{3.2.20}$$

$$[-k\sin\psi \mp (1+\varepsilon\sigma)\omega\cos\psi]_{x=0,1} = 0,$$

$$([-kX \mp (1+\varepsilon\sigma)X']_{x=0,1} = 0); \quad n = 1, 2, \ldots$$

and some others.

3.3. Approximation of Functions in Terms of the Approximate Basis

3.3.1. The problem of expansion in terms of an approximate basis. Let $f(y) \in C_0^{(2)}$ be a twice continuously differentiable function of $y \in [0,1]$ vanishing at the points $y = 0$ and $y = 1$. Then the Fourier series

$$f(y) = \sum_{n=1}^{\infty} f_n(\varepsilon)Y_n(y,\varepsilon), \quad f_n = (f, Y_n)_X, \tag{3.3.1}$$

is uniformly convergent. This convergence is due to the decay of the Fourier coefficients, $f_n(\varepsilon) \sim \nu_n^{-2}(\varepsilon)$, and the fact that the functions $\{Y_n(y,\varepsilon)\}$ are bounded and smooth.

Since it is practically possible to construct the functions Y_n with a finite error $O(\varepsilon^{p+1})$ (see Section 3.2)), it can be expected that in view of (3.3.1), (3.2.18), (3.2.19), the function $f(y) \in C_0^{(2)}$ admits the approximation

$$f(y) \sim f^{(p)}(y,\varepsilon) = \sum_{n=1}^{\infty} f_n^{(p)}(\varepsilon)Y_n^{(p)}(y,\varepsilon). \tag{3.3.2}$$

Now, it would be natural to ask whether the series (3.3.2) is convergent and the functions $f^{(p)}(y,\varepsilon)$ and $f(y)$ are close for $y \in [0,1]$ and $\varepsilon \in [0,\varepsilon_0]$. In this connection, the following result holds.

Theorem 3.3. *For an arbitrary function $f(y) \in C_0^{(2)}$ and sufficiently small $\varepsilon > 0$, the series (3.3.2) converges to the function $f^{(p)}(y,\varepsilon)$ in the norm of the Hilbert space L_2, and the following estimate of closeness holds:*

$$\|f(y) - f^{(p)}(y,\varepsilon)\|_{L_2} \le C\varepsilon^{p+1}, \quad C = \text{const.} \tag{3.3.3}$$

The proof of this result is based on the Cauchy condition for a sequence to be fundamental [47]. To that end, one considers the difference ($1 \le M < N < \infty$)

$$\sum_{n=M}^{N} f_n Y_n - \sum_{n=M}^{N} f_n^{(p)} Y_n^{(p)} \equiv \sum_{n=M}^{N} (f_n - f_n^{(p)}) Y_n^{(p)} + \sum_{n=M}^{N} f(Y_n - Y_n^{(p)}) \equiv S_1 + S_2 \quad (3.3.4)$$

and estimates in L_2 the segments of the series S_1, S_2. To estimate

$$\|S_1\|_{L_2}^2 = \sum_{n,m=M}^{N} (f_n - f_n^{(p)})(f_m - f_m^{(p)})(Y_n^{(p)}, Y_m^{(p)})_X \quad (3.3.5)$$

one utilizes elementary trigonometric formulas. By (3.2.17), we have

$$\begin{aligned} f_n - f_n^{(p)} &= -2(b_n \sin\varphi_n \sin^2(\tfrac{1}{2}\Delta\varphi_n^{(p)}), f)_X + (b_n \cos\varphi_n \sin\Delta\varphi_n^{(p)}, f)_X \\ &\quad -(\Delta b_n^{(p)} \sin\varphi_n \cos\Delta\varphi_n^{(p)}, f)_X - (\Delta b_n^{(p)} \sin\varphi_n \sin\Delta\varphi_n^{(p)}, f)_X, \end{aligned} \quad (3.3.6)$$

where $n \ge 1$, $p \ge 0$. Calculating these scalar products with the help of integration by parts and taking into account the estimates (3.2.17) for $\Delta\varphi_n^{(p)}$, $\Delta b_n^{(p)}$, and also the estimates

$$n\big(|\varphi_n'|^{-1} + |\varphi_n''|(\varphi_n')^{-2}\big) \le c, \quad |b_n'| + |\Delta\varphi_n^{(p)\prime}| \le c, \quad c = \text{const},$$

we obtain the following estimates for the coefficients (3.3.6):

$$|f_n - f_n^{(p)}| \le C n^{-1} \varepsilon^{p+1}. \quad n = 1, 2, \ldots, \quad C = \text{const}. \quad (3.3.7)$$

Similarly, using the inequalities

$$|\varphi_n^{(p)\prime} \pm \varphi_m^{(p)\prime}|^{-1} \le c|n \pm m|^{-1}, \quad n \ne m,$$

we obtain the following estimates for the coefficients in (3.3.5):

$$|(Y_n^{(p)}, Y_m^{(p)})_X| \le C((n+m)^{-1} + |n-m|^{-1}), \quad n \ne m. \quad (3.3.8)$$

From (3.3.7), (3.3.8), we immediately obtain the estimate

$$\|S_1\|_{L_2}^2 \le C\varepsilon^{2(p+1)} \left[\sum_{n=M}^{N} \frac{1}{n^2} + \sum_{n,m=M}^{N} \frac{1}{nm} \left(\frac{1}{n+m} + \frac{1}{|n-m|} \right) \right]. \quad (3.3.9)$$

Majorizing the double sum in (3.3.9) by the double integrals

$$\iint \frac{dx\,dy}{xy(x+y)}, \quad \iint \frac{dx\,dy}{xy|x-y|} \quad (x \ge 1,\ y \ge 1,\ |x-y| \ge 1),$$

we obtain the desired inequality

$$\|S_1\|_{L_2}^2 \le \varepsilon^{2(p+1)} A(M, N); \quad A \to 0, \quad M \to \infty, \quad N > M. \quad (3.3.10)$$

Here, we have $A(M, N) < \kappa$, with $\kappa > 0$ being arbitrarily small for sufficiently large $M = M(\kappa)$ and arbitrary $N > M$, i.e., the Cauchy condition for a sequence to be fundamental holds for $\|S_1\|_{L_2}^2$.

In order to estimate $\|S_1\|_{L_2}^2$ in (3.3.4), the following estimate of the Fourier coefficients can be used (this estimate holds in consequence of (3.1.7)):

$$|f_n| \le c\nu_n^{-2} \quad (\nu_n \sim n), \quad f(y) \in C_0^{(2)}, \quad y \in [0,1].$$

By (3.2.17) we obtain the inequality

$$||Y_n - Y_n^{(p)}||_{L_2} \le C\varepsilon^{p+1},$$

and therefore,

$$||S_2||_{L_2}^2 \le \varepsilon^{2(p+1)} B(M,N); \quad B \to 0, \quad M \to \infty, \quad N > M. \tag{3.3.11}$$

From (3.3.4), (3.3.10), (3.3.11) we obtain the desired result.

3.3.2. Uniform estimates. Let $f(x) \in C_{\ell-2}^{(\ell)}$, $\ell \ge 2$, i.e., the function f is ℓ times continuously differentiable and vanishes at the points $x = 0$, $x = 1$, together with all its derivatives of the orders $\le \ell - 2$. Suppose also that the functions $\sigma(x)$ and $\delta(x)$ are continuously differentiable $\ell - 1$ and $\ell - 2$ times, respectively. Then, for the Fourier coefficients $f_n(\varepsilon)$ with respect to the basis $\{X_n(x,\varepsilon)\}$ the following estimates hold:

$$f_n(\varepsilon) = a_n^*(\varepsilon)n^{-\ell}, \quad |a_n^*| \le a^* = \text{const} \quad (f_n = (f, X_n)_\mu). \tag{3.3.12}$$

The differences between the Fourier coefficients with respect to the exact basis and those with respect to the approximate basis satisfy the inequalities

$$|f_n(\varepsilon) - f_n^{(p)}(\varepsilon)| \le C\varepsilon^{p+1}, \quad C = \text{const} \quad (f_n^{(p)} = (f, X_n^{(p)})_\mu), \tag{3.3.13}$$

which are uniform in n and hold for all $\varepsilon \in [0, \varepsilon_0]$.

Further, consider the sequence of values $\varepsilon_n(\theta) = \varepsilon_0 \theta n^{-\ell/(p+1)}$ of the small parameter ε, where $\theta \in [0,1]$. Substituting $f_n(\varepsilon)$ defined by (3.3.12) into (3.3.13) for $\varepsilon = \varepsilon_n$, we obtain the estimates ($n = 1, 2, \ldots$)

$$|f_n^{(p)}(\varepsilon_n)| \le \left[C(\theta\varepsilon_0)^{p+1} + |a_n^*(\varepsilon_n)|\right] n^{-\ell} \le (C\varepsilon_0^{p+1} + a^*)n^{-\ell}. \tag{3.3.14}$$

According to (3.3.14), the coefficients $f_n^{(p)}(\varepsilon)$, f_n have the same decay rate with respect to n, and the following estimates hold:

$$\begin{gathered} f_n^{(p)}(\varepsilon) = a_n^{*(p)}(\varepsilon)n^{-\ell}, \quad |a_n^*(\varepsilon) - a_n^{*(p)}(\varepsilon)| \le d\varepsilon^{p+1}, \\ |f_n(\varepsilon) - f_n^{(p)}(\varepsilon)| \le d\varepsilon^{p+1}n^{-\ell}, \quad d = \text{const}. \end{gathered} \tag{3.3.15}$$

From (3.3.15), we obtain the estimates of closeness between the approximating series $f^{(p)}(x,\varepsilon)$ and the approximated function $f(x)$,

$$|f(x) - f^{(p)}(x,\varepsilon)| \le D\varepsilon^{p+1}, \quad f \in C_{\ell-2}^{(\ell)}, \quad p \ge 0, \quad \ell \ge 2, \quad D = \text{const}, \tag{3.3.16}$$

for $x \in [0,1]$ and $\varepsilon \in [0, \varepsilon_0]$.

The results obtained here justify the application of the asymptotic method of perturbations for the approximation of functions of class $C_{\ell-2}^{(\ell)}$ ($\ell \ge 2$) with respect to the approximate basis $\{X_n^{(p)}(x,\varepsilon)\}$.

3.4. Applications to Initial Boundary Value Problems

3.4.1. Approximate solution. The methods developed above can be used to construct approximate solutions of some hyperbolic problems describing vibrations of systems with distributed parameters. Such problems often occur in applications. Consider the following perturbed problem:

$$[1+\varepsilon\delta(x)]u_{tt} = [(1+\varepsilon\sigma(x))u_x]_x,$$
$$u(t,0,\varepsilon) = u(t,1,\varepsilon) = 0, \tag{3.4.1}$$
$$u(0,x,\varepsilon) = \varphi(x), \quad u_t(0,x,\varepsilon) = \psi(x).$$

Here, $u = u(t,x,\varepsilon)$, $t \in [0,T]$, $x \in [0,1]$, $\varepsilon \in [0,\varepsilon_0]$; the subscripts t and x indicate the derivatives in the time variable and the spatial variable; $\delta(x) \in C^{(1)}$, $\sigma(x) \in C^{(1)}$, $\varphi(x) \in C_0^{(2)}$, $\varphi(x) \in C_0^{(2)}$, $\psi(x) \in C^{(1)}$. System (3.4.1) is obtained as a reduction of the equations describing vibrations of an elastic system whose linear inertial and stiffness characteristics have slow variation with respect to length.

For small enough $\varepsilon_0 > 0$, a solution of problem (3.4.1) is constructed by the Fourier method in the form of an expansion in terms of the orthonormal eigenfunctions $X_n(x,\varepsilon)$ of the eigenvalue problem (3.1.1),

$$u(t,x,\varepsilon) = \sum_{n=1}^{\infty} F_n(t,\varepsilon)X_n(x,\varepsilon),$$
$$F_n(t,\varepsilon) = a_n(\varepsilon)\cos[\lambda_n(\varepsilon)t] + b_n(\varepsilon)\lambda_n^{-1}(\varepsilon)\sin[\lambda_n(\varepsilon)t], \tag{3.4.2}$$
$$a_n = (\varphi, X_n)_\mu, \quad b_n = (\psi, X_n)_\mu, \quad \mu = 1+\varepsilon\delta.$$

Let $\{X_n\}$ be the approximate basis constructed in Sections 3.1–3.3 with a given accuracy p with respect to ε (with the error $O(\varepsilon^{p+1})$), and let $u^{(p)}$ be an approximate analogue of the solution u,

$$u^{(p)}(t,x,\varepsilon) = \sum_{n=1}^{\infty} F_n^{(p)}(t,\varepsilon)X_n^{(p)}(x,\varepsilon). \tag{3.4.3}$$

3.4.2. Error estimates. Using the above results, in particular, estimates of the type (3.3.3), one establishes closeness of the solution u (3.4.2) to its $(p+1)$-th approximation $u^{(p)}$ (3.4.3).

Theorem 3.4. *For the solution $u(t,x,\varepsilon)$ of the form* (3.4.2) *and its approximation $u^{(p)}(t,x,\varepsilon)$ of the form* (3.4.3), *the following closeness estimate holds:*

$$\max_{t\in[0,T]} \|u - u^{(p)}\|_{L_2} \le C\varepsilon^{p+1}, \quad C = \text{const}, \tag{3.4.4}$$

where $\varepsilon \in [0,\varepsilon_0]$ and $\varepsilon_0 > 0$ is sufficiently small. The constant $C > 0$ can be constructively estimated in terms of the coefficients of problem (3.4.1).

Now, let $u_N^{(p)}(t,x,\varepsilon)$ be a finite segment of the Fourier series in terms of the approximate basis (3.4.3),

$$u_N^{(p)}(t,x,\varepsilon) = \sum_{n=1}^{N} F_n^{(p)}(t,\varepsilon)X_n^{(p)}(x,\varepsilon). \tag{3.4.5}$$

One can prove the following statement.

Theorem 3.5. *For a suitably chosen $N = N(\varepsilon)$ and sufficiently small $\varepsilon_0 > 0$, the following uniform estimate holds:*

$$\max_{t,x} |u - u_N^{(p)}| \le C\varepsilon^d, \quad d = (1 - \ell^{-1})(p+1), \quad C = \text{const}. \tag{3.4.6}$$

Here, $\ell \ge 2$ is the order of smoothness of the functions φ, ψ in (3.4.1) and $N(\varepsilon)$ has the form

$$N = N(\varepsilon) = N_0\varepsilon^{-\gamma}, \quad \gamma = (p+1)\ell^{-1}, \quad N_0 = \text{const}. \quad \varepsilon \in [0, \varepsilon_0].$$

The inequality (3.4.6) is obtained directly from the estimate

$$\max_{t,x} |u - u_N^{(p)}| \le D(N\varepsilon^{p+1} + N^{-\ell+1}), \quad D = \text{const}, \tag{3.4.7}$$

with the said $N = N(\varepsilon)$. The estimate (3.4.7) is established with the help of the relations

$$|u - u_N^{(p)}| \le S_1 + S_2 + S_3, \quad t \in [0, T], \quad x \in [0, 1], \quad \varepsilon \in [0, \varepsilon_0],$$

$$S_1 = \left|\sum_{n=1}^{N} (F_n - F_n^{(p)})X_n\right| \le C_1 N\varepsilon^{p+1}, \quad S_2 = \left|\sum_{n=1}^{N} F_n^{(p)}(X_n - X_n^{(p)})\right| \le C_2 N\varepsilon^{p+1},$$

$$S_3 = \left|\sum_{n=N+1}^{\infty} F_n X_n\right| \le C_3^* \sum_{n=N+1}^{\infty} \left(|a_n| + \frac{|b_n|}{\lambda_n}\right) \le C_3 N^{-\ell+1}.$$

These inequalities immediately yield the estimate (3.4.7), which implies (3.4.6).

Thus, for large enough $\ell \ge 2$, the solution $u(t, x, \varepsilon)$ (3.4.2) of problem (3.4.1) can be approximated with adequate precision by a function $u_N^{(p)}(t, x, \varepsilon)$ of the form (3.4.5) with a sufficiently small $N = N(\varepsilon)$ (the number of the preserved terms of the series (3.4.3)). For the first derivatives of the solution $u(t, x, \varepsilon)$ (3.4.2) in t and x, estimates similar to (3.4.6) hold for $\ell \ge 3$,

$$\begin{gathered} \max\nolimits_{t,x} |u_t - u_{Nt}^{(p)}| \le C\varepsilon^d, \quad \max\nolimits_{t,x} |u_x - u_{Nx}^{(p)}| \le C\varepsilon^d, \\ d = (1 - 2\ell^{-1})(p+1), \quad t \in [0, T], \quad x \in [0, 1], \quad \varepsilon \in [0, \varepsilon_0]. \end{gathered} \tag{3.4.8}$$

In general, for the derivatives of any order k,

$$\frac{\partial^k u}{\partial t^m \partial x^r}, \quad m + r = k, \quad 0 \le m, r \le k \le \ell - 2,$$

error estimates of the order of $O(\varepsilon^d)$ similar to (3.4.6), (3.4.8) can be obtained with the exponent

$$d = d(k) = [1 - (k+1)\ell^{-1}](p+1) > 0.$$

Our conclusion is that the basis $\{X_n\}$ can be used for the construction of approximate solutions of problem (3.4.1).

3.5. Exercises

Exercise 1. Consider problem (3.1.1) with the functions $\sigma(x)$, $\delta(x)$ being:

(a) $\sigma = 0, \ \delta = x^2$;

(b) $\sigma = 0, \ \delta = \sin(kx)$;

(c) $\sigma = 0, \ \delta = \exp(kx)$.

Exercise 2. Develop a scheme of the perturbation method for self-conjugate problems of the form

$$([\sigma_0(x) + \varepsilon\sigma(x)]u')' + \lambda^2[\delta_0(x) + \varepsilon\delta(x)]u = 0,$$

$$[\sigma_0(x) + \varepsilon\sigma(x)]u' \,|_{x=0,1} \mp k_{0,1}u \,|_{x=0,1} = 0,$$

under the assumption that the solutions

$$\lambda = \lambda_n, \quad u = u_n(x), \quad n = 1, 2, \ldots$$

of the unperturbed problem ($\varepsilon = 0$) are known.

Chapter 4

Generalized Sturm–Liouville Problem

Generalized eigenvalue problems arise in connection with many problems in mechanics, theory of vibrations and stability, optimal control, mathematical and theoretical physics, hydrodynamics, acoustics, dynamics of atmosphere and ocean, elasticity, etc. Examples of such problems will be considered below. So far no regular method has been proposed for solving generalized Sturm–Liouville problems.

4.1. Statement of the Generalized Sturm–Liouville Problem

4.1.1. Statement of the boundary value problem in differential form. Let $p = p(x, \lambda)$, $r = r(x, \lambda)$ be given functions defined on the set $0 \le x \le 1$, $\lambda \in \Lambda$. The properties of these functions will be specified below and are assumed to ensure that the corresponding constructions make sense. In the classical setting, the coefficient p does not depend on λ, while r linearly depends on λ.

The generalized Sturm–Liouville problem is formulated as follows: *Find the values of the parameter λ for which there exists a nontrivial solution of the first boundary value problem* [8]

$$(p(x,\lambda)u')' + r(x,\lambda)u = 0, \qquad u(0) = u(1) = 0. \tag{4.1.1}$$

We assume that

$$0 < p_1 \le p(x,\lambda) \le p_2 < \infty, \quad 0 < r_1 \le r(x,\lambda) \le r_2 < \infty, \quad 0 \le x \le 1. \tag{4.1.2}$$

Here and in what follows, the prime indicates the derivative in x. The functions p, r in (4.1.1) are assumed sufficiently smooth and separated from zero. The set of admissible values of λ is such that conditions (4.1.2) hold.

4.1.2. Basic definitions. By analogy with the classical case, the values of the parameter λ for which there exist nontrivial solutions are called eigenvalues and the solutions $u(x, \lambda)$ are called eigenfunctions. In general, the eigenvalues λ may be complex. From physical considerations, in applied problems it often makes sense to consider real values of λ (in particular, positive λ, which may be a squared frequency, a critical force, etc.) and the corresponding real-valued functions. In what follows, we deal with real λ and $u(x, \lambda)$. The properties of the solutions of the classical problems (see Chapters 1 and 2) have been investigated in great detail. The abundant results established for classical problems are at the basis of the theory of linear self-adjoint operators and functional analysis.

As a rule, the behavior of eigenvalues λ_n and eigenfunctions $u_n(x) = u(x, \lambda_n)$ $(n = 1, 2\dots)$ of the generalized eigenvalue problem (4.1.1) is quite different from the behavior of the solutions of the classical eigenvalue problems (see Subsection 4.2.1), and its investigation is a difficult task.

4.2. Some Sturm–Liouville Problems with Exact Solutions

4.2.1. Examples. Here, we consider some second-order equations (of Euler type) with variable coefficients. We are interested in cases allowing for exact analytical solutions whose "exotic" properties can be studied.

Example 1. Let $p = 1$, $r = (\lambda + x)^{-2}$ in (4.1.1). In the domain of admissible values $\Lambda = \{\lambda : \lambda < -1,\ \lambda > 0\}$, standard calculations yield

$$\begin{gathered}\lambda_n = [\exp(\gamma n) - 1]^{-1}, \quad \gamma = \frac{2\pi}{\sqrt{3}}; \quad \lambda_n + \lambda_{-n} = -1, \quad n = \pm 1, \pm 2, \dots, \\ u_n(x) = c_n\left(1 + \frac{x}{\lambda_n}\right)^{1/2} \sin\left[\frac{\pi}{\gamma}\ln\left(1 + \frac{x}{\lambda_n}\right)\right], \quad c_n = \text{const.}\end{gathered} \tag{4.2.1}$$

The eigenvalues λ_n (4.2.1) are concentrated in narrow regions near the points $\lambda = 0$ and $\lambda = -1$, and the following estimates hold

$$\begin{gathered}0 < \lambda_n \le c\exp(-\gamma n), \quad n \ge 1, \quad c \sim 1, \\ 0 < -(1 + \lambda_n) \le c\exp(\gamma n), \quad n \le -1.\end{gathered} \tag{4.2.2}$$

From (4.2.2), it follows that the values λ_n (4.2.1) rapidly (exponentially) approach one another as $|n| \to \infty$. For large enough $|n|$, the functions $u_n(x)$ (4.2.1) oscillate in an exotic manner and the oscillation rate can be arbitrarily high.

Example 2. For the function $r = \lambda^2(1 + \lambda x)^{-2}$, which is obtained from the above r (see Example 1) by the introduction of λ^{-1} and the transformation $\lambda^{-1} \to \lambda$, in the region of admissible values $\Lambda = \{\lambda :\ -1 < \lambda < \infty\}$, we obtain the solutions

$$\begin{gathered}\lambda_n = \exp(\gamma n) - 1, \quad (\lambda_n + 1)(\lambda_{-n} + 1) = 1; \quad n = \pm 1, \pm 2, \dots, \\ u_n(x) = c_n(1 + \lambda_n x)^{1/2} \sin\left[\frac{\pi}{\gamma}\ln(1 + \lambda_n x)\right], \quad c_n = \text{const.}\end{gathered} \tag{4.2.3}$$

The values λ_n (4.2.3), as well as the distances between them, have exponential growth as $n \to \infty$, while the functions $u_n(x)$ have fast oscillation in x. For $n \to -\infty$, we have $\lambda_n \to -1$ $(\lambda_n > -1)$ and the functions $u_n(x)$ have fast oscillation near the point $x = 1$.

4.2.2. Some basic general properties of solutions. There are some other elementary examples illustrating a variety of unusual properties of the solutions of the generalized problem (4.1.1). It should be mentioned that the spectrum of such problems may be discrete, discrete-continuous, finite, or empty. A specific feature of the eigenfunctions $u_n(x)$ is that their orthogonality condition has the form

$$\int_0^1 \{-[p(x, \lambda_n) - p(x, \lambda_m)]u_n'(x)u_m'(x) + [r(x, \lambda_n) - r(x, \lambda_m)]u_n(x)u_m(x)\}\, dx = 0, \tag{4.2.4}$$

where $\lambda_n \ne \lambda_m$. In the classical case, relations (4.2.4) coincide with the usual ones. Passing to the limit in (4.2.4), one obtains an analogue of the "generalized norm" of the eigenfunction $u_n(x)$,

$$\|u_n\|^2 = \left| \int_0^1 \left[-\frac{\partial p}{\partial \lambda}(x, \lambda_n) u_n'^2(x) + \frac{\partial r}{\partial \lambda}(x, \lambda_n) u_n^2(x) \right] dx \right|, \quad n = 1, 2 \ldots \tag{4.2.5}$$

Under certain conditions of fixed sign for the functions $\partial p/\partial \lambda$, $\partial r/\partial \lambda$, the expression (4.2.5) possesses the usual properties of the norm in the Sobolev space $W_2^{(1)}$ [30, 57]. Completeness of the countable set of the eigenfunctions $\{u_n(x)\}$ and the possibility of expanding any $f(x) \in W_2^{(1)}$ on the interval $[0, 1]$ in terms of this "basis" remains an open problem.

Basic results pertaining to generalized eigenvalue problems of the type (4.1.1), as well as more general problems in a complex domain, can be found in [25, 33, 44, 47], together with references to some fundamental works on these topics. Problem (4.1.1) is related to the theory of operator sheafs. General properties of operator sheafs were studied in the fundamental works of Keldysh, Steklov, Tamarkin et al. In particular, the conditions ensuring the existence of a real discrete spectrum were found. Calculation of eigenvalues and eigenfunctions with a given accuracy encounters difficulties of principal character.

4.3. Statement of an Auxiliary Variational Problem

4.3.1. Variational statement of the problem and its generalization. Our aim is to construct highly precise numerical-analytical solutions of the generalized Sturm–Liouville problem (4.1.1). We assume that conditions (4.1.2) are satisfied and the functions $p(x, \lambda)$, $r(x, \lambda)$ are sufficiently smooth in the region $0 \le x \le 1$, $\lambda \in \Lambda$, so that the constructions below make sense.

In order to find an eigenvalue and an eigenfunction of this problem, consider the following family of isoperimetric variational problems for continuously differentiable functions $U(x, \lambda)$ with the parameter λ:

$$\begin{gathered} J_\lambda[U] = \int_0^1 p(x, \lambda) U'^2 dx \to \min_U, \quad U(0) = U(1) = 0, \\ \Phi_\lambda[U] = \|U\|_r^2 = \int_0^1 r(x, \lambda) U^2 dx = 1, \quad \lambda \in \Lambda. \end{gathered} \tag{4.3.1}$$

For any fixed $\lambda \in \Lambda$, problem (4.3.1) is the classical one, and there exists a normalized continuously differentiable function $U_1(x, \lambda)$ realizing the absolute minimum of the functional J_λ : $J_\lambda[U_1] = \mu_1(\lambda) > 0$. The value $\mu_1(\lambda)$ is the smallest eigenvalue and $U_1(x, \lambda) = U(x, \mu_1(\lambda), \lambda)$ is the corresponding eigenfunction of the Sturm–Liouville problem with the parameter μ,

$$(p(x, \lambda) U')' + \mu r(x, \lambda) U = 0; \quad U(0) = U(1) = 0. \tag{4.3.2}$$

Problem (4.3.2) has countably many eigenvalues μ_m such that $\mu_{m+1} > \mu_m > 0\,(m = 1, 2 \ldots)$ and $c(\lambda) \le \mu_m(\lambda)/m^2 \le C(\lambda)$ (the functions $c(\lambda)$ and $C(\lambda)$ are positive and independent of m). The respective eigenfunctions $U_m(x, \lambda) = U(x, \mu_m(\lambda), \lambda)$ form a basis. Note that in order to determine subsequent eigenvalues μ_m and eigenfunctions U_m, $m \ge 2$, the variational problem (4.3.1) should be supplemented with the corresponding orthogonality conditions, i.e., the functions U should be orthogonal (with the weight $r(x, \lambda)$) to previous eigenfunctions $U_1, \ldots, U_{m-1}$,

$$\Psi_{\lambda j}[U] = (U_j, U)_r = \int_0^1 r(x, \lambda) U_j U \, dx = 0, \quad j = 1, 2, \ldots, m - 1. \tag{4.3.3}$$

4.3.2. Derivation and analysis of the determining relation. For the time being, assume that for any admissible $\lambda \in \Lambda$, the solutions of problem (4.3.2) or problems (4.3.1), (4.3.3) are known.

Then a solution of the original problem (4.1.1) can be obtained as follows. For any $m = 1, 2, \ldots$, let us consider the relation

$$\mu_m(\lambda) = 1, \quad \lambda \in \Lambda, \tag{4.3.4}$$

as an equation for the unknown λ and assume that $\mu_m(\lambda)$ is a smooth function of λ. The set of all real roots $\{\lambda_n\}$, $n = n(m)$, of equations (4.3.4) for all $m \geq 1$ determines the sought spectrum of problem (4.1.1). Simple examples show that this set may be either empty, finite, countable, continuous, or discrete-continuous. The behavior of the spectrum as a function of the index and other parameters of the system can be studied in detail only in rare cases, for instance, if analytical formulas for $\mu_m(\lambda)$ are known (see examples in Section 4.2, or problems close to the classical ones).

By standard operations we obtain the following expression for the derivative of $\mu_m(\lambda)$ in λ:

$$\mu'_m(\lambda) = \int_0^1 \left[\frac{\partial p}{\partial \lambda}(x, \lambda) U'^2_m(x, \lambda) - \mu_m(\lambda) \frac{\partial r}{\partial \lambda}(x, \lambda) U^2_m(x, \lambda)\right] dx. \tag{4.3.5}$$

For approximate or numerical calculations, it would be natural to assume that $\mu'_m(\lambda) \neq 0$ in a neighborhood of the sought value of λ. The derivative $\mu'_m(\lambda)$ (4.3.5) exists and is continuous in λ, provided that the functions p, r are continuously differentiable in $\lambda \in \Lambda$. The sign of μ'_m can be determined a priori, provided that the derivatives $\partial p/\partial \lambda$ and $\partial r/\partial \lambda$ are of the corresponding fixed sign.

For applications, it is crucial to know how to construct two-sided estimates of the eigenvalues λ_n and to obtain their highly precise approximations. In some cases, the existence of real eigenvalues λ_n of problem (4.1.1) can be established on the basis of rough estimates of the functions $p(x, \lambda), r(x, \lambda)$. Our investigation will be based on the variational statement (4.3.1), (4.3.3) of the problem and the application of the Rayleigh–Ritz method or the Rayleigh principle. Let us construct an upper bound for the set of values $\mu_1(\lambda)$, $\lambda \in \Lambda$. By the Rayleigh principle, we have

$$0 < \mu(\lambda) \leq \mu^*(\lambda) = \frac{J_\lambda[\psi]}{\Phi_\lambda[\psi]}, \quad \psi(0) = \psi(1) = 0. \tag{4.3.6}$$

Here, ψ is a continuously differentiable function chosen on the basis of general ideas about the first eigenfunction (absence of intermediate null-points, convexity, etc.). Suppose that such a bound $\mu^*(\lambda)$, $\lambda \in \Lambda$, has been constructed. Then an approximate value λ_1^0 can be obtained from equation (4.3.4),

$$\mu_1(\lambda) = 1; \quad \lambda^0 = \arg_\lambda[\mu^*(\lambda) - 1]; \quad \lambda, \lambda^0 \in \Lambda. \tag{4.3.7}$$

4.4. Closeness Criterion and the Theory of Perturbations

4.4.1. Some properties of the solution of the comparison problem. Let λ^0 be a root of equation (4.3.7). Inserting this root into equation (4.1.1), consider the following Cauchy problem:

$$(p(x, \lambda^0)v')' + r(x, \lambda^0)v = 0, \quad v(0) = 0, \quad v'(0) = 1. \tag{4.4.1}$$

Let us construct the solution of problem (4.4.1) by some numerical or analytical method. From (4.3.5), (4.3.6) and the second Sturm theorem about oscillations [36, 49], we obtain the following statement.

Theorem 4.1. *Suppose that $\partial p/\partial \lambda \leq 0$, $\partial r/\partial \lambda > 0$ or $\partial p/\partial \lambda < 0, \partial r/\partial \lambda \geq 0$. Then, for the first positive root ξ of the equation $U(x, \lambda^0) = 0$ the following inequalities hold:*

$$\xi < 1, \ \lambda^0 > \lambda \quad \text{or} \quad \xi > 1, \ \lambda^0 < \lambda; \quad \xi = \xi(\lambda^0) = \arg_x v(x, \lambda^0). \tag{4.4.2}$$

If $\partial p/\partial\lambda \geq 0, \partial r/\partial\lambda < 0$ *or* $\partial p/\partial\lambda > 0, \partial r/\partial\lambda \leq 0$), *then the inequalities opposite to* (4.4.2) *hold.*

This theorem follows from relation (4.4.3) proved below.

Let us introduce the parameter $\varepsilon = 1 - \xi$ and consider the strong inequality $|\varepsilon| \ll 1$ as a condition of closeness between λ^0 and λ_1. The above considerations show that the value $\lambda = \lambda^0$ and the function $v(x, \lambda^0)$ represent an exact solution of the generalized problem (4.1.1) on the known interval $0 \leq x \leq \xi(\lambda^0)$. Since the parameter ε is small, the perturbation procedure repeatedly used above allows us to refine the eigenvalue λ_1 and the corresponding eigenfunction.

Note that in order to obtain λ_1 (as well as other λ_n), one may take the following approach, which is quite obvious but requires much effort. Using the method of accelerated convergence, as we have done in Chapter 2, one can solve the boundary value problem (4.3.2) and thus obtain a highly precise description of the function $\mu_1(\lambda)$. Then, using interpolation, one solves the equation $\mu_1(\lambda) = 1$ (or the general equation $\mu_m(\lambda) = 1$). The smallest root λ_1 of this equation is taken as the first eigenvalue of the boundary value problem (4.1.1). In order to check whether λ is close to the exact eigenvalue, one should insert it into (4.4.1) and find the difference between the root of the equation $v(x, \lambda_1) = 0$ and $x = 1$. There is another, more simple, approach which does not require the construction of the function $\mu_1(\lambda)$ or the solution of the equation $\mu_1(\lambda) = 1$. To that end, one introduces the perturbed argument $y = x\xi$, so that (4.1.1) becomes a boundary value problem with an explicit small parameter.

4.4.2. Approximate solution of the perturbed problem. Let us take λ^0 and $v(x, \lambda^0)$ as the generating solution (the null-approximation) and then perform calculations as described in Chapter 2. As a result, we obtain a refined value λ_1 which takes into account the terms of the order ε (i.e., the error is $O(\varepsilon^2)$)

$$\begin{aligned} \lambda_{1(1)} &= \lambda^0 - \varepsilon\xi p(\xi, \lambda^0) v'^2(\xi, \lambda^0) d^{-1}(\lambda^0), \quad \varepsilon = 1 - \xi(\lambda^0), \\ d(\lambda^0) &= \int_0^\xi [-p'_\lambda(x, \lambda^0) v'^2(x, \lambda^0) + r'_\lambda(x, \lambda^0) v^2(x, \lambda^0)]\, dx. \end{aligned} \tag{4.4.3}$$

The "norm" $d(\lambda^0)$ can be expressed in another way through an auxiliary function $w(\xi, \lambda^0)$ which is found, together with $v(x, \lambda^0)$, as a solution of the Cauchy problem

$$(p(x, \lambda^0) w')' + r(x, \lambda^0) w = -\left(p^{-1}\frac{\partial p}{\partial\lambda}\right)' pv' + \left(rp^{-1}\frac{\partial p}{\partial\lambda} - \frac{\partial r}{\partial\lambda}\right) v, \quad w = w'(0) = 0,$$

$$d(\lambda_0) = p(\xi, \lambda^0) v'(\xi, \lambda^0) w(\xi, \lambda^0). \tag{4.4.4}$$

4.5. The Method of Accelerated Convergence for Generalized Sturm–Liouville Problems

Finding the solution of the Cauchy problem (4.4.1) and the root ξ and using formulas (4.4.3), (4.4.4), one can construct an accelerated convergence algorithm on the basis of the ideas repeatedly expounded above.

The recurrent formulas of that algorithm read as follows:

$$\begin{aligned} \lambda_{1(k+1)} &= \lambda_{1(k)} - \varepsilon_k \xi_k p(\xi_k, \lambda_{1(k)}) \frac{v'^2(\xi_k, \lambda_{1(k)})}{d(\lambda_{1(k)})}, \quad \xi_k(\lambda_{1(k)}) = \arg v(x, \lambda_{1(k)}), \\ \varepsilon_k &= 1 - \xi_k, \quad k = 0, 1, 2, \ldots; \quad \lambda^0 = \lambda_{1(0)}, \quad \varepsilon_0 = \varepsilon = 1 - \xi(\lambda_{1(0)}). \end{aligned} \tag{4.5.1}$$

After passing to the limit in (4.5.1) as $k \to \infty$, we obtain

$$\left.\frac{d\lambda_n}{d\xi}\right|_{\xi=1} = -p(1,\lambda_n)\frac{u'^2(1,\lambda_n)}{d(\lambda_n)}, \qquad n = 1, 2, \ldots \tag{4.5.2}$$

Without dwelling on the proof of this relation, we just mention that the convergence of the functions $v(x, \lambda_{1(k)})$ to $u(x, \lambda_1)$, as well as the convergence of their derivatives in x, is uniform for $0 \le x \le 1$ and has accelerated (quadratic) character with respect to the iteration number. Observe that with no loss of accuracy in ε_k, one can replace ξ_k in (4.5.1) by $\xi_k = 1$, except in the formula $\varepsilon_k = 1 - \xi_k$.

The accelerated convergence method can be modified in such a way that it becomes applicable to more general problems, in particular, problems for systems of equations, those with boundary conditions of the second or the third kind, or with periodic conditions. Computational experience shows that this method is highly effective in the sense that it can be easily implemented, requires little computer memory, is stable to computer failure, involves no roundoff error accumulation. This method is especially useful for mass highly precise calculations connected with parametric synthesis of systems described by generalized eigenvalue problems (see Chapter 9).

4.6. Model Problems

4.6.1. Test example for an integrable equation. In order to test our algorithm, consider some examples with calculations based on the method of accelerated convergence. Let $p \equiv 1$, $r(x, \lambda) = (\lambda + x^2)^{-2}$ in (4.1.1). We consider the region $0 \le x \le 1, \lambda \in \Lambda = \{\lambda : \lambda > 0, \lambda < -1\}$. For definiteness, let $\lambda > 0$. Rough estimates of λ_n can be obtained quite easily. Indeed, since $r(x, \lambda) \le 1/\lambda^2$, we have $0 < \lambda_n \le (\pi n)^{-1}$. With the help of some analytical tricks, it is possible to construct an exact solution of the form

$$u(x, \lambda) = C(\lambda + x^2)^{1/2} \sin\varphi(x, \lambda), \qquad \varphi(x, \lambda) = (1 + \lambda^{-1})^{1/2} \arctan(x\lambda^{-1/2});$$

$$\lambda_n = \mathrm{Arg}[\varphi(1, \lambda) - \pi n], \quad n = 1, 2, \ldots, \quad \lambda_n = (2n)^{-2} + O(n^{-4}), \quad n \gg 1; \tag{4.6.1}$$

$$\lambda_1 = 0.165643; \qquad \lambda_2 = 0.048674; \qquad \lambda_3 = 0.023214, \ldots$$

The accelerated convergence algorithm can be realized numerically with elementary facilities which ensure relative calculation errors of order 10^{-6} and have moderate calculation speed and minimal memory resources.

By the Rayleigh principle, using formulas (4.3.6), (4.3.7) with the test function $\psi(x) = \sin \pi x$, we obtain a rough lower bound, $\lambda_1^0 = \lambda_{1*} = 0.16$, for the first eigenvalue λ_1. Solving the Cauchy problem (4.4.1) and using (4.4.2), we obtain $\xi = 0.935549$, $\varepsilon = 0.064451$, and the refined $\lambda_{1(1)} = 0.165425$ close enough to λ_1 in (4.6.1). The next iteration yields the approximate value $\lambda_{1(2)} = 0.165656$, which is an upper bound and is also very close to λ_1. Using the two-sided estimates for λ_1, we calculate the mean value $\lambda_1 \sim 0.1655405$, which approximates the exact value with the relative error $\Delta\lambda_1/\lambda_1 \approx 5 \cdot 10^{-4}$. The small discrepancy between the values obtained and the exact ones is due to roundoff errors of these calculations. Note that the fast convergence of the algorithm exhausts the precision capacities of modern computers within several iterations. Thus, for $\varepsilon = 0.1 \div 0.01$, the number of iterations required for the calculations with the relative error of order $10^{-20} \div 10^{-30}$ should not exceed $k = 4 \div 6$. In practice, one usually performs calculations with relative errors of order $10^{-3} \div 10^{-5}$. Therefore, it suffices to consider 2 or 3 iterations (instead of $15 \div 20$ iterations by the shooting method).

For the sake of comparison, take the rough upper bound $\lambda_1^* = 0.17$ for λ_1. By the above method, we obtain $\xi = 1.052775$, $\varepsilon = -0.052775$, and the refined value $\lambda_{1(1)} = 0.165545$, which is a lower bound. The next iteration yields a refined lower bound $\lambda_{1(2)} = 0.165637$. Using the exact value λ_1 from (4.6.1), we find that the relative error satisfies the inequality $\Delta\lambda_1/\lambda_1 \le 4 \cdot 10^{-5}$. The next iteration ensures the relative error of order 10^{-6}.

4.6.2. Numerical example; two-sided estimates. Let us perform calculations in the case of a model problem for which no analytical solution is known. Let $p = 1$, $r = (\lambda + 0.1 \sin \pi x)^{-2}$. We consider the problem in the region $0 \le x \le 1$, $\lambda \in \Lambda = \{\lambda : \lambda > 0, \lambda < -0.1\}$. Let $\lambda > 0$. By the Rayleigh principle we obtain a rough bound for one of the eigenvalues, $\lambda_{1(0)} = 0.2$. Verification shows that this is a lower bound. The method of accelerated convergence yields the abscissa $\xi = 0,894047$, the parameter $\varepsilon = 0.105953$, and the refined eigenvalue approximation $\lambda_{1(1)} = 0.229819$. Using this, we obtain the abscissa $\xi_1 = 0.984047$ and the parameter $\varepsilon_1 = 0.015953$; the next refined lower bound is $\lambda_{1(2)} = 0.235203$. The third iteration yields the upper bound $\lambda_{1(3)} = 0.235283$, which allows us to find the relative error, $\Delta\lambda_1/\lambda_1 \le 2 \cdot 10^{-4}$. The fourth iteration ensures a relative error of order 10^{-5}. This example clearly demonstrates the quadratic convergence, $\varepsilon_1 = 0.015953 \sim \varepsilon^2$.

Similar calculations for the second eigenvalue easily allow us to obtain $\lambda_2 = 0.097163$ with the relative error $\Delta\lambda_2/\lambda_2 \le 4 \cdot 10^{-4}$.

These nontrivial test examples show the efficiency of the accelerated convergence method for solving generalized Sturm–Liouville problems. Calculations in these examples have been performed with the help of simplest programmable microcomputers.

4.7. Generalized Parametric Vibrations

4.7.1. Statement of the generalized periodic problem. Consider vibrations of a linear system described by the generalized periodic boundary value problem for the equation of Hill's type [13, 16, 19, 43, 61]

$$\ddot{u} + r(t, \lambda)u = 0, \qquad u(0) = u(1), \quad \dot{u}(0) = \dot{u}(1), \tag{4.7.1}$$

Here, the dots denote derivatives with respect to the time t $(0 \le t \le 1)$, $r(t, \lambda)$ is a piecewise continuous real-valued function, $r(t + 1, \lambda) \equiv r(t, \lambda)$. It is assumed that r has been periodically extended for all $|t| \ge 0$ as a piecewise continuous function. In particular, the function r may be smooth with respect to t. It is also assumed that r, as a function of the real parameter λ, is sufficiently smooth, for instance, continuously differentiable for $\lambda \in \Lambda$, where Λ is the set of admissible values of λ that consists of finitely many intervals.

Our aim is to find the solution of the generalized problem (4.7.1), i.e., find the eigenvalues λ and eigenfunctions $u(t, \lambda)$. Of special interest for applications is the case of discrete spectrum $\lambda \in \{\lambda_n\}$ and discrete set of vibration shapes $\{u_n(t)\}$, $n = 1, 2, \ldots$ As a rule, the coefficient r depends on other parameters of the system, denoted by γ (geometric, inertial, elastic, or other characteristics), i.e., $r = r(t, \lambda, \gamma)$, $\gamma \in \Gamma$. In such a case, it is required to construct curves or surfaces $\lambda_n(\gamma)$ and the respective vibration shapes (eigenfunctions) $u_n(t, \gamma)$.

In the classical periodic problem (see Section 2.9), the function r is linear with respect to λ, and one usually considers $r = \lambda + \gamma q(t)$, which corresponds to the Hill equation. For $q \equiv \cos 2\pi t$, we have the Mathieu equation, and for $q \equiv \operatorname{sign}(\cos 2\pi t)$, the Meissner equation [61].

In the vibration theory, motions described by equations with periodic coefficients are called parametric vibrations. Accordingly, we use to term "generalized parametric vibrations" [13] in reference to the motions described by boundary value problems of the type (4.7.1).

For periodic problems of the type (4.7.1) (in contrast to the Sturm–Liouville problem), there is usually no assumption that the function r should be positive in the domain of t, λ, γ under consideration. This is a great obstacle to the application of the oscillation theorems and the Sturm comparison theory and creates many difficulties in the way of constructing solutions and their analysis.

The investigation of the generalized periodic problem for equations of Hill's type (4.7.1) is very important for the theory of nonlinear vibrations and stability. Such problems arise in mechanics, electrical engineering, applied celestial mechanics, hydrodynamics, elasticity, and many other fields. Basic analytical results were obtained for the classical cases of the Hill equation — the equations of Mathieu and Meissner [43, 61]. Numerical-analytical methods for solving generalized periodic boundary value problems have not yet been developed to a desirable extent.

4.7.2. An example illustrating spectral properties. As mentioned above, the properties of generalized eigenvalues λ_n and eigenfunctions $u_n(t)$ may be essentially different from the properties of eigenvalues and eigenfunctions of classical problems. To demonstrate this fact, consider some simple problems of the form (4.7.1) which admit analytical solutions.

Example 1. Consider problem (4.7.1) with

$$r(t,\lambda) = (\lambda + t)^{-2}, \qquad \lambda \in \Lambda = \{\lambda : \lambda > 0, \lambda < -1\}, \qquad 0 \le t \le 1.$$

Analytical calculations show that the eigenvalues λ_n are determined by the roots of the following transcendental equation:

$$\begin{gathered} \cos(\sqrt{3}\ln\kappa) - \sqrt{3}\frac{\kappa^2-1}{\kappa^2+1}\sin(\sqrt{3}\ln\kappa) - \frac{2\kappa}{\kappa^2+1} = 0, \\ \xi(t) = \Bigl(1+\tfrac{t}{\lambda}\Bigr)^{1/2}, \qquad \kappa = \xi(1) = \Bigl(1+\tfrac{1}{\lambda}\Bigr)^{1/2}. \end{gathered} \tag{4.7.2}$$

The eigenvalues λ_n admit the following approximations:

$$\lambda_n \approx \left(\exp\left[\frac{2\pi}{\sqrt{3}}\left(n+\frac{1}{3}\right)\right]-1\right)^{-1}, \qquad |n| \gg 1.$$

The eigenfunctions $u_n(t)$ corresponding to λ_n have the form

$$\begin{gathered} u_n(t) = C_n\xi_n(t)[\kappa_n\sin\theta_n(1)\cos\theta_n(t) + (1-\kappa_n\cos\theta_n(1))\sin\theta_n(t)], \\ \xi_n(t) = \Bigl(1+\frac{t}{\lambda_n}\Bigr)^{1/2}, \quad \kappa_n = \xi_n(1), \quad \theta_n(t) = \sqrt{3}\ln\xi_n(t), \quad n = \pm1, \pm2, \dots \end{gathered} \tag{4.7.3}$$

Here, C_n are arbitrary constants, which are usually chosen on the basis of some additional conditions. From (4.7.2) it follows that the eigenvalues λ_n are concentrated near $\lambda = 0$ $(n \ge 1)$ and $\lambda = -1$ $(n \le -1)$, and exponentially approach these values as $|n| \to \infty$. The eigenfunctions $u_n(t)$ (4.7.3) oscillate in t for sufficiently large $|n|$, this oscillation may be arbitrarily fast and exhibits an exotic behavior.

Example 2. In a similar way, we examine problem (4.7.1) with $r = \lambda^2/(1+\lambda t)^2$, $\lambda > -1$. Replacing λ by λ^{-1} in (4.7.2) and (4.7.3), we obtain the following approximate expressions for the eigenvalues λ_n:

$$\lambda_n \approx \exp\left[\frac{2\pi}{\sqrt{3}}\left(n+\frac{1}{3}\right)\right] - 1, \qquad n = \pm1, \pm2, \dots \tag{4.7.4}$$

According to (4.7.2) and (4.7.4), we have $\lambda_n \to +\infty$ as $n \to \infty$, and $\lambda_n \to -1$ as $n \to -\infty$. The corresponding eigenfunctions for large $|n|$ also exhibit fast oscillation.

4.7.3. General properties of solutions of generalized periodic problems. The spectrum of the generalized periodic boundary value problem (4.7.1) may be discrete (finite or countable), discrete-continuous, or empty. The points of the discrete spectrum may be either simple or degenerate, i.e., the same eigenvalue may correspond to several linearly independent eigenfunctions.

Eigenfunctions corresponding to the points of the discrete spectrum of problem (4.7.1) satisfy the orthogonality conditions

$$\int_0^1 [r(t,\lambda_n) - r(t,\lambda_m)]u_n(t)u_m(t)dt = 0, \qquad r(t,\lambda_n) \not\equiv r(t,\lambda_m). \tag{4.7.5}$$

Passing to the limit in (4.7.5) for $\lambda_m \to \lambda_n$, we obtain the expression

$$N_n = \int_0^1 r'_\lambda(t,\lambda_n)u_n^2(t)dt. \tag{4.7.6}$$

Under the additional condition $r'_\lambda > 0$ for the derivative of r in λ, the expression (4.7.6) is equal to the squared norm of the function $u_n(t)$ with the weight $r'_\lambda(t,\lambda_n)$.

By a formal procedure, problem (4.7.1) can be represented as a continuous family of classical problems depending on the parameter $\lambda \in \Lambda$.

4.7.4. An extended setting of the problem and the procedure of its approximate solution. Consider the problem with the unknown μ, U

$$\begin{gathered} \ddot{U} + [\mu + r(t,\lambda)]U = 0, \quad 0 \le t \le 1, \quad U(0) = U(1), \quad U'(0) = U'(1), \\ \mu \in \{\mu_m(\lambda)\}, \quad U_m(t,\lambda) = U(t,\mu_m(\lambda),\lambda), \quad \lambda \in \Lambda, \quad m = 1,2,\dots \end{gathered} \tag{4.7.7}$$

Suppose that we have constructed a family of solutions of the auxiliary problem (4.7.7). Then, a solution of the original eigenvalue problem (4.7.1) can be obtained by the following formulas:

$$\lambda_n = \operatorname{Arg}\mu_m(\lambda), \quad \lambda \in \Lambda \quad (\mu_m(\lambda_n) = 0, \quad n = n(m)). \tag{4.7.8}$$

Thus, equating the eigenvalues μ_m to zero, we find the corresponding λ_n and the eigenfunctions of the original problem (4.7.1),

$$u_n(t) = U(t,0,\lambda_n). \tag{4.7.9}$$

This approach is connected with the construction of a discrete set of functions $\mu_m(\lambda)$ and the determination of their real roots λ_n. This construction for the classical problem (4.7.7) can be easily accomplished by the accelerated convergence method, as described in Section 2.9. Thus, after one or two iterations, we can find a bound $\lambda_0 \in \Lambda$ which allows us to introduce the small parameter ε. Without going into details, which are similar to those of Section 2.9, we describe the main steps of the accelerated convergence algorithm.

1. Having found the bound λ_0, insert it into equation (4.7.1) and find a numerical or analytical solution of two Cauchy problems ($i = 1, 2$)

$$v_i'' + r(t,\lambda_0)v_i = 0, \quad 1)\ v_1(0) = 0,\ v_1'(0) = 1; \quad 2)\ v_2(0) = 1,\ v_2'(0) = 0. \tag{4.7.10}$$

2. From the solutions $v_{1,2}(t,\lambda_0)$ of problems (4.7.10), construct the determinant Δ and find its root ξ closest to $t = 1$,

$$\Delta(t,\lambda_0) = v_1(t,\lambda_0) + v_2(t,\lambda_0) - 2 = 0, \qquad \xi = \xi(\lambda_0) = \arg_t \Delta(t,\lambda_0). \tag{4.7.11}$$

3. Using the formula of the general solution

$$v_0(t,\lambda_0) = C_1 v_1(t,\lambda_0) + C_2 v_2(t,\lambda_0), \tag{4.7.12}$$

find a periodic solution with period ξ.

4. Calculate the small parameter $\varepsilon = 1 - \xi$, $|\varepsilon| \ll 1$.

5. Calculate a refined value of λ_1,

$$\lambda_{1(1)} = \lambda_0 + \varepsilon\alpha(\xi,\lambda_0); \quad \alpha(\xi,\lambda_0) = -\frac{1}{N(\xi,\lambda_0)}[r(\xi,\lambda_0)v_0^2(\xi,\lambda_0) + \dot{v}^2(\xi,\lambda_0)],$$
$$N(\xi,\lambda_0) = \int_0^\xi r'_\lambda(t,\lambda_0)v_0^2(t,\lambda_0)\,dt. \tag{4.7.13}$$

It is assumed here that the "squared norm" differs from zero, i.e., $N(\xi,\lambda_0) \neq 0$. The case of $N(\xi,\lambda_0) = 0$ requires additional investigation.

Next, the refined value $\lambda_{1(1)}$ is inserted into (4.7.10) and the process is repeated again and again, so that we consecutively obtain the refinements $\lambda_{1(2)}$, $\lambda_{1(3)}$, etc. When the desired accuracy is reached, the process is stopped. The accuracy of calculations is tested on the basis of the criterion $\xi_k \to 1$, i.e., $\varepsilon_k \to 0$. Note that for a multiple root ξ of equation (4.7.11), the eigenvalue λ_1 may split in a small neighborhood of $\lambda = \lambda_0$. But this does not change the general scheme of calculations.

Under the above assumptions on $r(t,\lambda)$, the recurrent scheme (4.7.10)–(4.7.13) exhibits accelerated (quadratic) convergence with respect to the small parameter ε.

For practical calculations, the following remark may be useful. If we consider the interval $-1 \le t \le 1$ and $r(t,\lambda)$ is an even function of t, then the identity $u(t) \equiv u^+(t) + u^-(t)$ (here, $u^+(t) = (u(t) + u(-t))/2$ is an even solution and $u^-(t) = (u(t) - u(-t))/2$ is an odd solution) implies that the periodic problem (4.7.1) can be split into two generalized Sturm–Liouville problems. Thus, for $r(t,\lambda) = r(-t,\lambda)$, we have two boundary value problems (one of the first kind and another of the second) on the half-segment $0 \le t \le 1$,

$$\ddot{u} + r(t,\lambda)u = 0, \quad 1)\ u(0) = u(1) = 0, \quad 2)\ \dot{u}(0) = \dot{u}(1) = 0. \tag{4.7.14}$$

A meaningful example with calculations by this method is given in Chapter 12.

4.8. Generalized Boundary Value Problems with Spectral Parameter in Boundary Conditions

For some technical applications, it is necessary to find a solution of the generalized Sturm–Liouville problem with boundary conditions involving the spectral parameter. We consider a typical problem of this type: *Find eigenvalues and eigenfunctions of the boundary value problem*

$$(p(x,\lambda)u')' + r(x,\lambda)u = 0,$$
$$(k_0 - m_0\lambda)u(0) - p(0,\lambda)u'(0) = 0, \tag{4.8.1}$$
$$(k_1 - m_1\lambda)u(1) + p(1,\lambda)u'(1) = 0.$$

This modification of boundary conditions adds nothing new to the implementation of the accelerated convergence method. The scheme of calculations described in previous sections is absolutely the same. Therefore, we will only describe the main steps of the accelerated convergence algorithm and give the final formulas.

Suppose that we have found an approximate value λ_0. Following the standard scheme, we construct a solution of the Cauchy problem

$$
\begin{aligned}
&(p(x,\lambda_0)u')' + r(x,\lambda_0)u = 0,\\
&u(0) = C_1, \quad u'(0) = C_2, \quad u = U(x,\lambda_0).
\end{aligned}
\tag{4.8.2}
$$

The constants C_1 and C_2 are chosen from the boundary conditions at the point $x = 0$,

$$(k_0 - m_0\lambda_0)C_1 - p(0,\lambda)C_2 = 0. \tag{4.8.3}$$

One of these constants may be taken equal to unity. Using the solution $U(x,\lambda_0)$ of the Cauchy problem (4.8.2), we construct the auxiliary function

$$E(x,\lambda_0) = (k_1 - m_1\lambda_0)U(x,\lambda_0) + p(x,\lambda)U'(x,\lambda_0). \tag{4.8.4}$$

Next, we find the root $\xi = \xi(\lambda_0)$ of the function $E(x,\lambda_0)$ nearest to $x = 1$ and introduce the small parameter $\varepsilon = 1 - \xi$. Then we refine the eigenvalue λ_1,

$$
\begin{aligned}
\lambda_{1(1)} &= \lambda_0 - \varepsilon[p(\xi,\lambda_0)U'^2(\xi,\lambda_0) + \lambda_0 r(\xi,\lambda_0)U^2(\xi,\lambda_0)]N^{-1},\\
N &= d(\lambda_0) + m_0U^2(0,\lambda_0) + m_1U^2(\xi,\lambda_0).
\end{aligned}
\tag{4.8.5}
$$

Having found $\lambda_{1(1)}$, we repeat the process with $\lambda_{1(1)}$ instead of λ_0 and calculate another root ξ_1 and another parameter $\varepsilon_1 = 1 - \xi_1$, obtain a refined value $\lambda_{1(2)}$, etc.

All formulas of this section are obtained by standard calculations described in previous sections.

4.9. Exercises

Exercise 1. By analytic means, find a solution of the generalized Sturm–Liouville problem

$$u'' + (\lambda - x^2)^{-2}u = 0, \qquad u(0) = u(1) = 0. \tag{4.9.1}$$

Introducing a new unknown function v by the relation $u = v\sqrt{\lambda - x^2}$ and a new independent variable, $t = (\sqrt{\lambda} - x)(\sqrt{\lambda} + x)^{-1}$, transform equation (4.9.1) to an equation of Euler type, which can be solved analytically.

Exercise 2. Find a solution of problem (4.9.1) by the method of accelerated convergence. Calculate the first and the second eigenvalues. Compare the results with those obtained by analytical means. Take $\psi(x) = \sin(\pi x)$ and $\psi(x) = x(1 - x)$ as test functions.

Exercise 3. Using the method of accelerated convergence, find a solution of the generalized Sturm–Liouville problem

$$(\sqrt{\lambda + x}u')' + u = 0, \qquad u(0) = u(1) = 0.$$

Chapter 5

Asymptotics of Eigenvalues and Eigenfunctions of the Generalized Sturm–Liouville Problem for Higher Vibration Modes

For a wide class of systems, higher modes of free vibrations can be studied by a constructive approach based on the principle of separation of the so-called fast and slow motions (the method of averaging). This approach yields explicit expressions for eigenvalues and eigenfunctions with the desired accuracy with respect to inverse powers of the mode index.

5.1. General Notions Regarding the Asymptotic Behavior of Eigenvalues Corresponding to Higher Vibration Modes

5.1.1. Statement of the generalized problem. We are going to study free vibrations of mechanical systems described by the following generalized eigenvalue problem [2, 8]:

$$\begin{gathered}(p(x,\lambda)u')' + f(x,\lambda)u = 0, \qquad u(0) = u(L) = 0;\\ 0 < C_1 \le p, f \le C_2 < \infty, \quad 0 \le x \le L < \infty.\end{gathered} \tag{5.1.1}$$

The coefficients p and f are determined by the mechanical properties of the system under consideration (see Section 5.6). These functions are assumed sufficiently smooth in the domain of x, λ and possess certain additional properties. The parameter λ characterizes vibration frequency or is a physical parameter of the system. In what follows, we consider this problem for arbitrarily large λ (i.e., $\lambda \to \infty$). Relations (5.1.1) are written in dimensionless variables. The problem is to find real eigenvalues λ_n and eigenfunctions $u_n(x)$ for sufficiently large $n \gg 1$ [24, 33, 46–48].

For the sake of definiteness, we consider problem (5.1.1) with the boundary conditions of the first kind. It is also possible to consider more general boundary conditions, namely, those of the third kind at one or both end-points (see Section 5.5), or periodic boundary conditions [13]. The generalized eigenvalue problem (5.1.1) belongs to the class of problems studied in [22, 25, 33, 43, 46, 47?].

The main results were obtained for the self-conjugate problem with $f \equiv \lambda r(x) - q_0(x)$. Usually, one takes $p = r \equiv 1$. The theory, as well as analytical and numerical methods for the construction of the solution λ_n, $u_n(x)$ (variational methods, the finite element method, the method of accelerated convergence, etc.), was developed in [4–6, 18–24, 27–30, 33, 37, 39, 41–48, 53, 55, 58, 60–62, 68]. The numerical methods are effective for relatively small $n \sim 1$, i.e., for lower vibration modes. From the theoretical standpoint, as well as for applications, it is important to know the behavior of the solution for $n \gg 1$, in particular, its asymptotic behavior as $n \to \infty$.

Asymptotic formulas for λ_n and $u_n(x)$ were constructed in [24], where the leading terms of the expansions are given. For applications, it would be useful to have more precise approximations which allow for practical calculations with arbitrarily small absolute errors for relatively small $n \sim 1$.

We propose an efficient numerical-analytical method which can be used for the construction of highly precise approximate solutions of the generalized boundary value problem (5.1.1) with the functions p, f satisfying the following conditions:

$$\begin{gathered} p \equiv p(x), \quad f \equiv \omega^2 r(x) - \omega q(x,\varepsilon), \quad \lambda = \omega^2, \quad \varepsilon = \omega^{-1}, \\ 0 < C_1 \le p, r \le C_2 < \infty, \quad 0 < \varepsilon \le \varepsilon_0, \quad \varepsilon_0 \ll 1, \quad \omega \gg 1. \end{gathered} \tag{5.1.2}$$

Functions of the form (5.1.2) often occur in applications (see Example 5.6.4). The parameter ω characterizes frequency, and ε is the period of vibrations. The parameter ω is to be determined. The functions p, r, q smoothly depend on x, and q satisfies the Lipschitz condition in ε. Note that by the nonsingular linear transformation $u^* = u\sqrt{p}$, equation (5.1.1) is reduced to an equation with $p \equiv 1$.

5.1.2. Classical results. The classical Sturm–Liouville problem is a special case of problem (5.1.2) with $q \equiv \varepsilon q_0(x)$, i.e., $q(x,0) \equiv 0$. For this case, the following fundamental result is known [24] (see Subsection 1.2.3):

$$\begin{gathered} \lambda_n = (\pi n)^2 \left(\int_0^L \nu_0(x)\,dx \right)^{-2} + O(1), \quad \nu_0(x) = \left[\frac{r(x)}{p(x)} \right]^{1/2}; \\ u_n(x) = \left(\frac{2}{L} \right)^{1/2} [\mu_0(x)]^{1/2} \sin\left(\frac{\pi n}{L} \int_0^x \nu_0(y)\,dy \right) + O\left(\frac{1}{n} \right), \\ \mu_0(x) = [r(x)p(x)]^{-1/2}, \quad (u_n, u_m)_r = \sigma_{nm} + O(n^{-1} + m^{-1}). \end{gathered} \tag{5.1.3}$$

The approximation (5.1.3) does not take into account the function $q_0(x)$, and therefore, it is desirable to have more precise formulas.

Highly precise asymptotic expansions for the standard Sturm–Liouville problem (see Subsection 1.2.3) have been discussed previously, but the formulas found in literature are nonconstructive, since they are not expressed in terms of the original coefficients, have implicit character, and are based on solutions of transcendental equations, in general.

Next, we describe a constructive method for obtaining an approximate solution of the generalized boundary value problem (5.1.1) (5.1.2). This approach is based on the method of averaging, the method of accelerated convergence, and asymptotic expansions. Our approach can be extended to wider classes of problems, as shown below. Note that in the problem under consideration, the functions $f(x,\lambda)$ and $\lambda r(x) - q_0(x)$ are asymptotically close for $\lambda \to \infty$; the absolute values of these functions may grow and be of the order $O(\sqrt{\lambda})$.

5.2. Application of Asymptotic Methods of the Theory of Nonlinear Vibrations

5.2.1. "Amplitude–phase" variables. Consider the generalized boundary value problem (5.1.1), (5.1.2) for asymptotically large values of the parameter λ. It is assumed that these relations are written in dimensionless form, i.e., λ, ω, ε, L are numerical parameters, $\lambda, \omega \gg 1$, $L \sim 1$. We introduce the argument $s = \omega x$ varying on the asymptotically large interval $0 \le s \le L\varepsilon^{-1}$. Then, (5.1.1) turns into a problem for a vibration system with the slowly varying parameter $x = \varepsilon s$.

Let us write the boundary value problem (5.1.1)–(5.1.2) in the standard "amplitude-phase" variables a, φ [19, 48]

$$u = a\sin\varphi, \quad \frac{du}{ds} = a\nu_0(x)\cos\varphi, \quad a > 0.$$

As result, we obtain

$$\begin{gathered} \tfrac{da}{ds} = \varepsilon a F(x,\varphi,\varepsilon), \quad \tfrac{d\varphi}{ds} = \nu_0(x) + \varepsilon N(x,\varphi,\varepsilon), \\ \varphi(0) = 0, \quad \varphi(L\varepsilon^{-1}) = \pi n, \end{gathered} \tag{5.2.1}$$

where

$$F \equiv \tfrac{1}{2}g(x,\varepsilon)\sin 2\varphi - \tfrac{1}{2}h(x)(1+\cos 2\varphi), \qquad g \equiv q(x,\varepsilon)\mu_0(x)$$

$$N \equiv -\tfrac{1}{2}g(x,\varepsilon)(1-\cos 2\varphi) + \tfrac{1}{2}h(x)\sin(2\varphi), \quad h \equiv -(\ln\mu_0(x))'.$$

The derivative in the "fast" variable $s = \omega x$ characterizes the change rate of the phase φ: $d\varphi/ds \approx \nu_0 \sim 1$; and x is the "slow" variable. Note that the right-hand sides of equations (5.2.1) are π-periodic in φ, and the variable φ can be separated. For this variable, we have a boundary value problem on the asymptotically large interval $0 \le s \le L\varepsilon^{-1}$ whose length is determined by $n \gg 1$. The parameter $\varepsilon = \omega^{-1}$ is not given and is found from the boundary condition. Having calculated the function $\varphi(s,\varepsilon)$ and the values $\varepsilon_n = \varepsilon(n)$, we obtain the unknown function $a_n = a(s,\varepsilon_n)$ in terms of a simple quadrature (see Section 5.4).

5.2.2. Approximation of the phase. The fast phase $\varphi(s,\varepsilon)$ on the interval $0 \le s \le L\varepsilon^{-1}$ is constructed in approximation by the standard averaging method. According to this method, the function φ is replaced by a close function ψ, called the "averaged" phase, which satisfies a certain equation with the desired error in ε. Thus, we seek φ in the form [2]:

$$\begin{aligned} \varphi &= \psi + \varepsilon V(x,\psi,\varepsilon) = \psi + \varepsilon V_1(x,\psi,\varepsilon) + \varepsilon^2 V_2 + \cdots + \varepsilon^k V_k + \cdots, \\ \tfrac{d\psi}{ds} &= \nu_0(x) + \varepsilon M(x,\varepsilon) = \nu_0(x) + \varepsilon\nu_1(x,\varepsilon) + \varepsilon^2\nu_2 + \cdots + \varepsilon^k\nu_k + \cdots. \end{aligned} \tag{5.2.2}$$

Here, V_i, ν_i are unknown functions regularly depending on ε through $g(x,\varepsilon)$. The approximation order k in (5.2.2) is limited only by the smoothness of the function N with respect to x, since N is analytic in φ. Taking into account the terms of the order $O(\varepsilon^k)$ yields errors of the order $O(\varepsilon^k)$ on the interval $0 \le s \le L\varepsilon^{-1}$. The unknown functions V_i, ν_i are constructed in a standard way, with the π-periodicity in ψ taken into account. The averaged derivative $d\psi/ds$ can be represented in explicit analytical form,

$$\begin{aligned} \frac{d\psi}{ds} &= \nu^{(k)}(x,\varepsilon) + O(\varepsilon^{k+1}), \quad \nu^{(k)}(x,\varepsilon) \equiv \sum_{i=0}^{k} \varepsilon^i \nu_i(x,\varepsilon); \\ \nu_0(x) &= \left(\frac{r(x)}{p(x)}\right)^{1/2}, \quad \nu_1(x,\varepsilon) = \langle N\rangle = -\frac{1}{2}g(x,\varepsilon), \\ \nu_2 &= -\left\langle\frac{\partial V_1}{\partial x}\right\rangle + \left\langle\tfrac{\partial N}{\partial\psi}V_1\right\rangle = -\frac{1}{4}H' - \tfrac{\nu_0}{8}(H^2+G^2), \quad H \equiv \tfrac{h}{\nu_0}, \quad G \equiv \tfrac{g}{\nu_0}, \\ \nu_3 &= -\left\langle\tfrac{\partial V_2}{\partial x}\right\rangle + \frac{1}{2}\left\langle\frac{\partial^2 N}{\partial\psi^2}V_1^2\right\rangle + \left\langle\tfrac{\partial N}{\partial\psi}V_2\right\rangle \\ &= -\tfrac{1}{32}HG' - \tfrac{3}{32}H'G - \tfrac{\nu_0^2}{32}G(H^2+G^2) + \tfrac{1}{8}\left(\tfrac{G'}{\nu_0}\right)', \end{aligned} \tag{5.2.3}$$

etc. The angular brackets here and in what follows denote the integral mean value of the corresponding π-periodic functions of ψ. These functions have elementary trigonometric structure and their mean values can be calculated explicitly. The functions $V_i(x,\psi,\varepsilon)$ are also defined explicitly as follows:

$$\begin{aligned} V_1(x,\psi,\varepsilon) &= \tfrac{1}{\nu_0}\int_0^{\psi}(N-\langle N\rangle)d\varphi = \frac{1}{4\nu_0(x)}\left[g(x,\varepsilon)\sin 2\varphi + h(x)(1-\cos 2\varphi)\right], \\ V_2(x,\psi,\varepsilon) &= \tfrac{1}{\nu_0}\int_0^{\psi}\left(\left\langle\frac{\partial V_1}{\partial x}\right\rangle - \frac{\partial V_1}{\partial x} + \frac{\partial N}{\partial \psi}V_1 - \left\langle\frac{\partial N}{\partial \psi}V_1\right\rangle\right)d\varphi - \frac{\nu_1}{\nu_0}V_1 \\ &= -\tfrac{G'}{8\nu_0}(1-\cos 2\psi) + \tfrac{1}{8}\left(H^2+G^2+\tfrac{H'}{\nu_0}\right)\sin 2\psi \\ &\quad + \tfrac{1}{16}HG(1-\cos 4\psi) - \tfrac{1}{32}(H^2-G^2)\sin 4\psi. \end{aligned} \tag{5.2.4}$$

Subsequent coefficients ν_i, V_{i-1} in the asymptotic expansions (5.2.2) can be found in a similar way [19, 48].

Using (5.2.3), (5.2.4), we can write approximations of the sought variables ψ, φ with the absolute error $O(\varepsilon^3)$ for all $0 \le s \le L\varepsilon^{-1}$:

$$\begin{aligned} \psi^{(k)}(s,\varepsilon) &= \int_0^s \nu^{(k)}(\varepsilon y,\varepsilon)\,dy, \qquad k=0,1,2,\ldots; \qquad s = x\varepsilon^{-1}, \\ \varphi^{(3)}(x,\varepsilon) &= \psi^{(3)}(s,\varepsilon) + \varepsilon V_1(x,\psi^{(2)}(s,\varepsilon),\varepsilon) + \varepsilon^2 V_2(x,\psi^{(1)}(s,\varepsilon),\varepsilon). \end{aligned} \tag{5.2.5}$$

Here, $\nu^{(k)}$ are defined by (5.2.3). It is unnecessary to substitute higher approximations of the phase ψ into the coefficients $V_{1,2}$, and one can take $V_2(x,\psi^{(1)},0)$ and drop the terms of the order $O(\varepsilon)$ in $\psi^{(1)}(s,\varepsilon)$. Recall that the dependence of the functions ν_i and V_i on ε is due to the possible nonsmoothness of $g(x,\varepsilon)$ in (5.1.2), which is only assumed Lipschitz continuous in ε. Note that $V_i(x,\pi n,\varepsilon)\equiv 0$, i.e., the boundary conditions for $\psi^{(k)}$ and $\varphi^{(k)}$ coincide. Using the approximate expressions obtained above for ψ, φ, one can calculate the eigenvalues ε_n, as well as $\omega_n = \varepsilon_n^{-1}$ and $\lambda_n = \omega_n^2$, with the desired accuracy. As mentioned above, the quantities ε_n characterize small periods of free vibrations and ω_n are high frequencies for $n \gg 1$.

Similar expansions can be used for problem (5.1.1) with the coefficients p and f of a more general structure than in (5.1.2). For instance, one can take

$$f \equiv \omega^K r(x,\varepsilon) + \omega^{K-1}q(x,\varepsilon), \quad p \equiv \omega^{K-2}w(x,\varepsilon), \quad K \ge 2, \quad \omega = \sqrt{\lambda} = \varepsilon^{-1}, \tag{5.2.6}$$

with the usual assumptions on the sign of the functions r, w in (5.2.6).

5.3. Determination of Eigenfrequencies and Vibration Phases

5.3.1. Introduction of intermediate parameters. Approximate values of the spectral parameters ε_n, $\omega_n = \varepsilon_n^{-1}$, $\lambda_n = \omega_n^2$ of problem (5.1.1), (5.1.2) are calculated on the basis of the boundary condition for ψ at $s = L/\varepsilon$ in (5.2.1) and formulas (5.2.5). To facilitate calculations, we introduce a new parameter, $\pi n\varepsilon$, which remains bounded on eigenvalues as $n\to\infty$. For the unknown z we obtain the following equation with the error $O(\varepsilon_n^4) = O(n^{-4})$:

$$\begin{aligned} z &= \alpha_0 + \alpha_1\zeta z + \alpha_2(\zeta z)^2 + \alpha_3(\zeta z)^3 + O((\zeta z)^4), \qquad \zeta = (\pi n)^{-1} \ll 1, \\ \alpha_i &= \alpha_i(\zeta z) = \int_0^L \nu_i(x,\zeta z)\,dx, \qquad i=0,1,2,3; \qquad \varepsilon = \zeta z. \end{aligned} \tag{5.3.1}$$

An approximate solution $z_n = z(\zeta)$ with the error $O(\zeta^4)$ is constructed by the method of asymptotic expansions [19, 48]. Estimates for the absolute and the relative approximation errors for z_n coincide. It is convenient to find the values of z_n by successive approximations with respect to powers of the parameter ζ. In particular, for $k \le 3$, we have

$$z_n^{(3)} = \alpha_0 + \alpha_1^{(2)}\zeta z_n^{(2)} + \alpha_2^{(1)}(\zeta z_n^{(1)})^2 + \alpha_3^{(0)}(\zeta z_n^{(0)})^3, \quad |z_n - z_n^{(3)}| = O(\zeta^4),$$
$$z_n^{(0)} = \alpha_0, \quad z_n^{(1)} = \alpha_0(1+\alpha_1(0)\zeta), \quad z_n^{(2)} = \alpha_0(1+\alpha_1(\zeta z_n^{(1)})\zeta(1+\alpha_1(0)\zeta)) + \alpha_2(0)\zeta^2\alpha_0^2),$$
$$\alpha_1^{(2)}(\zeta) = \alpha_1(\zeta z_n^{(2)}(\zeta)), \quad \alpha_2^{(1)}(\zeta) = \alpha_2(\zeta z_n^{(1)}(\zeta)), \quad \alpha_3^{(0)} = \alpha_3(0). \tag{5.3.2}$$

Simple estimates show that the intermediate approximations $z_n^{(k)}$, $k = 0, 1, 2$, differ from $z_n^{(3)}$ by $O(\zeta^{k+1})$. Now, having constructed $z_n^{(k)}$ (5.3.2), we easily find the eigenvalues $\varepsilon_n^{(k)}$, and also $\omega_n^{(k)}$, $\lambda_n^{(k)}$.

5.3.2. Finding the original quantities. According to (5.3.1), we have $\varepsilon_n^{(k)} = \zeta z^{(k)}(\zeta) = z_n^{(k)}/(\pi n)$, for $k = 0, 1, \ldots$ The parameter $\varepsilon_n^{(k)}$ is found with the absolute error $O(\zeta^{k+2})$, while the relative error is $O(\zeta^{k+1})$, just as for $z_n^{(k)}$. It is of principal importance to construct approximations for the value $\omega_n = \varepsilon_n^{-1}$,

$$\omega_n^{(k)} = \frac{\pi n}{z_n^{(k)}}, \quad \varepsilon_n^{(k)} = \frac{z_n^{(k)}}{\pi n}, \quad \lambda_n^{(k)} = \omega_n^{(k)2}. \tag{5.3.3}$$

The value $\omega_n^{(k)}$ approximates the corresponding free vibration frequency with the absolute error $O(\zeta^k)$. For $k = 0$, we obtain the expression $\omega_n^{(0)} = \pi n/\alpha_0$ corresponding to the classical formula (5.1.3), since $\lambda_n^{(k)} = \omega_n^{(k)2}$. Naturally, the absolute error obtained when calculating $\lambda_n^{(k)}$ is $O(\zeta^{k-1})$, in general. Obviously, the relative error $O(\zeta^{k+1})$ remains the same as for $z_n^{(k)}$, $\varepsilon_n^{(k)}$, $\omega_n^{(k)}$.

Thus, we have obtained the following principal result. Formulas (5.3.1)–(5.3.3) give explicit analytic expressions for refined approximations of ε_n, ω_n, λ_n. These formulas are very simple and reduce to simple integrals of known functions, which can often be calculated analytically. These formulas can be used for fairly precise calculations, even if the mode index n is not very large. These expressions can be utilized for applied problems and yield acceptable errors for $n \ge 1$. For $n = 2, 3, \ldots$, their absolute error does not exceed $10^{-2} \div 10^{-3}$, provided that the functions p, r, q are sufficiently smooth (see examples in Section 5.6). When dealing with specific problems, it is useful to carry out preliminary studies based on the numerical analysis of lower vibration modes with $n = 1, 2, \ldots$ and compare the results with the asymptotic expressions (5.3.1)–(5.3.3) (Example 5.6.4).

The free vibration phases $\varphi_n(x)$ can be approximated according to the following scheme. Using (5.2.3), (5.2.5), (5.3.3), we obtain expressions for the averaged frequencies $\nu_n^{(k)}(x)$ and the averaged phases $\psi_n^{(k)}(x)$

$$\nu_n^{(k)}(x) = \sum_{i=0}^{k}(\varepsilon_n^{(k)})^i \nu_i(x, \varepsilon_n^{(k)}) = \sum_{i=0}^{k}(\varepsilon_n^{(k-i)})^i + O(\zeta^{k+1}),$$
$$\psi_n^{(k)}(x) = \omega_n^{(k)} \int_0^x \nu_n^{(k)}(y)\,dy, \quad k = 0, 1, 2, \ldots; \quad \zeta = \frac{1}{\pi n}, \quad n \gg 1. \tag{5.3.4}$$

The constructions and the estimates from Sections 5.2, 5.3 allow us to estimate the convergence of $\nu_n^{(k)}$, $\psi_n^{(k)}$ to the exact values,

$$|\nu_n(x) - \nu_n^{(k)}(x)| \le C\zeta^{k+1}, \quad |\psi_n(x) - \psi_n^{(k)}(x)| \le C\zeta^k, \quad 0 \le x \le L. \tag{5.3.5}$$

The free vibration phases $\varphi_n(x)$ are approximated with any given accuracy in ζ with the help of (5.2.2), (5.3.3),

$$\varphi_n^{(k)}(x) = \psi_n^{(k)}(x) + \sum_{i=1}^{k} (\varepsilon_n^{(k)})^i V_i(x, \psi_n^{(k)}(x), \varepsilon_n^{(k)}), \qquad |\varphi_n - \varphi_n^{(k)}| \le C\zeta^k. \tag{5.3.6}$$

In (5.3.4), (5.3.6), the terms of the order $O(\zeta^k)$ may be dropped. Recall that by construction the functions V_i (5.2.4) satisfy the condition $V_i(x, \pi n, \varepsilon) \equiv 0$.

Rigorous proof of the convergence of these approximations to the exact solution as $k \to \infty$ is problematic even for a system whose coefficients are analytic in x, because these expansions have an asymptotic character and, in general, are divergent as $k \to \infty$.

It is known that higher accuracy of calculations based on the method of averaging or that of accelerated convergence and connected with the introduction of fast and slow variables requires higher smoothness of the coefficients of the equations or the utilization of smoothing techniques. This is due to the growing order of the derivatives in the slow variable x. In the case under consideration, the function N (5.2.1) is a trigonometric polynomial and is analytic in φ. However, even if the right-hand sides are analytic in x, one cannot guarantee convergence of these procedures as $k \to \infty$, because there is "resonance" between the fast and the slow variables φ and x.

5.3.3. Procedure of successive approximations. Combining the method of averaging with that of successive approximations, one can develop a procedure of refining the solution $\varphi(x, \varepsilon)$ of equation (5.2.1) with any given accuracy. This procedure requires no analyticity. Let us briefly describe the corresponding scheme and discuss its properties.

First, we pass from the variable φ to ψ with the help of the known function $V_1(x, \psi, \varepsilon)$ (5.2.2). Differentiation in s and algebraic transformations bring us to the following boundary value problem:

$$\begin{aligned}
&\frac{d\psi}{ds} = \nu^{(1)}(x, \varepsilon) + \varepsilon^2 M(x, \psi, \varepsilon), \qquad \psi(0) = 0, \qquad \psi(L\varepsilon^{-1}) = \pi n; \\
&\varphi = \psi + \varepsilon V_1(x, \psi, \varepsilon), \quad V_1 \equiv \tfrac{1}{4}[\nu_0(x)]^{-1}\left[g(x, \varepsilon)\sin 2\varphi + h(x)(1 - \cos 2\varphi)\right], \\
&M \equiv \left[\tfrac{1}{\varepsilon}(N(x, \psi + \varepsilon V_1, \varepsilon) - N(x, \psi, \varepsilon)) - (N - \langle N\rangle)\langle N\rangle\right]\left[1 + \tfrac{\varepsilon}{\nu_0}(N - \langle N\rangle)\right]^{-1}.
\end{aligned} \tag{5.3.7}$$

For small enough $\varepsilon > 0$, the function M (5.3.7) is analytic in ψ in some strip and is continuous in x. This function is Lipschitz continuous in ε, and this fact is used when constructing a solution of the boundary value problem, i.e., when finding ε_n, ω_n, λ_n. In order to calculate $\psi(x, \varepsilon)$ on the interval $0 \le x \le L$, we solve the Cauchy problem (5.3.7) by the method of successive approximations,

$$\psi^{(k+1)}(x, \varepsilon) = \psi^{(1)}(x, \varepsilon) + \varepsilon \int_0^x M(y, \psi^{(k)}(y, \varepsilon), \varepsilon)\, dy, \qquad k = 1, 2, \ldots \tag{5.3.8}$$

The function $\psi^{(1)}(x, \varepsilon)$ in (5.3.8) is an integral of $\nu^{(1)}$ (see (5.2.5)). Successive approximations (5.3.8) are absolutely and uniformly convergent for sufficiently small $\varepsilon > 0$, and their limit is a function $\psi(x, \varepsilon)$ which is continuously differentiable in x. The difference $\psi - \psi^{(1)} = O(\varepsilon)$, $0 \le x \le L$ and the function ψ satisfy the Lipschitz condition in ε. This property allows us to construct approximations of the eigenvalues ε_n, as well as ω_n, λ_n, by a procedure of successive approximations which is similar to the algorithm (5.3.1)–(5.3.3). It should be observed that the calculation of (5.3.8) is a very difficult task, because it requires highly precise integration of a rapidly oscillating function. Having constructed $\psi_n(x) = \psi(x, \varepsilon_n)$, we find the phases $\varphi_n(x)$ from (5.3.7).

The procedure of consecutive replacements $\varphi \to \psi$ (5.3.7) may be continued in a recurrent manner, and this would bring us to the known scheme of the accelerated convergence method. However, fast decrease of errors, which is the attractive feature of that method, can only be ensured under additional smoothness assumptions, just as for the method of averaging. There are also obstacles of computational character connected with catastrophic complexity of the analytical expressions obtained for $M^{(k)}(x, \psi^{(k)}, \varepsilon)$ as the number of iterations grows, $k = 1, 2, \ldots$ $(M^{(1)} \equiv M)$. The situation worsens if smoothing techniques are used.

The above procedure based on averaging and asymptotic expansions is fairly efficient for practical calculations. In what follows, the parameters ε_n, ω_n, λ_n and the functions $\varphi_n(x)$ are assumed known with a given error with respect to the parameter $\zeta = (\pi n)^{-1}$. As shown above, the smallness of the error is ensured either by increasing n (the index of the vibration mode or the resonance zone) or by increasing the number of iterations k, or by both.

5.4. Finding Amplitudes and Shapes of Free Vibrations

5.4.1. Approximate calculation of higher mode amplitudes. After the parameters ε_n and the phases $\varphi_n(x)$ have been determined, the amplitudes $a_n(x)$ are found in terms of elementary quadratures, as indicated in Section 5.2. Thus,

$$a_n(x) = a_n^0 \exp\left[\int_0^x F(y, \varphi_n(y), \varepsilon)\, dy\right], \qquad a_n^0 = \text{const}. \tag{5.4.1}$$

Here, a_n^0 is an arbitrary constant which is chosen from additional conditions, say, the condition of normalization with the weight $f'_\lambda(x, \lambda_n)$. However, highly precise numerical integration of the rapidly oscillating function F in (5.4.1) (just as the integration of M in (5.3.8)) is a very difficult task, since the integration step should be very small, $\Delta y \ll \varepsilon_n \ll 1$ for $n \gg 1$. It seems more efficient to use the analytical procedure of averaging with subsequent utilization of numerical methods for the integration of regular (smooth) functions (see Sections 5.2, 5.3).

Let us introduce an averaged variable b corresponding to a and construct the transformation of the variables $(a, \varphi) \to (b, \psi)$, which is close to the identical transformation and is such that the equations for b, ψ do not contain ψ up to a given order in ε. Such phase replacement $\varphi \to \psi$ was independently constructed in Section 5.2 (see (5.2.1)). The replacement $a \to b$ depends on ψ, x, ε and has the form

$$\begin{aligned} &a = b(1 + \varepsilon U(x, \psi, \varepsilon)), \qquad U = U_1(x, \psi, \varepsilon) + \varepsilon U_2 + \cdots + \varepsilon^{k-1} U_k + \cdots, \\ &\frac{db}{ds} = \varepsilon b d(x, \varepsilon), \qquad d = d_1(x, \varepsilon) + \varepsilon d_2 + \cdots + \varepsilon^{k-1} d_k + \cdots. \end{aligned} \tag{5.4.2}$$

The functions U_i, d_i are determined similarly to V_i, ν_i (5.2.2)–(5.2.4) in terms of the known expressions for V_i, ν_i. These elementary operations yield the desired representations. In particular, for $d_{1,2,3}$ and $U_{1,2}$ we obtain the explicit analytical expressions

$$\begin{aligned} &d_1 = \langle F \rangle = -\tfrac{1}{2} h(x), \\ &d_2 = \langle F U_1 \rangle + \left\langle \frac{\partial F}{\partial \varphi} V_1 \right\rangle - d_1 \langle U_1 \rangle - \left\langle \frac{\partial U_1}{\partial x} \right\rangle = \frac{\nu_0}{8} H(x) G(x, \varepsilon) - \tfrac{1}{4} G'(x, \varepsilon), \\ &d_3 = \left\langle F U_2 + \frac{\partial F}{\partial \varphi}(U_1 V_1 + V_2) + \frac{1}{2}\frac{\partial^2 F}{\partial \varphi^2} V_1^2 + d_2 U_1 + d_1 U_2 - \frac{\partial U_2}{\partial x} \right\rangle, \\ &U_1 = \frac{1}{\nu_0} \int_0^\psi (F - \langle F \rangle)\, d\varphi = -\frac{1}{2\nu_0(x)} N(x, \psi, \varepsilon), \\ &U_2 = \frac{1}{\nu_0} \int_0^\psi \left(F U_1 - \frac{\partial F}{\partial \varphi} V_1 - d_2 - d_1 U_1 - \frac{\partial U_1}{\partial x} - \nu_1 \frac{\partial U_1}{\partial \varphi} \right) d\varphi. \end{aligned} \tag{5.4.3}$$

Note that the functions FU_1, $(\partial F/\partial\varphi)V_1$, $\partial U_1/\partial\varphi$ and $\partial U_2/\partial\varphi$ have zero mean values. By analogy with (5.4.3), we perform analytical calculations of subsequent expansion coefficients (5.4.2). For the kth approximation it is unnecessary to find U_k (see above).

Further, using the replacement formulas (5.4.2), we obtain the sought approximation of the amplitude $a_n^{(k)}(x)$ (with the error $O(\zeta^k)$)

$$a_n^{(k)}(x) = a_n^0 \exp\Big(\int_0^x d^{(k)}(y,\varepsilon_n^{(k)})\,dy\Big)\Big(1+\varepsilon_n^{(k)}U^{(k-1)}(x,\psi_n^{(k-1)},\varepsilon_n^{(k-1)})\Big),$$
$$d^{(k)} = d_1 + \varepsilon d_2 \cdots + \varepsilon^{k-1}d_k, \qquad U^{(k-1)} = U_1 + \varepsilon U_2 + \cdots + \varepsilon^{k-2}U_{k-1}. \tag{5.4.4}$$

5.4.2. Finding eigenfunctions corresponding to higher modes. Substituting the known functions $a_n^{(k)}(x)$ (5.4.4) and $\varphi_n^{(k)}(x)$ into the replacement formulas (5.3.1), we obtain the desired kth approximations of the eigenfunctions and their derivatives

$$u_n^{(k)}(x) = a_n^{(k)}(x)\sin\varphi_n^{(k)}(x), \quad |u_n - u_n^{(k)}| \le C\zeta^k, \quad 0\le x\le L,$$
$$u_n^{(k)\prime}(x) = \omega_n^{(k)}\nu_0(x)a_n^{(k)}(x)\cos\varphi_n^{(k)}(x), \quad |u_n' - u_n^{(k)\prime}| \le C\zeta^{k-1}. \tag{5.4.5}$$

The functions $u_n^{(k)}(x)$, $u_m^{(k)}(x)$ (5.4.5) satisfy the approximate orthogonality condition for n, $m \gg 1$,

$$\frac{1}{\lambda_n^{(k)}-\lambda_m^{(k)}}\int_0^L u_n^{(k)}(x)u_m^{(k)}(x)\Big(f(x,\lambda_n^{(k)}) - f(x,\lambda_m^{(k)})\Big)\,dx = \delta_{nm} + O(\zeta_*^k), \tag{5.4.6}$$

where $\zeta_* = \pi^{-1}\max(n^{-1},m^{-1})$. In (5.4.6), it is assumed that $n\sim m$; and it has been taken into account that $\omega_n^{(k)}$ and $\omega_m^{(k)}$ may differ substantially, $\omega_n^{(k)} - \omega_m^{(k)} = O(1)$.

In the second approximation, which refines the expression (5.1.3) by one order with respect to powers of $\zeta = (\pi n)^{-1}$, we obtain the following explicit analytical expressions for $a_n(x)$, $u_n(x)$:

$$a_n^{(2)}(x) = A_n^0\mu_0^{1/2}(x)(1+\varepsilon_n^{(0)}I_-(x)), \quad A_n^0 = \text{const},$$
$$u_n^{(2)}(x) = A_n^0\mu_0^{1/2}(x)(\sin\psi_n^{(2)}(x) + \varepsilon_n^{(0)}I_+(x)\sin\psi_n^{(1)}(x)), \tag{5.4.7}$$
$$I_\pm = \tfrac{1}{8}\int_0^x h(y)G(y,0)\,dy \pm \frac{1}{4}(G(x,0)-G(0,0)).$$

In (5.4.7), the terms of the order $O(\zeta^2)$ have been discarded. Note that the expressions for subsequent approximations are very cumbersome and can be obtained with the help of symbolic calculations software.

An approximate solution of equations (5.2.1) for a given $\varepsilon = \varepsilon_n$ can be obtained by a procedure similar to (5.3.7), (5.3.8). This procedure combines methods based on transforming the variables (the method of averaging and that of accelerated convergence) and the method of successive approximations. However, the implementation of this procedure is very difficult, because it requires numerical integration of rapidly oscillating functions (see above).

Of basic interest for applications is the calculation of frequencies of free vibrations with high precision. These quantities are not much affected by the perturbing factors which are difficult to take into account, such as nonlinearities or dissipation of different physical nature, drift of parameters, etc. The eigenfrequencies are important stable characteristics of vibration systems.

5.5. Other Types of Boundary Value Problems

The scheme proposed above for the determination of higher mode frequencies and vibration shapes can be directly extended to boundary value problems with other types of boundary conditions.

5.5.1. Boundary conditions of the second kind. Consider equation (5.1.1) with the boundary condition of free ends, $u'(0) = u'(L) = 0$. In this case, the phase φ in (5.2.1) should satisfy the boundary conditions

$$\varphi(0) = \tfrac{1}{2}\pi, \quad \varphi(L\varepsilon^{-1}) = \tfrac{1}{2}\pi(2n+1), \quad n = 1, 2, \dots$$

Note that it is convenient to seek a solution in the form

$$u = a\cos\varphi, \qquad \frac{du}{ds} = -a\nu_0 \sin\varphi. \tag{5.5.1}$$

Then, the boundary conditions preserve their form (5.2.1), but the equations for a, φ change somewhat. Omitting repetitions, we just give the formulas for the coefficients in the replacements of the type (5.2.2)–(5.2.4), (5.4.2), (5.4.3) for the second approximation

$$\begin{gathered}\nu_0 = \sqrt{r/p}, \quad \nu_1 = -\tfrac{1}{2}g, \quad \nu_2 = \tfrac{1}{4}H' - \tfrac{1}{8}\nu_0(H^2 + G^2),\\ V_1 = \tfrac{1}{2}\nu_0^{-1}F(x, \psi, \varepsilon), \quad V_1(x, \tfrac{1}{2}\pi(2n+1), \varepsilon) \equiv 0,\\ d_1 = -\tfrac{1}{2}h, \quad d_2 = \tfrac{1}{4}G' - \tfrac{1}{8}\nu_0 HG,\\ U_1 = -\tfrac{1}{2}\nu_0^{-1}N(x, \psi, \varepsilon), \quad U_1(x, \tfrac{1}{2}\pi(2n+1), \varepsilon) \equiv 0.\end{gathered} \tag{5.5.2}$$

By similarity with Section 5.3, one finds approximations for the eigenvalues of the parameters ε_n, ω_n, λ_n. The formulas for the desired functions remain practically the same. Similar calculations can be done for the case of boundary conditions corresponding to one end-point being fixed and the other free.

5.5.2. General boundary conditions of the third kind. Let us briefly examine the asymptotic behavior of higher mode frequencies and vibration shapes for system (5.1.1), (5.1.2) in the general case of elastic fixation at the end-points, $pu' \mp k_{0,L}u = 0$ for $x = 0, L$, where $k_{0,L}$ are the coefficients characterizing concentrated elastic forces at the end-points. From physical considerations, it may be concluded that for $\omega \to \infty$, the main contribution to the boundary conditions is due to the term characterizing distributed elastic forces at the points $x = 0, L$, i.e., $u'(0) = u'(L) = 0$ for $k_{0,L} = O(1)$. Therefore, just as in the case of free ends, it is convenient to change the variables by (5.5.1) and represent the original problem in the form

$$\begin{gathered}\frac{da}{ds} = \varepsilon a F\left(x, \varphi - \frac{\pi}{2}, \varepsilon\right), \quad a(0) = a^0 \neq 0,\\ \frac{d\varphi}{ds} = \nu_0(x) + \varepsilon N\left(x, \varphi - \frac{\pi}{2}, \varepsilon\right), \quad x = \varepsilon s,\\ \varphi(0) = \varphi^0 = \arctan(\varepsilon k_0 \mu(0)), \quad \varphi(L\varepsilon^{-1}) = \varphi^L = \pi n + \arctan(\varepsilon k_L \mu_0(L)).\end{gathered} \tag{5.5.3}$$

From (5.5.3), it follows that for $k_{0,L} \sim 1$, the concentrated elastic force at the end-points affects the solution only in the second approximation with respect to ε. For the unknown coefficients ν_i, V_i we obtain the following expressions:

$$\begin{gathered}\nu_0 = \sqrt{r/p}, \quad \nu_1 = -\tfrac{1}{2}g, \quad \nu_2 = \tfrac{1}{4}H'\cos 2\varphi^0 - \tfrac{1}{4}G'\sin 2\varphi^0 - \tfrac{1}{8}\nu_0(H^2 + G^2),\\ V_1 = -\tfrac{1}{4}\left[H(\cos 2\varphi^0 - \cos 2\psi) + G(\sin 2\psi - \sin 2\varphi^0)\right].\end{gathered} \tag{5.5.4}$$

The leading term of the approximation of the frequency ω_n and its approximations of the second and the third orders are obtained from formulas similar to (5.3.1)–(5.3.3),

$$\varepsilon_n^{(k)} = \frac{1}{\omega_n^{(k)}}, \quad \lambda_n^{(k)} = \omega_n^{(k)2}; \quad \omega_n^{(0)} = \frac{\pi n}{\alpha_0}, \quad \omega_n^{(1)} = \frac{\pi n}{\alpha_0} - \frac{\alpha_1(0)}{\alpha_0}, \tag{5.5.5}$$
$$\omega_n^{(2)} = \frac{\pi n}{\alpha_0} - \frac{\alpha_1(\varepsilon_n^{(0)})}{\alpha_0} - \frac{\alpha_2(0)}{\pi n} + \frac{\delta\varphi^L - \varphi^0}{\alpha_0}, \qquad \delta\varphi^L = \varphi^L - \pi n.$$

It should be kept in mind that φ^0, $\delta\varphi^L \sim \zeta$, according to (5.5.3). Using (5.5.4), (5.5.5), we obtain (as in the case of (5.3.4), (5.3.6)) approximations for the phases $\varphi_n(x)$ and then approximations for the amplitudes $a_n(x)$ and the vibration shapes $u_n(x)$ with the desired accuracy in ζ (see Section 5.4).

5.5.3. Remarks about generalizations. The above approach, which combines the method of averaging with that of asymptotic expansions, can be extended to generalized boundary value problems with mixed boundary conditions (for instance, the Hill problem with periodic boundary conditions), to problems with boundary conditions depending on the spectral parameter λ, and to problems with more general state equations, in particular, systems with variable coefficients which may also depend on λ in some way. For the investigation of elastic and parametric vibrations, it is important to develop asymptotic methods of nonlinear mechanics for interacting systems, as well as asymptotic methods for partial differential equations, in particular, those concerned with the asymptotic behavior of frequencies and shapes of free vibrations of membranes whose border has a complex structure.

The constructive approach developed above for the approximation of higher mode frequencies and vibration shapes with given accuracy, together with the highly efficient numerical-analytical method of accelerated convergence for the approximation of lower frequencies and vibration shapes, can be used for detailed investigations of a wide class of mechanical systems.

5.6. Calculations for Some Specific Mechanical Systems

Let us consider some examples which may be interesting for applications, in particular, for the analysis of vibrations of elastic systems or parametric vibrations with complex loading. We will calculate higher eigenvalues for some vibration systems, using the approach described in Sections 5.2, 5.3, and 5.5.

5.6.1. Longitudinal vibrations of an inhomogeneous rectilinear beam. The eigenfrequencies and shapes of free vibrations of a beam are obtained by solving the following Sturm–Liouville problem [40, 64]:

$$(ES(x)u')' + \lambda dS(x)u = 0, \qquad 0 \le x \le l; \tag{5.6.1}$$
$$1)\ u(0) = u(l) = 0; \qquad 2)\ u'(0) = u'(l) = 0.$$

Here, l is the beam's length, E is the Young modulus, d is the volume density, $S(x)$ is the area of the cross-section, λ is the separation constant (squared frequency) to be determined. For definiteness, we consider boundary conditions of the first or the second kind in (5.6.1) By suitable transformations, the functions u, S, the argument x, and the parameter λ can be made dimensionless, and therefore, we may assume that $E = d = l = 1$. We are going to study problem (5.6.1) for $\lambda \gg 1$, i.e., $\lambda dl^2/E \gg 1$ in the original variables.

In terms of (5.1.1), (5.1.2), we have $r = p \equiv S(x)$, $q \equiv 0$, and writing the two problems (5.6.1) in the form (5.2.1), (5.5.2), we have $\nu_0 \equiv 1$, $g \equiv 0$, $h = S'/S$. By (5.2.3), (5.5.2), we get

$$\nu_{1,3} \equiv 0, \qquad \nu_2^{1,2} = \mp\tfrac{1}{4}h' - \tfrac{1}{8}h^2,$$

where the subscripts correspond to the boundary conditions 1 and 2 in (5.6.1).

Using (5.3.1)-(5.3.3), (5.5.2), we obtain the eigenfrequencies with the error $O(\zeta^3)$,

$$\omega_n^{1,2} = \pi n + \delta\omega_n^{1,2}, \qquad \delta\omega_n^{1,2} = \frac{1}{4\pi n}\left[\pm \left.\frac{S'(x)}{S(x)}\right|_0^1 + \frac{1}{2}\int_0^1 \left(\frac{S'(x)}{S(x)}\right)^2 dx\right]. \tag{5.6.2}$$

Hence, we see that the relative contribution of the inhomogeneity is of the order of ζ^2, while its absolute contribution is of the order of ζ. This contribution is due to two factors, one local (at the points at $x = 0$ and $x = 1$) and the other nonlocal (integral).

Consider some special cases of $S(x)$.

Let $S(x) = 1 + \theta(2x - 1)$ be a linear function of x, where $|\theta| < 1$. Then the value $S(1/2) = 1$ is independent of θ, and the volume of the beam is $V = 1$. Substituting $S(x)$ into (5.6.2), we obtain the correction terms

$$\delta\omega_n^1 = -\frac{1}{2\pi n}\theta^2(1-\theta^2)^{-1}, \qquad \delta\omega_n^2 = \frac{3}{2\pi n}\theta^2(1-\theta^2)^{-1}. \tag{5.6.3}$$

Expressions (5.6.2), (5.6.3) allow us to make the following qualitative conclusion: in the case of fixed ends, linear inhomogeneity in x makes the eigenvalues decrease; and in the case of free ends, it makes them increase.

Now, consider vibrations of a beam with a circular cross-section whose radius linearly depends on the linear coordinate x, $R(x) = R_0 + \theta(2x - 1)$, where $|\theta| < R_0$. The area of the cross-section is $S(x) = \pi R^2(x)$, and $S(1/2) = \pi R_0$ is independent of θ. According to (5.6.2), we obtain the correction terms

$$\delta\omega_n^1 = 0, \qquad \delta\omega_n^2 = 4(\pi n)^{-1}\theta^2(R_0^2 - \theta^2)^{-1}, \qquad |\theta| < R_0. \tag{5.6.4}$$

Hence, we see that in the second approximation with respect to ζ, the frequency of longitudinal vibrations of the conical beam does not depend on θ in the case of fixed edges; and this frequency is an increasing function of $\theta \neq 0$ for the beam with free edges. Assuming that the volume of the beam is constant, $V = \pi\mathfrak{B}^2 = \text{const}$, we obtain correction terms of the form

$$\delta\omega_n^1 = 0, \qquad \delta\omega_n^2 = (\pi n)^{-1}\gamma^2\left(1 - \tfrac{1}{3}\gamma^2\right)^{-1}, \qquad \gamma^2 = 4\theta^2\beta^{-2} < 3. \tag{5.6.5}$$

Thus, variation of the cone parameter θ for a family of beams of constant volume leads us to similar qualitative conclusions regarding the correction terms for higher mode frequencies. Formulas (5.6.2) can be used for the calculation of the correction term $\delta\omega_n^{1,2}$ for arbitrary differentiable functions $S(x)$. These corrections may be of an arbitrary sign or equal to zero, depending on the local factor; the integral term is always positive.

5.6.2. Vibrations of an inhomogeneous string. Eigenfrequencies and shapes of transverse vibrations of a stretched string are determined by the classical boundary value problems for the equation

$$Tu'' + \lambda\rho(x)u = 0, \qquad \rho(x) = dS(x), \qquad 0 \le x \le l. \tag{5.6.6}$$

For definiteness, we consider the same boundary conditions as in (5.6.1). Here, T is the constant string tension, $\rho(x)$ is its linear density, l is its length. Without the loss of generality, we can take $T = l = d = 1$. The functions u, ρ, the argument x, and the parameter λ are assumed dimensionless.

In terms of (5.1.1), (5.1.2), we have $r \equiv \rho(x)$, $p \equiv 1$, $q \equiv 0$. Writing (5.6.6) in the form (5.2.1), we get $\nu_0 = \sqrt{\rho(x)}$, $g \equiv 0$, $h = \frac{1}{2}\rho'(x)/\rho(x)$, where $\rho(x) = S(x)$. Just as in example (5.6.1), formula (5.2.3) yields

$$\nu_{1,3} \equiv 0, \qquad \nu_2^{1,2} = \mp\tfrac{1}{4}H' - \tfrac{1}{8}\nu_0 H^2.$$

Using (5.3.1)–(5.3.3), (5.5.2), we find third-order approximations of the eigenfrequency $\omega_n^{1,2}$ (the relative and the absolute errors are $O(\zeta^4)$ and $O(\zeta^3)$, respectively),

$$\omega_n^{1,2} = \frac{\pi n}{\alpha_0} + \delta\omega_n^{1,2}, \qquad \delta\omega_n^{1,2} = -\frac{\alpha_2^{1,2}}{\pi n},$$
$$\alpha_0 = \int_0^1 \sqrt{\rho(x)}dx, \qquad \alpha_2^{1,2} = \mp\frac{1}{4}(H(1) - H(0)) - \frac{1}{8}\int_0^1 \nu_0(x)H^2(x)dx. \tag{5.6.7}$$

Let us examine the correction terms $\delta\omega_n^{1,2}$ (5.6.7) for different $\rho(x)$. Just as in (5.6.2), the integral term is always positive, while the local term may have any sign or be equal to zero. For particular types of $\rho(x)$ (i.e., $S(x)$) the quantities α_0, $\alpha_2^{1,2}$ in (5.6.7) are found by numerical or analytical methods. Next, we consider some special cases.

For the linear function $\rho(x) = 1 + \theta(2x - 1)$, $|\theta| < 1$, the expressions for α_0, $\delta\omega_n^{1,2}$ are obtained in analytical form (5.6.7)

$$\alpha_0 = 2\left(1 + \tfrac{1}{3}\theta^2\right)\left[(1+\theta)^{3/2} + (1-\theta)^{3/2}\right]^{-1},$$
$$\delta\omega_n^1 = -\tfrac{5}{16}D(\pi n)^{-1}, \qquad \delta\omega_n^2 = \tfrac{7}{16}D(\pi n)^{-1}, \tag{5.6.8}$$
$$D = 4\theta^2\left(1 + \tfrac{1}{3}\theta^2\right)(1-\theta^2)^{-3/2}\left[(1+\theta)^{3/2} + (1-\theta)^{3/2}\right]^{-1}.$$

From (5.6.8), the following qualitative conclusions (similar to those for the beam vibration model; see (5.6.3)) can be made. Linear variation of the density $\rho(x)$ decreases frequencies of free vibrations of a string with fixed ends. If transverse displacements are allowed for the string ends, then the correction term is positive. Note that for this family of strings, the volume (mass) is constant $V(\theta) \equiv 1$ $(m(\theta) \equiv 1)$.

Suppose now that the string has a circular cross-section whose radius linearly depends on the coordinate, and the mass of the string is constant, $m(\theta) = \pi$ (is independent of θ). Then the following constraint appears: $R_0^2 = 1 - \theta^2/3$. As in the case of (5.6.8), we get

$$\alpha_0 = \left(1 - \tfrac{1}{3}\theta^2\right)^{1/2}, \qquad \delta\omega_n^1 = -\frac{D}{2\pi n}, \qquad \delta\omega_n^2 = \frac{3D}{2\pi n},$$
$$D = 4\theta^2\left(1 - \tfrac{1}{3}\theta^2\right)^{1/2}\left(1 - \tfrac{4}{3}\theta^2\right)^{-2}, \qquad \theta^2 < \tfrac{3}{4}. \tag{5.6.9}$$

From (5.6.9), it follows that $\delta\omega_n^1 < 0$, just as in (5.6.3), (5.6.8), but opposite to (5.6.4), (5.6.5). The frequency corrections in the case of free ends are again positive. Formulas (5.6.7) can be used to determine the leading terms and the corrections for eigenfrequencies in the case of arbitrary smooth $\rho(x)$. Note that the refining corrections may have an arbitrary sign due to the substantial effect of the local factor (at $x = 0$, $x = 1$).

5.6.3. Asymptotics of eigenvalues of the Hill problem. Consider the problem of highly precise calculation of periodic solutions of the equation

$$u'' + [\lambda - q(x)]u = 0; \qquad u(0) = u(1), \quad u'(0) = u'(1). \tag{5.6.10}$$

Here, $q(x)$ is a sufficiently smooth periodic function, $q(x+1) \equiv q(x)$, which may depend on some other parameters. The problem is to find real λ_n corresponding to sufficiently high vibration modes, $n \gg 1$. Problem (5.6.10) is obtained as a result of reduction of numerous problems occurring in the theory of vibrations, elasticity, and mathematical physics (see, for instance, [16, 19, 24, 28, 33, 42, 43, 46, 48, 61, 62, 67, 68]).

Suppose that $q(x) \equiv q(-x)$ is an even function. This case often occurs in applications. Then problem (5.6.10) is equivalent to two Sturm–Liouville problems, one with the boundary conditions of the first kind and the other with those of the second kind. The solutions of these two problems are odd and even functions, respectively. Note that together with the resonance frequencies $1/n$, there exist resonance frequencies $2/n$, the latter including the former for n even. According to the notation (5.1.1), (5.1.2), we have a classical problem with the perturbation of the order ε^2, where $\varepsilon = \omega^{-1}$, $\lambda = \omega^2$. By elementary calculations similar to those of Section 5.2, we obtain the desired expressions for the functions ν_i corresponding to the fifth approximation of the frequency with respect to powers of ε (the relative error is $O(\varepsilon^6)$)

$$\begin{gathered}\nu_0 = 1, \quad \nu_2 = -\tfrac{1}{2}q(x), \quad \nu_4 = \tfrac{1}{8}(\mp q''(x) - q^2(x)), \quad \nu_1 = \nu_3 = \nu_5 = 0; \\ \nu^{(5)}(x,\varepsilon) = \nu_0 + \varepsilon^2\nu_2(x) + \varepsilon^4\nu_4(x),\end{gathered} \tag{5.6.11}$$

where the sign "+" before $q''(x)$ corresponds to the boundary conditions of the first kind and the sign "−" to those of the second kind. The eigenfrequencies with the absolute error $O(\zeta^5)$ are calculated on the basis of (5.6.11) by the procedure described in Section 5.3. As a result, we obtain the desired representations

$$\omega_n^{(5)} = \pi n - \frac{\alpha_2}{\pi n} - \frac{\alpha_2^2 + \alpha_4}{(\pi n)^3}, \qquad \alpha_{2,4} = \int_0^1 \nu_{2,4}(x)dx. \tag{5.6.12}$$

From (5.6.11), (5.6.12), it follows that the difference between the asymptotic behavior of the eigenvalues ω_n for the boundary conditions of the first and the second kinds remains unmanifest in the terms up to the order ζ^3, although ν_4 contains the function q'' with different signs for different boundary conditions. This happens, because the integral of $q''(x)$ is equal to zero due to the smoothness and the periodicity of $q(x)$. It is only if higher-order terms are taken into account that different asymptotic behavior of the eigenvalues becomes manifest. Further, if the function q has zero mean value, then $\alpha_2 = 0$ and the correction term is positive,

$$\delta\omega_n^{1,2} = \tfrac{1}{8}\langle q^2\rangle(\pi n)^{-3}, \qquad \langle q\rangle = 0,$$

where $\langle q^2\rangle$ is the mean value of q^2 (see the Mathieu equation, the Hill equation, etc. [19, 43, 46, 61]). In the general case, with the function $q(x)$ possibly non-even, a special procedure can be developed for finding the asymptotic behavior of the eigenvalues λ_n of the periodic boundary value problem (see Sections 2.9, 4.7 and 5.3, 5.5). In the case of discontinuous $q(x)$ (the Meissner equation [43, 61]), formulas (5.6.11) hold for the approximations of the order ζ^3.

5.6.4. Spatial vibrations of a satellite. Consider a dynamically symmetric satellite whose polar axis oscillates with an arbitrary amplitude in the plane of its circular orbit. At some instant, its motions are subjected to small perturbations, and we want to examine small angular deviations of the satellite axis from the orbit plane. The corresponding periodic boundary value problem in dimensionless form reads as follows (for the sake of definiteness, we consider an oblate body):

$$\begin{gathered}u'' + 16\mathbf{K}^2(k)\left[(\chi + k\,\mathrm{cn}\,\theta)^2 - k^2\,\mathrm{sn}^2\,\theta\right]u = 0, \\ u = u(x), \quad 0 \le x \le 1, \quad u(0) = u(1), \quad u'(0) = u'(1); \\ k^2 = 2h_0\chi^2, \quad \chi = (3(\alpha-1))^{-1/2}, \quad 0 \le k < 1, \quad 1 < \alpha \le 2.\end{gathered} \tag{5.6.13}$$

Here, u is the angular deviation of the polar axis, the prime indicates the derivative in the argument $x = w/(2\pi)$, w is an independent variable characterizing plane vibrations; $\theta = 4\mathbf{K}(k)x$ is the argument of the Jacobi elliptic functions cn, sn; $\mathbf{K}(k)$ is the complete elliptic integral of the first kind. The above equation contains two constant parameters: the modulus k and the parameter χ; h_0 is dimensionless energy of plane vibrations; α is the ratio of the polar and the equatorial moments of inertia.

In the case of an elongated body ($0 \le \alpha < 1$), the equation of small vibrations (5.6.13) should be modified.

Since the coefficient of u is an even function of x, the periodic problem (5.6.13) is equivalent to two Sturm–Liouville problems (see Subsection 5.6.3). Our aim is to describe the asymptotic behavior of eigenvalues $\chi_n(k)$ for sufficiently large n (the index of the resonance zone) and $0 \le k < 1$. In the notation of (5.1.1), (5.1.2), we have

$$\lambda = \omega^2 = (4\mathbf{K}\chi)^2, \quad \varepsilon = \omega^{-1}, \quad \beta = 4\mathbf{K}k,$$
$$p = r \equiv 1, \quad q = -2\beta \operatorname{cn}\theta - \varepsilon\beta^2(\operatorname{cn}^2\theta - \operatorname{sn}^2\theta). \tag{5.6.14}$$

Let us find the eigenvalues $\omega_n(\beta)$ of problem (5.6.13), (5.6.14) in the second approximation with respect to $\zeta = (\pi n)^{-1}$, i.e., with the relative error $O(\zeta^3)$ for $\beta \sim 1$ and k "not too close" to $k = 1$. In this approximation, the eigenvalues $\omega_n^{1,2}(\beta)$ of both boundary value problems coincide. Using the results of Sections 5.2, 5.5 and taking into account (5.6.14), we get

$$\nu_0 = 1, \quad \nu_1 = \beta\operatorname{cn}\theta + \frac{1}{2}\varepsilon\beta^2(\operatorname{cn}^2\theta - \operatorname{sn}^2\theta), \quad \nu_2 = -\frac{1}{2}\beta^2\operatorname{cn}^2\theta,$$
$$\omega_n(\beta) = \pi n + \frac{1}{2\pi n}\beta^2\langle\operatorname{sn}^2\theta\rangle, \quad \langle\operatorname{sn}^2\theta\rangle = \frac{1}{k^2}\left(1 - \frac{\mathbf{E}(k)}{\mathbf{K}(k)}\right), \tag{5.6.15}$$
$$\chi_n(k) = \frac{\pi n}{4\mathbf{K}(k)} + \frac{2}{\pi n}(\mathbf{K}(k) - \mathbf{E}(k)).$$

Here, $\mathbf{E}$ is the complete elliptic integral of the second kind.

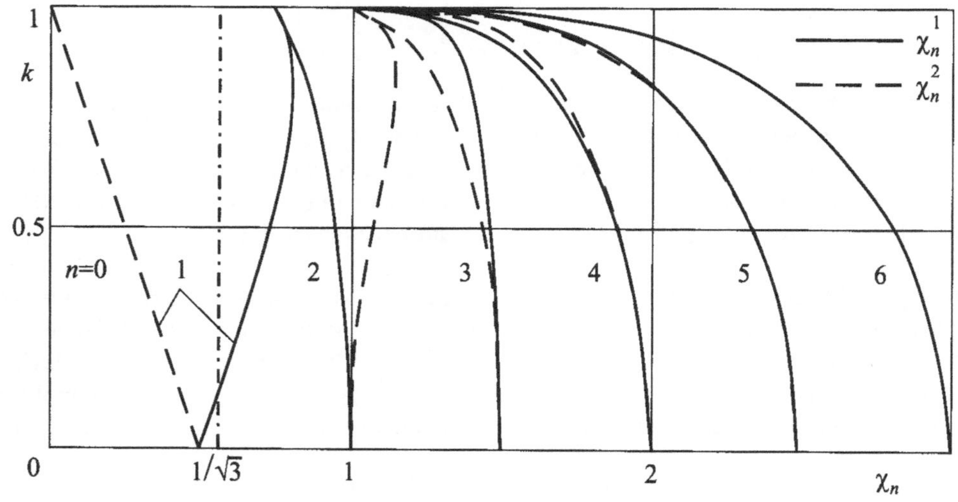

Fig. 5.1

From (5.6.15), we obtain the bound $\chi_n = (n/2)(1 - k^2/4)$ for $k^4 \ll 1$. For $k \to 1$, the leading term of the expressions for χ_n monotonically decreases, since $\mathbf{K} \to \infty$.

Numerical results for $\chi_n^{1,2}$, $0 \le k \le 0.999$, are represented in Fig. 5.1 (see Section 12.2); the indices 1 and 2 correspond to odd and even eigenfunctions, respectively. These curves (obtained in Section 12.2 and [13] by the highly precise method of accelerated convergence) show that the expressions (5.6.15) are fairly precise for $n = 2$, provided that $0 \le k \le 0.2$. There is some divergence of the curves χ_2^1 and χ_2^2. For $n = 3$, we have acceptable precision for $0 \le k \le 0.4$. As the index n of the resonance zone grows, the precision becomes higher and the range of k becomes wider. The curves $\chi_n^1(k)$ and $\chi_n^2(k)$ diverge less and less and are deflected according to the asymptotic behavior of the leading term in (5.6.15). It is interesting to observe the following qualitative result:

$$\chi_n^{1,2}(k) \to 1, \quad \alpha_n^{1,2}(k) \to \tfrac{4}{3}, \quad k \to 1, \quad n > 2.$$

Note that the case of an elongated body and rotation regimes can be studied in a similar manner on the basis of the corresponding modified equations.

Our conclusion is that calculations for meaningful specific problems demonstrate the efficiency of our asymptotic approach to the investigation of higher vibration modes of mechanical systems. In the case of smooth coefficients, the calculations are fairly precise even for $n \sim 1$ being "not too large".

5.7. Exercises

Exercise 1. Using (5.6.2), calculate the frequency corrections $\delta\omega_n^{1,2}$ and examine their behavior depending on the parameters. Consider the following classes of the function $S(x)$ characterizing the area of the beam cross-section.

a) $S(x) = (1 + \alpha x)^k$, $-1 < \alpha < \infty$, $|k| < \infty$. Pay attention to the case $k = \pm 2$.

b) $S(x) = 1 + \beta(1 + \alpha x)^k$, $S > 0$, $0 \le x \le 1$. Consider the case $k = 2$, $\alpha = -2$.

c) $S(x) = 1 + \beta \sin(\pi N x)$, $N = 1, 2, \ldots$, $|\beta| < 1$. Examine the case $N \gg 1$, $|\beta| \ll 1$.

Exercise 2. Using formulas (5.6.7), find the frequency corrections $\delta\omega_n^{1,2}$ for cases a), b), c) (see Exercise 1) of the cross-sectional area $S(x)$ of the string. Study their behavior depending on the parameters by analytical and numerical methods.

Exercise 3. Using (5.2.3), (5.3.3), (5.5.4), (5.5.5), calculate the frequency corrections $\delta\omega_n$ in the first approximation for a homogeneous beam in the presence of elastic environment characterized by the function $q(x)$.

a) $q(x) = \beta(1 + \alpha x)^k$. Consider the case $\beta \sim 1$, $|\alpha| \ll 1$.

b) $q(x) = \gamma + \beta \sin(\alpha x + \theta)$. Consider the cases: $|\beta| \ll 1$, $\alpha \sim 1$ and $\beta \sim 1$, $\alpha \gg 1$.

Chapter 6

Solutions of Fourth-Order Self-Conjugate Problems. Oscillation Properties

Some important classes of dynamical systems occurring in physics and mechanics are described by higher-order (4th, 6th, 8th, etc.) equations. For many problems, separation of the time and the spatial variables results in eigenvalue problems of Sturm-Liouville type. Theoretical and numerical methods of solving such problems for higher-order equations whose coefficients have considerable variation have not been developed to a desirable extent or are altogether absent. In Chapters 6, 7, and 8, we suggest constructive approaches to the solution of self-conjugate fourth-order eigenvalue problems with different types of boundary conditions. Our numerical-analytical methods possess great generality and can be extended to arbitrary self-conjugate problems with a scalar argument, and also to some systems of partial differential equations.

6.1. Statement of a Self-Conjugate Fourth-Order Boundary Value Problem

6.1.1. Statement of the problem in differential form. Some remarks. Consider the following eigenvalue problem (of Sturm–Liouville type) in dimensionless variables [3, 11, 12, 28, 45, 64]:

$$(p(x)u'')'' - \lambda r(x)u = 0, \qquad 0 \le x \le l, \tag{6.1.1}$$

$$u(0) = u'(0) = u(l) = u'(l) = 0. \tag{6.1.2}$$

Here, $\lambda > 0$ is the eigenvalue and $u(x) \not\equiv 0$ is the eigenfunction to be determined. The range of the argument x, i.e., the length $l > 0$, is not fixed and can be varied in the process of solution. It is assumed that the coefficients $p(x)$, $r(x)$ are continuous and extended to a larger interval $(l, \xi]$; moreover,

$$0 < p^- \le p(x) \le p^+ < \infty, \qquad 0 < r^- \le r(x) \le r^+ < \infty.$$

It is also possible to consider the case of discontinuous $p(x)$, $r(x)$ with finitely many jumps of the first kind; see Subsection 6.1.3.

One may also consider equations and boundary conditions more general than those in (6.1.1) (6.1.2). Such problems have important applications in mathematical physics and the theory of elasticity.

The problem is to find the sequence of eigenvalues $\lambda = \lambda_n$ (frequencies $\omega_n = \sqrt{\lambda_n}$) and eigenfunctions $u = u_n(x)$ (vibration shapes). Of main interest from the standpoint of theory and applications are the lower vibration modes, $n = 1, 2, \ldots$ It is known that problem (6.1.1), (6.1.2) has a countable set of eigenvalues (discrete spectrum) and the corresponding eigenfunctions (vibration shapes), which form an orthonormal basis in the weighted Lebesgue space L^2 with the weight $r(x)$,

$$\begin{gathered}\lambda \in \{\lambda_n\}, \quad 0 < \lambda_1 < \lambda_2 < \cdots < \lambda_n < \cdots, \quad \lambda_n \sim n^4, \\ u_n(x) = u(x, \lambda_n), \quad (u_n, u_m)_r = \|u_n\|^2 \delta_{nm}; \quad n, m = 1, 2, \ldots\end{gathered} \tag{6.1.3}$$

Here, δ_{nm} is the Kronecker symbol, $(\cdot,\cdot)_r$ is the scalar product in L^2 with the weight $r(x)$, and $\|\cdot\|_r$ is the corresponding norm. The norm of an eigenfunction can be calculated in finite form with the help of the *sensitivity function* $v = \partial u / \partial \lambda$ as follows:

$$\begin{gathered}\|u_n\|_r^2 = (u_n, u_n)_r = \int_0^l u_n^2(x) r(x)\, dx = \left[(pu_n'')' v_n - (pu_n'') v_n'\right]_{x=l}, \\ (p(x)v'')'' + \lambda_n r(x) v = -r(x) u_n(x), \quad v(0) = v'(0) = v''(0) = v'''(0) = 0.\end{gathered} \tag{6.1.4}$$

The function $v = v_n(x)$ is the solution of the linear Cauchy problem (6.1.4) and is constructed numerically by joint integration of equations (6.1.4) and (6.1.1) (see below).

6.1.2. Statement of the problem in variational form. Problem (6.1.1), (6.1.2) can be easily formulated as an isoperimetric variational problem [24],

$$\begin{gathered}J[u] = \frac{1}{2} \int_0^l p(x) u''^2\, dx \to \min, \quad I[u] = \|u\|_r^2 = \int_0^l r(x) u^2\, dx = 1, \\ u(0) = u'(0) = u(l) = u'(l) = 0; \\ \lambda \in \{\lambda_n\}, \quad u_n(x) = u(x, \lambda_n), \quad (u_n, u_m)_r = \delta_{nm}, \quad n = 1, 2, \ldots\end{gathered} \tag{6.1.5}$$

Here, λ is the double Lagrange multiplier for this variational problem, λ_1 is equal to the double global minimum of the quadratic functional J (6.1.5). Every subsequent eigenvalue λ_n, $n \geq 2$, is equal to the minimum of J on the subspace formed by u orthogonal to the preceding eigenfunctions, $(u, u_k)_r = 0$, $k = 1, 2, \ldots, n-1$. The variational statement of the problem has been used for the development of functional methods aimed at approximating the eigenvalues λ_n and the eigenfunctions $u_n(x)$. There exist algorithms for the calculation of effective upper bounds λ_n^+. The construction of highly precise lower bounds λ_n^- encounters great computational difficulties. The existing theoretical methods of Weinstein–Aronszajn [28] seem unsuitable for practical purposes, because of their formal character and rather cumbersome formulas hardly usable for the construction of algorithms.

Observe that the following rough two-sided estimates hold for the eigenvalues of problem (6.1.1), (6.1.2):

$$\begin{gathered}\lambda_n^- \leq \lambda_n \leq \lambda_n^+, \quad \lambda_n^\pm = \left(\frac{\gamma_n}{l}\right)^4 \frac{p^\pm}{r^\mp}, \quad \gamma_n = \operatorname{Arg}(\cos\gamma \cosh\gamma - 1), \quad n = 1, 2, \ldots; \\ \gamma_1 = 4.7300, \quad \gamma_2 = 7.8532, \quad \gamma_3 = 10.9956, \quad \gamma_4 = 14.1372, \quad \gamma_5 = 17.2788; \\ \gamma_n = \left(n + \frac{1}{2}\right)\pi + O(e^{-\pi n}), \quad n \gg 1.\end{gathered} \tag{6.1.6}$$

These estimates follow directly from (6.1.5)). The bounds $\lambda_n^\pm$ (6.1.6) for λ_n (6.1.3) are fairly precise, if the functions $p(x)$ and $r(x)$ have small variation on the interval $x \in [0, l]$, i.e., if the differences $p^+ - p^-$ and $r^+ - r^-$ are relatively small. In such a case, the method of perturbations can be used (see Chapter 8) to obtain analytical refinements of the solutions. But the numerical calculations based on that method are ineffective. In the general case of the problem considered here, the estimates (6.1.6) can be used for the construction of the initial approximation in recurrent schemes (of successive approximations, the method of accelerated convergence, etc.). These algorithms will be fairly effective when combined with the procedure of continuation in the parameters of the system (parametric synthesis) or in the parameters introduced artificially to improve convergence (see Chapters 7 and 8). The methods and algorithms described in scientific literature, i.e., the methods based on the Rayleigh–Ritz principle and finite element approximations, are essentially insufficient for highly precise mass on-the-fly calculations and the description of properties of the solutions depending on n and the parameters of the system [28, 45].

6.1.3. Introduction of natural physical variables. Equation (6.1.1) is not very convenient for the investigation by numerical-analytical methods. For a twice continuously differentiable $p(x)$, this equation can be easily reduced to the standard fourth-order linear equation or written as a system of first-order equations (in Cauchy's form). However, it seems more natural to write this equation in terms of auxiliary variables z, μ, which have a certain mechanical meaning. Thus, problem (6.1.1), (6.1.2) can be transformed to

$$\begin{gathered} u' = \theta, \quad \theta' = -\frac{z}{p(x)}, \quad z' = \mu, \quad \mu' = -\lambda r(x) u; \\ u(0) = \theta(0) = u(l) = \theta(l) = 0 \qquad \Big(u'' = -\frac{z}{p(x)}, \quad z'' = -\lambda r(x) u\Big). \end{gathered} \tag{6.1.7}$$

In the theory of elastic vibrations, these new variables have a certain physical meaning, namely, z is the moment of elastic forces and μ is the cross-cutting force. The problem in the form (6.1.7) does not require that the coefficient $p(x)$ be differentiable, it suffices that $p(x)$ be continuous (or even piecewise continuous).

It is possible to give a mechanical interpretation of various types of self-conjugate boundary value problems. Apart from the condition of fixed ends (6.1.2) or (6.1.7), one may consider other boundary conditions corresponding to self-conjugate problems. For instance, the condition $u = z = 0$ describes hinged fixation, the condition $z = \mu = 0$ corresponds to free edges, and $\theta = \mu = 0$ to the fixation of the tangential direction. These boundary conditions may be realized on one or both edges. As a general case of boundary conditions, we consider the so-called conditions of elastic fixation with respect to the displacement and elastic fixation with respect to the rotation of the tangent at the end-points of the rod,

$$[(1-\kappa_x)\mu \mp \kappa_x u]_{x=0,l} = 0, \quad [(1-\sigma_x)z \mp \sigma_x \theta]_{x=0,l} = 0, \quad 0 \le \kappa_{0,l}, \sigma_{0,l} \le 1. \tag{6.1.8}$$

The normalized coefficients $\kappa_{0,l}$ and $\sigma_{0,l}$ determine the relative effect of the rigidity of the elastic fixation of the edges with respect to the displacement and with respect to the rotation of the tangent line. The case (6.1.2), (6.1.7) of absolutely rigid fixation corresponds to "infinite values" of the rigidity coefficients, i.e., $\kappa_{0,l} \to 1$, $\sigma_{0,l} \to 1$. The above special cases of boundary conditions are limit cases of (6.1.8) corresponding to the extremal values of the parameter $\kappa_{0,l}$ or (and) $\sigma_{0,l}$. The case of infinite displacement rigidity or (and) infinite rotation rigidity can only be realized approximately, because the material of the base to which the edges are attached possesses bounded (often not very large) rigidity.

Note that the equation of vibrations of a rod (6.1.1) or (6.1.7) can take into account the influence of an external elastic medium (Winkler's foundation). From the standpoint of theory and applications, it is important to generalize the above model to the case of vibrations of beams (thick rods) and flexural-torsional vibrations of curvilinear strongly inhomogeneous rods.

6.1.4. Scheme of solution. The standard procedure of finding the eigenvalues and the eigenfunctions of problem (6.1.7) consists in constructing a general solution of the equations depending on the parameter λ and then separating the solutions satisfying the boundary conditions at the points $x = 0$, $x = l$. The necessary and sufficient condition for a solution to be nontrivial is that the determinant of the matrix of the fundamental system should be equal to zero. Hence, we obtain the characteristic equation for the eigenvalues λ_n. The general theory claims that the said determinant is an entire function of λ with countably many roots $\{\lambda_n\}$ on the real semi-axis $\lambda > 0$, and these roots satisfy the condition $\lambda_n \sim n^4$, $n = 1, 2, \ldots$ These eigenvalues satisfy the two-sided estimates (6.1.6).

In connection with problem (6.1.7), it is required to construct two families of solutions (λ is the parameter of each family) of the Cauchy problem with the following initial values at $x = 0$:

$$\begin{gathered} 1)\quad u(0) = \theta(0) = \mu(0) = 0, \quad z(0) = 1; \\ 2)\quad u(0) = \theta(0) = z(0) = 0, \quad \mu(0) = 1; \\ u(x,\lambda) = c_1 u_1(x,\lambda) + c_2 u_2(x,\lambda), \quad \theta(x,\lambda) = c_1\theta_1(x,\lambda) + c_2\theta_2(x,\lambda), \\ z(x,\lambda) = c_1 z_1(x,\lambda) + c_2 z_2(x,\lambda), \quad \mu(x,\lambda) = c_1\mu_1(x,\lambda) + c_2\mu_2(x,\lambda). \end{gathered} \tag{6.1.9}$$

The desired solution (u, θ, z, μ) is defined to within a constant coefficient, which is usually found from the normalization condition (6.1.4). We can take that coefficient equal to the constant $c_1 \neq 0$ (or $c_2 \neq 0$) in (6.1.9) and consider the ratio $c_2/c_1 = \eta$ (or $c_1/c_2 = \eta$) and the parameter λ as the sought quantities. The equations for λ, η are provided by the boundary conditions at $x = l$,

$$u_1(l,\lambda) + \eta u_2(l,\lambda) = 0, \quad \theta_1(l,\lambda) + \eta\theta_2(l,\lambda) = 0. \tag{6.1.10}$$

When solving system (6.1.10), the sets $\lambda \in \Lambda$, $\eta \in H$ for every λ_n, η_n can be roughly estimated with the help of relations (6.1.6). Standard numerical methods for obtaining closer approximations of the sought quantities involve a search through a large number of alternatives. This number is of the order 10^{2N}, where 10^{-N} is the desired relative error. The construction of a solution and its analysis become more difficult with the growth of n, and also if the parameters of system (6.1.7) have substantial variation. The difficulties are due to the existence of oscillating, rapidly growing or decaying solutions, in particular, exponential solutions.

Let us briefly consider the case of more general boundary conditions (6.1.8). According to the standard approach, we have to construct four families of solutions (the fundamental matrix)

$$\begin{gathered} 1)\ u(0,\lambda) = 1, \quad 2)\ \theta(0,\lambda) = 1, \quad 3)\ z(0,\lambda) = 1, \quad 4)\ \mu(0,\lambda) = 1; \\ u(x,\lambda) = \Sigma c_i u_i, \quad \theta(x,\lambda) = \Sigma c_i\theta_i, \quad z(x,\lambda) = \Sigma c_i z_i, \quad \mu(x,\lambda) = \Sigma c_i\mu_i. \end{gathered} \tag{6.1.11}$$

The other components of the solution (the components complementary to the above ones) are taken equal to zero (as in the case of (6.1.9)). In (6.1.11), the summation is over i from 1 to 4. Using the boundary conditions in (6.1.8), we obtain the following system of equations for the unknown λ, c_i:

$$\begin{gathered} (1-\kappa_0)c_4 - \kappa_0 c_1 = 0, \quad (1-\sigma_0)c_3 - \sigma_0 c_2 = 0; \\ \Sigma c_i\left[(1-\kappa_l)\mu_i + \kappa_l u_i\right]_{x=l} = 0, \quad \Sigma c_i\left[(1-\sigma_l)z_i + \sigma_l\theta_i\right]_{x=l} = 0. \end{gathered} \tag{6.1.12}$$

Dividing these relations by $c_j \neq 0$, we obtain four equations for λ, η_i ($\eta_j = 1$). In the general situation, we have $0 < \kappa_{0,l}, \sigma_{0,l} < 1$. Now, from the first two equations, the constants c_3, c_4 are expressed through c_2, c_1 and then substituted into the last two equations. Dividing these equations by c_1 (or c_2), we obtain equations of the form (6.1.10). Passing to the limit as $\kappa_{0,l}, \sigma_{0,l} \to 1$ or $\kappa_{0,l}, \sigma_{0,l} \to 0$, we obtain different types of boundary conditions.

6.2. The Method of Sagittary Function. Sturm's Theorems

6.2.1. Construction of the characteristic equation and the sagittary function. The characteristic equation for the eigenvalues of the parameter λ for problem (6.1.7) has the form

$$
\begin{aligned}
&S(\lambda, l) = 0, \quad \lambda = \lambda_n(l); \quad S(\lambda, x) \equiv u_1(x,\lambda)\theta_2(x,\lambda) - u_2(x,\lambda)\theta_1(x,\lambda), \\
&0 \le x \le l, \quad 0 < l < \infty, \quad \lambda > 0; \quad S(\lambda, x) > 0, \quad 0 < x \ll l, \quad \lambda \sim 1.
\end{aligned} \tag{6.2.1}
$$

As mentioned in Section 6.1, the boundary value problem admits a nontrivial solution of the form (6.1.9), if and only if the determinant of the linear system for c_1, c_2 at $x = l$ is equal to zero, and therefore, relation (6.2.1) should hold. Obviously, this relation is also obtained by the elimination of the unknown η in system (6.1.10). It is common practice to fix the parameter l ($l = 1$) and assume S to be a function of only one argument, λ. Below, we describe some methods and computation algorithms based on the notion of *sagittary function* $S(\lambda, x)$ of two variables λ and x. This function can be used for the investigation of basic properties of solutions of eigenvalue problems. These properties are of the kind considered in Sturm's theorems and their corollaries. The meaning of the term "sagittary function" will become clear from what follows.

The characteristic equation and the sagittary function in the case of boundary conditions (6.1.8) are defined in a similar manner,

$$
\begin{gathered}
S(\lambda, l) = 0, \qquad \lambda = \lambda_n(l); \\
S(\lambda, x) \equiv \kappa_0(\sigma_0 d_1 + (1-\sigma_0)d_2) - (1-\kappa_0)(\sigma_0 d_3 + (1-\sigma_0)d_4), \\
d_1 = M_3 Z_4 - M_4 Z_3, \quad d_2 = M_2 Z_4 - M_4 Z_2, \quad d_3 = M_1 Z_3 - M_3 Z_1, \\
d_4 = M_1 Z_2 - M_2 Z_1, \quad d_i = d_i(\lambda, x), \\
M_i = (1-\kappa_l)\mu_i + \kappa_l u_i, \quad Z_i = (1-\sigma_l) z_i + \sigma_l \theta_i, \quad i = 1,2,3,4.
\end{gathered} \tag{6.2.2}
$$

The function $S(\lambda, x)$ defined by (6.2.2) can be represented in a simpler form, if we use two families of linearly independent solutions satisfying the following boundary conditions at $x = 0$:

$$
\begin{gathered}
1)\ u(0) = 1-\kappa_0, \quad \theta(0) = 0, \quad z(0) = 0, \quad \mu(0) = \kappa_0, \quad 0 < \kappa_0 < 1; \\
2)\ u(0) = 0, \quad \theta(0) = 1-\sigma_0, \quad z(0) = \sigma_0, \quad \mu(0) = 0, \quad 0 < \sigma_0 < 1; \\
u = c_1 u_1 + c_2 u_2, \quad \theta = c_1\theta_1 + c_2\theta_2, \quad z = c_1 z_1 + c_2 z_2, \quad \mu = c_1\mu_1 + c_2\mu_2.
\end{gathered} \tag{6.2.3}
$$

Condition (6.1.8) at $x = 0$ is automatically fulfilled for the solution (6.2.3) for arbitrary λ, c_1, c_2. Now, the characteristic function and the sagittary function can be written in the form

$$
\begin{aligned}
&S(\lambda, l) = 0, \qquad \lambda = \lambda_n(l); \\
&S(\lambda, x) \equiv ((1-\kappa_l)\mu_1(x,\lambda) + \kappa_l u_1(x,\lambda))\,((1-\sigma_l)z_2(x,\lambda) + \sigma_l\theta_2(x,\lambda)) \\
&\quad - ((1-\kappa_l)\mu_2(x,\lambda) + \kappa_l u_2(x,\lambda))\,((1-\sigma_l)z_1(x,\lambda) + \sigma_l\theta_1(x,\lambda)).
\end{aligned} \tag{6.2.4}
$$

In analytical or numerical calculations, one can choose either form of the solution, as dictated by considerations of convenience. Thus, in the first case (6.1.11), the fundamental system is constructed independently of the parameters κ_0, σ_0, which are taken into account on subsequent stages of the determination of the sagittary function and the characteristic equation (6.2.2). When constructing linearly independent solutions in the second case (6.2.3), one has to take into account κ_0, σ_0, but the number of the Cauchy problems to be solved is reduced two times.

From now on, the function $S(\lambda, x)$ is assumed known. It can be constructed either analytically (in rare cases) or by solving the corresponding Cauchy problems. For this purpose, one has either to calculate the functions u_i, θ_i, z_i, μ_i in consecutive order for a fixed λ and save their values in memory, or to solve jointly two or four Cauchy problems and calculate $S(\lambda, x)$ by finite algebraic formulas. The function S may be defined as a solution of the said Cauchy problems coupled with a nonlinear equation. In the case (6.1.7), the function S (6.2.1) satisfies the relations

$$\begin{gathered} S' = \frac{z_1 u_2 - z_2 u_1}{p(x)}, \quad S(\lambda, 0) \equiv 0, \quad 0 \le x \le l, \\ S(\lambda, x) = O(x^4), \quad S > 0, \quad 0 < x \ll l, \quad \lambda \sim 1. \end{gathered} \tag{6.2.5}$$

Relations similar to (6.2.5) hold for the sagittary function in the case of other boundary conditions considered in Section 6.1. The function $S(\lambda, x)$ allows us to describe basic properties of the solution of the original problem and formulate results similar to Sturm's oscillation theorems and their corollaries obtained previously for second-order equations of the form $(pu')' + \lambda r u = 0$ with the corresponding boundary conditions. These properties are helpful in calculations of approximate solutions (see Sections 6.9 and 6.4).

6.2.2. Oscillation properties of the sagittary function. For a fixed $l > 0$, the function $S(\lambda, l)$ oscillates with respect to $\lambda > 0$. For sufficiently large $\lambda = \lambda(l) > 0$, the function $S(\lambda, x)$ oscillates with respect to x for $0 \le x \le l$ and has arbitrarily many null-points. If $r/p \ge c > 0$ for $x > 0$, then the function $S(\lambda, x)$ oscillates in x independently of $\lambda > 0$. Oscillation also takes place, if $r/p \to 0$ as $x \to \infty$, so that $r/p \sim x^{-\gamma}$, $0 < \gamma \le 4$. It should be observed that both solutions $(u_1(x, \lambda), \theta_1(x, \lambda))$, $(u_2(x, \lambda), \theta_2(x, \lambda))$ (6.1.9) of the above system do not oscillate in the above sense.

Suppose that for a fixed $l > 0$, we have found $\lambda > 0$ such that $S(\lambda, l) = 0$ and the function $S(\lambda, x)$ has no intermediate null-points with respect to x. Then, the corresponding $\lambda = \lambda_1(l)$ is the first eigenvalue of problem (6.1.7). If there are $n - 1 \ge 1$ intermediate null-points, then $\lambda = \lambda_n(l)$ is the nth eigenvalue and $0 < \lambda_1 < \lambda_2 < \cdots < \lambda_n$. Using the properties of the function $S(\lambda, x)$, one can establish some statements which can be used for the construction of approximate numerical solutions of the problem. These statements follow from a relation (established in Chapter 7) between the magnitude of λ and the length l of the interval on which the Sturm–Liouville problem for the fourth-order equation is solved [3, 12]. The theorems formulated below may also be proved on the basis of the fact that $S(\lambda, x)$ is an entire function of λ [24, 28, 33, 45, 67]. For any fixed x, this function has finitely many real roots λ_n. With the decrease of x, all roots of $S(\lambda, x)$ (i.e., eigenvalues) increase.

Theorem 6.1. *Suppose that for $\lambda = \lambda^*$, the function $S(\lambda^*, x)$ has n ($n = 1, 2, \ldots$) intermediate roots in x, $0 < x \le l$. Then, for $\lambda = \lambda^{**} > \lambda^*$, the function $S(\lambda^{**}, x)$ has at least n roots on the said interval.*

Moreover, as $\lambda \to \infty$, the number of the roots on the given interval $0 \le x \le l$ tends to infinity.

Theorem 6.2. *Suppose that for $\lambda = \lambda^*$, we have $S(\lambda^*, x) = 0$ for $x = x_1$ and $x = x_2$, where $0 < x_1 < x_2 \le l$. Then, for $\lambda = \lambda^{**} > \lambda^*$, there exists a point $x = x_3$, $x_1 < x_3 < x_2$, such that $S(\lambda^{**}, x_3) = 0$.*

For the construction of an approximate solution of the eigenvalue problem (6.1.7), one can utilize the following statement regarding the null-points of the sagittary function $S(\lambda, x)$ for $\lambda = \lambda^*, \lambda^{**}$.

Theorem 6.3. *Let $x = x_0$ be a common null-point of the functions $S(\lambda^*, x)$ and $S(\lambda^{**}, x)$. Then, for $\lambda^{**} > \lambda^*$, the next null-point $x = x_1$ of $S(\lambda^*, x)$ and the next null-point $x = x_2$ of $S(\lambda^{**}, x)$ satisfy the inequality $x_2 < x_1$.*

This theorem implies that all roots $x_j > 0$ of the function $S(\lambda, x)$ are shifted to the left with the increase of λ and are shifted to the right with the decrease of λ. This property establishes a relation between the sought parameter λ and the length l of the interval and can be used in calculations. This property allows us to find upper bounds λ_n^+ and lower bounds λ_n^- for the eigenvalues λ_n, and this is very important for evaluating the actual precision of approximations. In particular, for the first eigenvalue $\lambda_1(l)$, the following statement can be made about the bounds $\lambda_1^\pm$ and the corresponding null-points $x = \xi_1^\pm$ of the function $S(\lambda_1^\pm, x)$.

Theorem 6.4. *Let λ_1^+ be an upper bound for the first eigenvalue, i.e., $\lambda_1^+ \geq \lambda_1(l)$. Then, $S(\lambda_1^+, \xi^+) = 0$, where $\xi^+ \leq l$. Similarly, if there is a lower estimate $\lambda_1^- \leq \lambda_1(l)$, then $S(\lambda_1^-, \xi^-) = 0$, where $\xi^- \geq l$ (for $\xi^- > l$, the functions $p(x)$, $r(x)$ are assumed to be smoothly extended to the interval $(l, \xi^-]$). If the values $\lambda^\pm$ are sufficiently close, then l, $\xi^\pm$ can be made arbitrarily close, and $S(\lambda_1^+, l) < 0$, $S(\lambda_1^-, l) > 0$.*

The inequality $S(\lambda_1^-, x) > 0$ for $0 < x < l$ follows from simple estimates of the functions $u_{1,2}(x, \lambda_1^-)$, $\theta_{1,2}(x, \lambda_1^-)$, according to (6.1.9), (6.2.5) (problem (6.1.7)). Naturally, for $n \geq 2$, the sagittary function $S(\lambda_n^-, x)$ may change sign up to $(n-1)$ times, and the function $S(\lambda_n^+, x)$ changes sign at least n times. In the general case, cumbersome calculations are needed to determine the sign of the function S for $0 < x \ll l$ and find the number of its alterations on the interval $0 < x < l$. It is convenient to perform these calculations on the basis of numerical integration of the above Cauchy problems.

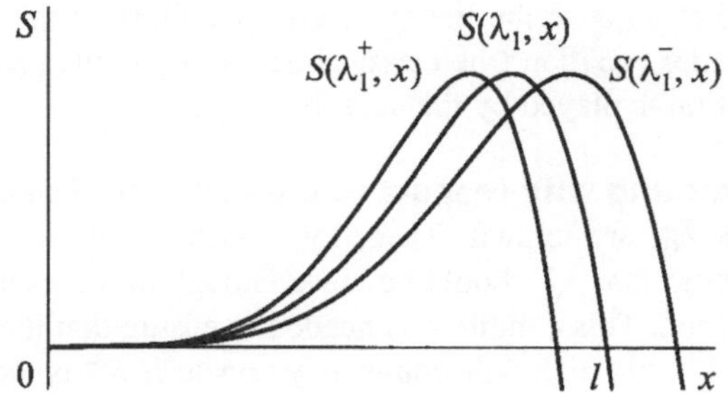

Fig. 6.1

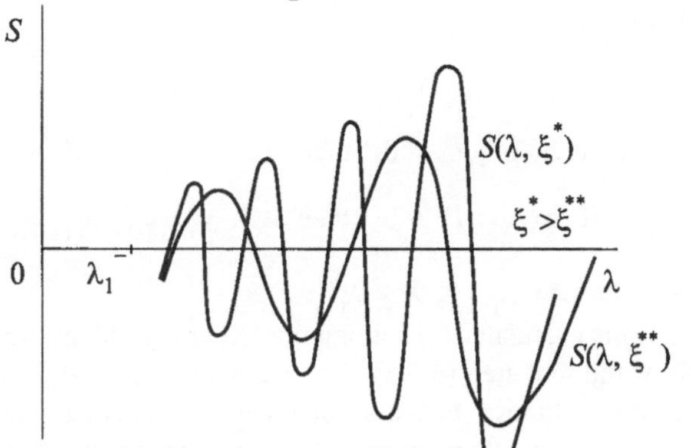

Fig. 6.2

A qualitative picture of the behavior of the function $S(\lambda, x)$ is given in Figures 6.1, 6.2. The "separating" property of the function S can be naturally used in numerical methods aimed at consecutively refining approximations of the desired solution. Among such methods, we should mention the shooting method, the method of successive approximations, the methods of accelerated convergence of Newton's type, etc. This explains the term sagittary function in reference to S (from the Latin *sagitta* – arrow). Note that the shooting method is preferable on the initial stage of solving the problem with the help of preliminary estimates in terms of the bounds $\lambda_n^{\pm}$ (6.1.6) or on the basis of variational estimates, with the relations (6.1.5) taken into account. A fast convergent method, like that of Newton, can be used to refine these values (see Chapter 7). This combined approach is quite effective for parametric synthesis, for instance, for shape optimization.

6.3. Computation Algorithms of the Shooting Method Based on the Sagittary Function

The above properties of the sagittary function $S(\lambda, x)$ (6.2.1) allow us to construct two relatively simple recurrent algorithms, which resemble the shooting method and do not require lengthy calculations. The major part of the calculations is connected with solving two Cauchy problems (6.1.7), (6.1.9) for a given λ, which has been found on the preceding step of the iterative procedure. The sagittary function $S(\lambda, x)$ is used for refining the value of λ used on the next step of the algorithm. This involves an evaluation of the accuracy of the solution (error) with respect to both the "ordinate" (i.e., $|S(\lambda, l)|$) and the abscissa ξ (i.e., $\delta = |l - \xi|$, where $\xi = \arg_x S(\lambda, x)$).

First, consider the problem of finding the first eigenvalue $\lambda = \lambda_1(l)$, dropping the subscript for the sake of brevity. Let us describe the standard operations of the shooting method similar to those for the second-order equation (the classical Sturm–Liouville problem), in which case the role of the sagittary function is played by the solution $u(x, \lambda)$.

6.3.1. Algorithm of shooting with respect to the ordinate. Suppose that sufficiently close upper and lower bounds $\lambda_0^{\pm}$ are known. Then, by Theorem 6.4, we have $S(\lambda_0^+, l) < 0$ and $S(\lambda_0^-, l) > 0$. It is important that λ_0^+ should be close enough to the exact λ, and this is achieved by the Rayleigh–Ritz method. This condition is needed to ensure that there are no additional null-points of the function $S(\lambda^+, x)$ (such null-points may appear if λ^+ is too large (Theorem 6.1)).

The first step of the algorithm consists in calculating the average $\lambda_{(1)}$ and constructing: 1) a refined upper bound $\lambda_{(1)}^+$ or 2) a refined lower bound $\lambda_{(1)}^-$ on the basis of the separating property of the sagittary function S,

$$\begin{gathered}\lambda_{(1)} = \tfrac{1}{2}(\lambda_0^+ + \lambda_0^-);\\ 1)\ \text{if}\ S(\lambda_{(1)}, l) < 0,\ \text{then}\ \lambda_{(1)}^+ = \lambda_{(1)},\ \lambda_{(1)}^- = \lambda_0^-;\\ 2)\ \text{if}\ S(\lambda_{(1)}, l) > 0,\ \text{then}\ \lambda_{(1)}^- = \lambda_{(1)},\ \lambda_{(1)}^+ = \lambda_0^+.\end{gathered} \tag{6.3.1}$$

We obtain the refined estimate: $\lambda_{(1)}^- \leq \lambda \leq \lambda_{(1)}^+$.

The value $S(\lambda_{(1)}, l)$ is calculated by solving the Cauchy problems, as indicated in Sections 6.1 and 6.2. As a result of the first step (6.3.1), the uncertainty interval $\delta_1 = \lambda^+ - \lambda^-$ within which λ is sought is reduced two times. On the second step, we calculate the average $\lambda_{(2)}$ and, using the sign of $S(\lambda_{(2)}, l)$, determine its position with respect to the exact value $\lambda(l)$. In this way, the uncertainty interval for λ is reduced 4 times, etc. On the i-th step, we have

$$\begin{gathered}\lambda_{(i)} = \tfrac{1}{2}(\lambda_{(i-1)}^+ + \lambda_{(i-1)}^-);\\ 1)\ \text{if}\ S(\lambda_{(i)}, l) < 0,\ \text{then}\ \lambda_{(i)}^+ = \lambda_{(i)},\ \lambda_{(i)}^- = \lambda_{(i-1)}^-;\\ 2)\ \text{if}\ S(\lambda_{(i)}, l) > 0,\ \text{then}\ \lambda_{(i)}^- = \lambda_{(i)},\ \lambda_{(i)}^+ = \lambda_{(i-1)}^+.\end{gathered} \tag{6.3.2}$$

As a result, we obtain the refined estimate

$$\lambda^-_{(i)} \le \lambda \le \lambda^+_{(i)}, \quad \delta_i = (\lambda^+_{(i)} - \lambda^-_{(i)}) = \frac{1}{2^i}(\lambda^+_0 - \lambda^-_0), \quad i = 1, 2, \ldots$$

The calculations by the scheme (6.3.2) are continued until the desired relative error is attained. In the meantime, one can control the errors using the ordinates $S(\lambda^\pm_{(i)}, l)$, as well as the solutions $u(l, \lambda^\pm_{(i)})$, $\theta(l, \lambda^\pm_{(i)})$. The quantity $\nu_i = \delta_i/(2\lambda_{(i)})$ can be taken as a measure of relative closeness between the approximate solution and the exact one.

Together with the elementary procedure of bisecting the uncertainty interval, other methods of splitting the interval may be used, for instance, that based on the golden section. In order to accelerate the convergence, the uncertainty interval δ_{i-1} can be split into parts proportional to the residuals $S^\pm_{i-1}$ of the function S, i.e.,

$$\begin{aligned} \lambda_{(i)} &= \lambda^-_{(i-1)} + \delta_{i-1}\frac{s^-_{i-1}}{s_{i-1}} = \lambda^+_{(i-1)} - \delta_{i-1}\frac{s^+_{i-1}}{s_{i-1}}, \\ s^\pm_{i-1} &= |S(\lambda^\pm_{(i-1)}, l)|, \quad s_{i-1} = S(\lambda^-_{(i-1)}, l) - S(\lambda^+_{(i-1)}, l). \end{aligned} \tag{6.3.3}$$

The procedure of refining the sought solution (6.3.3) will converge faster, if S is a linear non-constant function of λ near the eigenvalue.

The above scheme of the shooting method can be used for the problem with the general boundary conditions of elastic fixation (6.1.8). In this case, one has to determine the sign of S for $0 < x \ll l$. The sagittary function $S(\lambda, x)$ is constructed in the same way as in Sections 6.1, 6.2 (see (6.2.2)–(6.2.4)) and has the same property of separating the null-points ξ depending on the parameter λ (see Theorems 6.1–6.4). It should be observed that the root $\lambda_0 = 0$ is simple for the problem with one end fixed and the other end free. If both ends are fixed, then the root $\lambda_0 = 0$ has double multiplicity. For $\lambda = \lambda_1$, the above statements are valid.

Approximations of subsequent eigenvalues $\lambda_n(l)$ and eigenfunctions $\{u_n(x,l),\ \theta_n(x,l),\ z_n(x,l),\ \mu_n(x,l)\}$ for $n = 2, 3, \ldots$ are constructed in the same way as above. It should be kept in mind that the function $S(\lambda_n, x)$ has $(n-1)$ intermediate roots on the interval $0 \le x \le l$, and one should take into account the sign of the function $S(\lambda^\pm_n, x)$ near the points $x = l$; in particular, for $n = 2$, we have $S(\lambda^+_2, l) > 0$ and $S(\lambda^-_2, l) < 0$.

6.3.2. Algorithm of shooting with respect to the abscissa. There is another algorithm of the shooting method which is more clear, spares more computational resources, and is more stable. This algorithm is based on finding discrepancies with respect to the abscissa, i.e., the roots $\xi^\pm_{(i)}$ of the function $S(\lambda^\pm_{(i)}, x)$ for given $\lambda = \lambda^\pm_{(i)}$. The procedure of consecutive refinements of the first eigenvalue (as well as other eigenvalues) utilizes the results of Theorems 6.3, 6.4. In contrast to the procedure (6.3.1), (6.3.2), finding the null-points $x = \xi^\pm_{(i)}$ allows us to judge with certainty the existence of a solution of the boundary value problem and its closeness to the exact solution corresponding to $x = \xi = l$. As a measure of relative closeness between the approximation and the exact solution one can use either ν_i or $\chi_i = \zeta_i/(2l)$, where $\zeta_i = \xi^-_i - \xi^+_i$ is the difference of the abscissa discrepancies. The scheme considered here does not require the determination of the sign of the function $S(\lambda^\pm_{(i)}, x)$ and is based on finding the null-points $x = \xi^\pm_i > 0$ with the desired index.

Thus, in order to find the first eigenvalue $\lambda = \lambda_1(l)$, one should find positive roots $\xi^\pm_i$. On the first step of the recurrent procedure, one should perform operations similar to (6.3.1), namely,

$$\begin{aligned} &\lambda_{(1)} = \tfrac{1}{2}(\lambda^+_0 + \lambda^-_0), \quad \arg_x S(\lambda_{(1)}, x) = \xi_1 > 0; \\ &1)\ \text{if } \xi_1 < l, \ \text{ then } \lambda^+_{(1)} = \lambda_{(1)}, \ \lambda^-_{(1)} = \lambda^-_0; \\ &2)\ \text{if } \xi_1 > l, \ \text{ then } \lambda^-_{(1)} = \lambda_{(1)}, \ \lambda^+_1 = \lambda^+_0. \end{aligned} \tag{6.3.4}$$

We obtain the refined estimate

$$\lambda^-_{(1)} \le \lambda \le \lambda^+_{(1)}, \quad \delta_1 = \tfrac{1}{2}(\lambda^+_0 - \lambda^-_0) = \tfrac{1}{2}\delta.$$

As a result of (6.3.4), the uncertainty of the admissible values of λ is reduced two times. On subsequent steps, this uncertainty decreases in geometrical progression with ratio $1/2$. The ith step consists of the operations (see (6.3.2))

$$\begin{aligned}
&\lambda_{(i)} = \tfrac{1}{2}(\lambda^+_{(i-1)} + \lambda^-_{(i-1)}); \quad \arg_x S(\lambda_{(i)}, x) = \xi_i > 0;\\
&1)\ \text{if } \xi_i < l, \ \text{then } \lambda^+_{(i)} = \lambda_{(i)},\ \lambda^-_{(i)} = \lambda^-_{(i-1)};\\
&2)\ \text{if } \xi_i > l, \ \text{then } \lambda^-_{(i)} = \lambda_{(i)},\ \lambda^+_{(i)} = \lambda^+_{(i-1)};
\end{aligned} \tag{6.3.5}$$

where $i = 1, 2, \dots$. As a result, we obtain the refined estimate

$$\lambda^-_{(i)} \le \lambda \le \lambda^{(+)}_{(i)}, \quad \delta_i = \tfrac{1}{2}\delta_{i-1} = 2^{-i}\delta, \quad \delta_0 = \delta.$$

Together with finding the relative error with respect to the parameter λ, we find abscissa discrepancies for the functions S, u, θ. It is also possible to split the uncertainty interval $[\lambda^-_{(i)}, \lambda^+_{(i)}]$ in another way, in particular, using the gold section rule (see above). If the function $\lambda(l)$ can be linearized in a small neighborhood of the exact value, then (by analogy with the algorithm (6.3.3)) the uncertainty interval can be split in proportion to the discrepancies $l - \xi^+_{i-1}$, $\xi^-_{i-1} - l$, i.e.,

$$\begin{aligned}
&\lambda_{(i)} = \lambda^-_{(i-1)} + \delta_{i-1}\frac{\varepsilon^-_{i-1}}{\varepsilon_{i-1}} = \lambda^+_{(i-1)} - \delta_{i-1}\frac{\varepsilon^+_{i-1}}{\varepsilon_{i-1}},\\
&\varepsilon^\pm_{i-1} = \left|1 - \frac{\xi^\pm_{i-1}}{l}\right|, \qquad \varepsilon_{i-1} = \frac{\xi^-_{i-1} - \xi^+_{i-1}}{l}.
\end{aligned} \tag{6.3.6}$$

The refinement scheme (6.3.6) possesses much faster convergence, if $\lambda(l)$ essentially depends on l, so that $-\infty < \lambda'(l) \le -c < 0$ $(c \sim 1)$.

The algorithm (6.3.4)–(6.3.6) can be used for constructing approximations of subsequent eigenvalues λ_n and functions $\{u_n(x,l),\ \theta_n(x,l),\ z_n(x,l),\ \mu_n(x,l)\}$ for $n = 2, 3, \dots$ The abscissa $\xi^\pm_i$ coincides with the nth positive null-point of the function $S(\lambda^\pm_i, x)$. In the case of general boundary conditions of the form (6.1.8), the algorithm remains the same.

The above procedures of the shooting method are quite effective for obtaining preliminary moderately precise estimates of solutions. These procedures entail no accumulation of roundoff errors (as in the method of successive approximations) and are stable in the case of computer failure. The major part of computations is connected with the integration of the Cauchy problem for system (6.1.7). If the coefficients $p(x)$, $r(x)$ have intricate structure and the range of the parameters of the system is wide, the shooting method may happen to be ineffective for highly precise mass calculations. In such cases, it is better to solve the problem of parametric synthesis by a more complex method of accelerated convergence described in Chapter 7. The case of rapidly oscillating or sharply changing coefficients $p(x)$, $r(x)$ requires further investigation on the basis of numerical experiments.

6.4. Examples

In order to demonstrate computational efficiency of the method based on the sagittary function, we consider some problems of the type (6.1.1), (6.1.2) or (6.1.7) with sharply varying smooth functions $p(x)$ and $r(x)$. In the theory of elastic vibrations, these functions have standard expressions in terms of the moment of inertia and the area of the cross-section of the beam.

6.4.1. A model test example. In order to test the above method, we perform calculations for a model problem for an equation of the type (6.1.1) or a system of the type (6.1.7) which admit analytical solutions in terms of elementary functions. Consider the problem with the boundary conditions (6.1.2) or (6.1.7) and the coefficients

$$p(x) = p_0(b+ax)^2, \quad r(x) = r_0(b+ax)^{-2}, \quad b+ax \neq 0. \tag{6.4.1}$$

Thus, we have the Euler equation, whose solution can be represented as a power function $(b + ax)^k$, where k is a complex parameter determined by the algebraic equation $k^2(k-1)^2 = \lambda/a^4$, which can be easily solved in radicals. The expression obtained for the general solution is rather cumbersome and contains power functions, as well as trigonometric and logarithmic functions. It is hardly possible to obtain analytical expressions of the roots of the characteristic equation (6.2.1); these roots have to be constructed numerically for specific values of the parameters.

Without the loss of generality, two of the three parameters a, b, l, as well as the constants p_0, r_0 in (6.4.1), may be assumed equal to unity. For definiteness, consider the case $ab > 0$. Then, we can take $a = b = 1$, $l > 0$, and the solution (6.1.9) for $\lambda > 1/16$ takes the form

$$\begin{gathered} u = c_1u_1(x,\lambda) + c_2u_2(x,\lambda) = \sqrt{1+x}\,(c_1w_1 + c_2w_2), \quad \theta = u', \\ w_1 = \sin(f^-h) - \frac{f^-}{f^+}\sinh(f^+h), \quad w_2 = \cos(f^-h) - \cosh(f^+h), \\ f^\pm = \left(\sqrt{\lambda} \pm \tfrac{1}{4}\right)^{1/2}, \quad h = \ln(1+x), \quad 0 \le x \le l, \end{gathered} \tag{6.4.2}$$

and $u \equiv 0$ for $\lambda \le 1/16$. Note that the hyperbolic functions can be represented in terms of power functions of the form $(1+x)^{\pm f^\pm}$. From (6.4.2), it follows that the functions $u_{1,2}$, $\theta_{1,2}$ do not oscillate. However, the sagittary function $S(\lambda, x)$ (6.2.1) oscillates both in x and λ. Indeed, we have

$$S(\lambda,x) = 2f^-\cos(f^-h)\cosh(f^+h) - \tfrac{1}{2}(f^+)^{-1}\sin(f^-h)\sinh(f^+h) - 2f^-. \tag{6.4.3}$$

The characteristic equation for any $x = l > 0$ admits countably many roots $\lambda_n(l)$, which can be found numerically by the algorithms described in Sections 6.2, 6.3. In particular, for $l = 1$, we have the following "exact" value of λ_1 and the bound for λ_n:

$$\lambda_1 = 2181.355; \qquad \sqrt{\lambda_n} = \left(\frac{\pi}{\ln 2}\right)^2\left(n + \frac{1}{2}\right)^2 - \frac{1}{4} + O\left(\frac{1}{n}\right) \quad \text{if} \quad n \gg 1. \tag{6.4.4}$$

Let us use the numerical algorithm of Section 6.3 to find λ_1. The two-coordinate (trigonometric) approximation by the Rayleigh–Ritz method yields the value $\lambda_1^+ = 2338.442$, which diverges considerably from the exact value (6.4.4). By the sagittary function method, using (6.3.4)–(6.3.6), we obtain the abscissa $\xi^+ = 0.9770$. Having found ξ^+, we obtain the lower bound $\lambda^- = (\xi^+)^4\lambda^+ = 2131.2312$, according to the results described in Chapter 7. The value $\lambda_{(1)}$ (6.3.4) yields a refined upper bound, since $\xi_{(1)} < 1$. Seven iterations yield the bounds $\lambda_1^\pm$ quite close to λ_1 (6.4.4), namely,

$$\lambda_{1(7)}^+ = 2181.4093, \quad \lambda_{1(7)}^- = 2181.3478, \quad \lambda_{1(8)} = 2181.3785, \quad \frac{\Delta\lambda_1}{\lambda_1} \sim 10^{-5}.$$

Similarly, one calculates desirably precise approximations of subsequent eigenvalues λ_n, for $n \ge 2$ and other $l > 0$.

In the case of the boundary conditions $u = z = 0$ at $x = 0, 1$, six iterations of the algorithms described in Section 6.3 yield the following bounds for the first eigenvalue:

$$\lambda^+_{1(6)} = 432.3410, \quad \lambda^-_{1(6)} = 432.2820, \quad \lambda_{1(7)} = 432.3150, \quad \frac{\Delta\lambda_1}{\lambda_1} \sim 10^{-4}.$$

The above model example illustrates basic theoretical features and demonstrates the efficiency of the computational algorithm. Moreover, it has been shown that in rare cases of analytical integrability (usually in terms of special functions) one has to utilize numerical methods. Thus, the numerical-analytical algorithm based on the sagittary function seems preferable to the algorithms described in Sections 6.2, 6.3.

6.4.2. Comparison with the results of S. Gould. For the sake of comparison with other approaches, consider equation (6.1.1) with $p(x) = 1 + 2\sin^2 \pi x$, $r(x) = 1$, and boundary conditions $u = z = 0$ at the points $x = 0$, $x = l$. This eigenvalue problem was solved for $n = 1, 3$ by the Weinstein–Aronszajn method in [28]: $\lambda_1 = 2.36388\pi^4$, $\lambda_3 = 149.6520\pi^4$ with the relative error $\Delta\lambda/\lambda \sim 10^{-6}$. The sagittary function algorithm, after six iterations, yields the following estimates for $\lambda_{1,3}$:

$$2.36387\pi^4 \le \lambda_{1(6)} \le 2.36389\pi^4, \qquad 149.6517\pi^4 \le \lambda_{3(6)} \le 149.6531\pi^4.$$

It should be observed that the Weinstein–Aronszajn method involves cumbersome formulas and can hardly be used for effective calculations. The above calculations show that the sagittary function method has essential advantages for the class of problems under consideration.

6.4.3. Parametric synthesis for conical beams. Consider a problem with applications in mechanics and technology. Let us examine free vibrations of a conical beam. The geometrical and the mass-inertial characteristics of the beam are described by the relations

$$\begin{gathered} R(x) = R_0\left(1 - \frac{\alpha x}{l}\right), \quad G(x) = \pi R^2(x), \quad I(x) = \tfrac{1}{4}\pi R^4(x), \\ V_\alpha = V_1(3 - 3\alpha + \alpha^2), \quad 0 \le \alpha \le 1, \quad p(x) = EI(x), \quad r(x) = dG(x). \end{gathered} \tag{6.4.5}$$

Here, R is the radius of the cross-section, α is the parameter characterizing the cone angle, G is the area of the cross-section, I is the moment of inertia of the cross-section, V_α is the volume of the cone, V_1 corresponds to $\alpha = 1$ ($V_1 = \frac{1}{3}\pi R_0^2 l$), E is the Young modulus of the material, d is its volume density. All parameters in (6.4.5), except α, are fixed. Introducing dimensionless parameters and the argument $x \in [0, 1]$, we can write equation (6.1.1), (6.4.5) with the rigidity coefficient $p(x) = (1 - \alpha x)^4$ and the linear density $r(x) = (1 - \alpha x)^2$. Thus, we consider free vibrations of a family of conical beams of fixed length, the radius of the beam at $x = 0$ is fixed, but its volume may vary. The case $\alpha = 0$ corresponds to a homogeneous cylindrical beam, and $\alpha = 1$ corresponds to a tapering conical beam for which equation (6.1.1) and system (6.1.7) have singularities at the points $x = l(= 1)$ ($p(1) = r(1) = 0$) . Note that for $\alpha \to 1$, one has to watch that $\xi_i^- \alpha < 1$.

The investigation of free vibrations of conical beams with various types of boundary conditions is important for applications. Consider the following three types of boundary conditions:

1) hinged support of both edges

$$u(0) = u''(0) = u(1) = u''(1) = 0; \tag{6.4.6}$$

2) rigid fixation (clamping) of the left edge and hinged fixation of the right edge

$$u(0) = u'(0) = u(1) = u''(1) = 0; \tag{6.4.7}$$

3) both edges clamped

$$u(0) = u'(0) = u(1) = u'(1) = 0. \tag{6.4.8}$$

Calculation results for the first eigenvalue $\lambda_1(\alpha)$ corresponding to the fundamental vibration mode are represented by solid lines in Fig. 6.3. Curves 1, 2, and 3 correspond to the boundary conditions (6.4.6)–(6.4.8). The calculations were performed for $0 \le \alpha \le 0.95$ by the sagittary function method with the relative error $\Delta\lambda/\lambda_1 \sim 10^{-4}$. Note that for $\alpha = 0.95$, the coefficients $p(x)$ and $r(x)$ have large variation ($1.6 \cdot 10^5$ and 400 times, respectively) as x varies from $x = 0$ to $x = 1$. Highly precise calculations can be performed by the methods of Chapter 7 for $\alpha = 0.999$ and larger values of $\alpha \to 1$. The results obtained admit a mechanical interpretation.

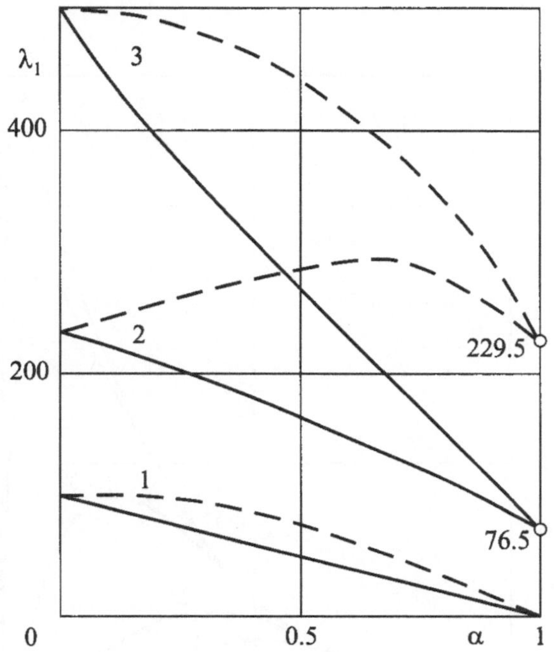

Fig. 6.3

Calculation results and curve 1 show that for $\alpha \to 1$, the first eigenvalue (the fundamental vibration frequency) of the beam with hinged support (6.4.6) tends to zero. This corresponds to free rotation of the beam about the hinge axis at $x = 0$; the edge $x = 1$ exhibits no elastic reaction, since the radius tends to zero. The function $\lambda_1(\alpha)$ is nearly linear, and $\lambda_1(0) = \pi^4$, $\lambda_1(1) = 0$.

In case 2, with the left edge $x = 0$ clamped and the right edge $x = 1$ subjected to hinged fixation (6.4.7), the first eigenvalue (frequency) is much larger (curve 2 is located much higher than curve 1). For $\alpha \to 1$, the value $\lambda_1(\alpha)$ approaches $\lambda_1(1) \approx 76.5$ from above, which corresponds to a conical beam with cantilever fixation [64]. Curve 2 has also a relatively simple structure, but its convexity is more prominent.

In the case (6.4.8) of both ends clamped (curve 3), the eigenvalues $\lambda_1(\alpha)$ are much larger. However, for $\alpha \uparrow 1$, there is fast convergence to $\lambda_1(1) \approx 76.5$ corresponding to a conical beam with cantilever fixation, just as in the case (6.4.7). The function $\lambda_1(\alpha)$ is nearly linear. The eigenfunctions $u_1(x, \lambda)$ normalized with the weight $r(x)$ for $\alpha_i = 0, 0.5, 0.9, 0.99, 0.999$ are represented in Fig. 6.4. Note that all these curves have zero slope at $x = 0$ and $x = 1$.

In a similar way, using the algorithm of Section 6.3 and the properties of the sagittary function described in Section 6.2, one calculates subsequent eigenvalues $\lambda_n(\alpha)$ (vibration frequencies) for the conical beam, as well as the eigenvalues in the case of other types of boundary conditions. This method can be successfully used for the optimization of the cross-section of beams. In particular, if the volume of a conical beam is fixed, $V = 1$, then the eigenvalues can be calculated by formulas

$$\lambda_n^{(1)}(\alpha) = \lambda_n(\alpha)\big(1 - \alpha + \tfrac{1}{3}\alpha^2\big)^{-1}, \qquad n = 1, 2, \ldots$$

The functions $\lambda_1^{(1)}(\alpha)$ are represented in Fig. 6.3 by dashed lines. Since the volume of the beam does not decrease, the eigenfrequencies $\lambda_1^{(1)}$ are much larger than for beams whose volume decreases. Moreover, the dashed curve 2 has a clear maximum, which shows that the conical parameter $\alpha \approx 0.7$ is optimal in the sense that it ensures the maximal first eigenvalue for beams of constant volume.

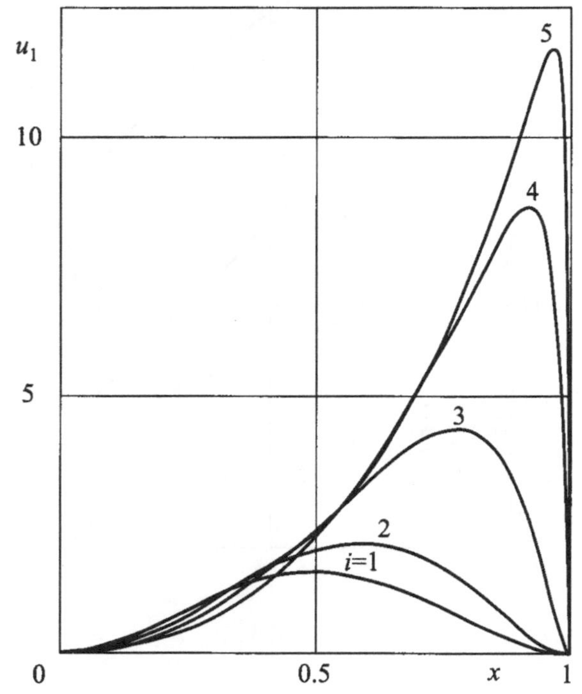

Fig. 6.4

Chapter 7

The Method of Accelerated Convergence for Eigenvalue Problems for Fourth-Order Equations

In this chapter, we develop a rapidly convergent method for solving fourth-order self-conjugate problems. As a basic mathematical model we take the equation of transverse vibrations of a thin inhomogeneous rod subjected to arbitrary boundary conditions. Just as for second-order equations, we establish a differential relation between the eigenvalue and the length of the interval. High efficiency of the algorithm is demonstrated by examples. The problem of parametric synthesis is solved for conical beams whose flexural rigidity may have variation of order up to 10^{12} from one edge to the other, while its linear density variation may be of an order up to 10^6.

7.1. Two-Sided Estimates for Lower Mode Eigenvalues

7.1.1. Differential and variational statements of the problem. First, we formulate the eigenvalue problems in brief form, as we have done in Section 6.1. The original problem is to study small transverse vibrations of a thin rectilinear inhomogeneous beam described by the standard initial boundary value problem [24, 28, 45, 64]. By the methods of mathematical physics, this problem can be reduced to a problem of Sturm–Liouville type for eigenvalues and eigenfunctions. For definiteness, consider the problem of free vibrations of a beam with both edges clamped,

$$\begin{gathered}(p(x)u'')'' - \lambda r(x)u = 0;\\ u(0) = u'(0) = u(1) = u'(1) = 0.\end{gathered} \tag{7.1.1}$$

Here, x is the normalized Euler coordinate ($0 \le x \le 1$) and $\lambda > 0$ is the spectral parameter. The eigenvalues λ_n and the corresponding eigenfunctions $u_n(x) = u(x, \lambda_n)$, $n \ge 1$, are the quantities to be found. The function $p(x)$ in (7.1.1) can be interpreted in terms of mechanics as flexural rigidity, and $r(x)$ is the linear density of the beam (per unit length). The Young modulus and the volume density of the material are assumed constant, and the masses are distributed in such a way that the neutral line of the beam is straight. In order to avoid technical difficulties, we assume that the function $p(x)$ is twice continuously differentiable. Moreover, it is essential that the functions p and r be bounded and separated from zero, $0 < c_{(1)} \le p,\ r \le c_{(2)} < \infty$. It should be mentioned that equation (7.1.1) is just a tribute to tradition [22, 28, 33]. This form of the equation is not very convenient for calculations, because one has to assume additional smoothness of the function $p(x)$. It is more convenient to introduce the variables $z = -pu''$, $w = z'$ characterizing the moment of elastic forces and the cutting force on the cross-section x. Then, the boundary value problem (4.1.1) can be written in the form

$$u' = k, \quad k' = -\frac{z}{p(x)}, \quad z' = w, \quad w' = -\lambda r(x)u;$$
$$u(0) = k(0) = 0, \quad u(1) = k(1) = 0. \tag{7.1.2}$$

In calculations, one actually uses the standard system of equations (7.1.2) which involves no derivatives of $p(x)$. Note that even in the case of constant p, r, a solution of problem (7.1.1) or (7.1.2) obtained analytically in terms of elementary functions is rather cumbersome, because one has to find roots of a transcendental equation, and this can only be done by numerical methods. For variable $p(x)$, $r(x)$, in rare cases when the problem can be solved analytically in terms of special functions, it is very difficult to find the roots λ_n of the corresponding characteristic equation. Modern programmable microcomputers and personal computers can be successfully used for the implementation of a new very efficient numerical-analytical approach to problem (7.1.1) or (7.1.2). This approach is based on a generalization of the well-known Rayleigh–Ritz method, which has been created, thoroughly described, and justified mainly for second-order equations, as mentioned above. Much less attention has been given to fourth-order equations of the type (7.1.1). The Rayleigh–Ritz method yields good results for lower eigenvalues (in particular, the first eigenvalue) and eigenfunctions. For large $\lambda_n^{1/4} \gg 1$ $(n \gg 1)$, asymptotic methods cannot be used for the construction of approximations, because the solutions exhibit non-oscillating behavior (see Chapter 6).

Let us formulate a variational problem equivalent to the Sturm–Liouville problem (7.1.1). The problem is to minimize an integral functional on the class of twice-differentiable functions $u(x)$ satisfying the boundary conditions (7.1.1) and an additional normalization condition, i.e.,

$$J[u] = \int_0^1 p(x)u''^2(x)\,dx \to \min, \quad \Phi[u] = \| u \|^2 = \int_0^1 r(x)u^2(x)\,dx = 1;$$
$$u(0) = u'(0) = u(1) = u'(1) = 0. \tag{7.1.3}$$

The solution of the variational problem (7.1.3) corresponding to the absolute minimum of the functional is the first eigenfunction $u_1(x)$ of the Sturm–Liouville problem (7.1.1), and the absolute minimum $J[u_1] = \lambda_1$ is the first eigenvalue.

In order to determine the second eigenfunction $u_2(x)$ and the second eigenvalue λ_2, one has to solve the variational problem on a more narrow class of functions $u(x)$ orthogonal with the weight $r(x)$ to the function $u_1(x)$. In order to find subsequent $u_n(x)$, λ_n, $n \geq 2$, one has to solve problem (7.1.3) with the orthogonality conditions

$$\Psi_k[u] = \int_0^1 r(x)u_k(x)u(x)\,dx = 0, \quad k = 1, 2, \dots \tag{7.1.4}$$

We are going to construct approximations of the first eigenvalue λ_1 and the eigenfunction $u_1(x)$ of problem (7.1.3), for these quantities are especially interesting for theory and applications. In applications, approximate solutions are often constructed by the well-known Rayleigh–Ritz method, whose special case is the Rayleigh principle.

7.1.2. Construction of upper bounds. The Rayleigh principle immediately yields an upper bound λ_1^* for the first eigenvalue λ_1 in terms of quadratures,

$$0 < \lambda_1 \leq \lambda_1^* = \frac{J[\psi_1]}{\Phi[\psi_1]}; \quad \psi_1(0) = \psi'(0) = \psi_1(1) = \psi_1'(1) = 0. \tag{7.1.5}$$

Here, $\psi_1(x)$ is any test function which is twice continuously differentiable and satisfies the boundary conditions (7.1.5). To have a possibly closer approximation, this test function should take into account the properties of $p(x)$, $r(x)$. The equality $\lambda_1^* = \lambda_1$ holds if and only if

$\psi_1(x) = au_1(x)$, $a = \text{const}$. Thus, the precision of the estimate (7.1.5) depends on the closeness of the function $\psi_1(x)$ to $u_1(x)$. Intuitively, it is clear that $\psi_1(x)$ should resemble the first vibration shape of the beam. Computational experience shows that even a very rough approximation $\psi_1(x)$ of the eigenfunction $u_1(x)$ (for instance, the standard symmetrical function $\psi_1(x) = x^2(1-x)^2$ or $\psi_1(x) = \sin^2 \pi x$) quite often provides a good upper bound λ_1^* for λ_1. This fact allows us to develop an efficient numerical-analytical procedure of the perturbation method for the numerical construction of λ_1, $u_1(x)$ with any given accuracy (usually in the first approximation).

Note that in general, the estimate (7.1.5) does not allow us to decide whether λ_1^* is close to λ_1. An indirect confirmation of the closeness may be obtained by using multi-coordinate (two- or three-coordinate) approximations in the Rayleigh–Ritz method. A direct confirmation of the closeness may be obtained by finding a lower bound λ_{1*} for λ_1. The construction of lower bounds for fourth-order equations by traditional methods is even more difficult than that for second-order equations. Standard approaches involve cumbersome formulas and require lengthy calculations of multiple integrals of functions having a complex structure and defined numerically.

It seems important to implement the following plan, which is similar to that stated in Chapter 2 in connection with the Sturm–Liouville problem.

1. Find a reasonable criterion of closeness between λ_1 and its upper bound λ_1^* obtained by the Rayleigh–Ritz method with not too many coordinate test functions $\psi_i(x)$.
2. Construct a lower bound λ_{1*} for λ_1. This bound, together with λ_1^* determines the range of possible values of λ_1.
3. Develop a constructive algorithm for refining the bound λ_1^*.
4. Find effective procedures for refining the bound λ_1^* and study the problems of justification and convergence.
5. Find approaches to the calculation of subsequent eigenvalues λ_n and eigenfunctions $u_n(x)$, $n \geq 2$.
6. Discuss the construction of solutions of other self-conjugate problems, in particular, problems with other types of boundary conditions.

7.1.3. Relation between the upper bound and the length of the interval. Consider the problem of approximating the first eigenvalue λ_1 and the first eigenfunction $u_1(x)$. For the sake of brevity, we drop the subscript $n = 1$ ($\lambda_1 \to \lambda$, $\lambda_1^* \to \lambda^*$, $u_1 \to u$, etc.), which is unlikely to cause misunderstanding.

Let λ^* be a known upper bound for the first eigenvalue obtained by the Rayleigh–Ritz method or the Rayleigh principle with some test function $\psi(x)$. For λ^*, we construct an "eigenfunction" $v(x, \lambda^*)$ in the following way. Consider two auxiliary Cauchy problems for equation (7.1.1.) (or system (7.1.2))

$$\begin{gathered}
(p(x)v_i'')'' - \lambda^* r(x) v_i = 0, \quad i = 1, 2, \quad 0 \le x \le \xi^* \le 1; \\
v_1(0) = v_1'(0) = v_1'''(0) = 0, \quad v_1''(0) = 1 \quad \text{(1st problem)}; \\
v_2(0) = v_2'(0) = v_2''(0) = 0, \quad v_2'''(0) = 1 \quad \text{(2nd problem)}.
\end{gathered} \tag{7.1.6}$$

It is assumed that the solutions $v_i(x, \lambda^*)$ of problems (7.1.6) are known and have been constructed by some analytical or numerical method. Obviously, these functions are linearly independent. On the basis of the functions v_i, we construct a two-parameter family of solutions

$$\begin{gathered}
v(x, \lambda^*) = c_1 v_1(x, \lambda^*) + c_2 v_2(x, \lambda^*), \\
v(0, \lambda^*) = v'(0, \lambda^*) = 0.
\end{gathered} \tag{7.1.7}$$

Here $c_{1,2}$ are constants chosen from the condition that v and v' should vanish at some point $x = \xi \in [0, 1]$,

$$v(\xi,\lambda^*) = v'(\xi,\lambda^*) = 0. \tag{7.1.8}$$

System (7.1.8) admits a nontrivial solution if and only if its determinant is equal to zero,

$$\begin{gathered}\Delta(\xi,\lambda^*) = v_1(\xi,\lambda^*)v_2'(\xi,\lambda^*) - v_1'(\xi,\lambda^*)v_2(\xi,\lambda^*) = 0;\\ \Delta(0,\lambda^*) = \Delta'(0,\lambda^*) = \Delta''(0,\lambda^*) = \Delta'''(0,\lambda^*) = 0.\end{gathered} \tag{7.1.9}$$

From (7.1.9), it follows that $\Delta = \frac{1}{12}\xi^4 + O(\xi^5)$ for small ξ. This relation is obtained from the definition of Δ and conditions (7.1.6) for the functions $v_{1,2}(x)$ and their derivatives at $x = 0$. Equation (7.1.9) has countably many roots ξ_k, $k = 1, 2, \ldots$ Let $\xi^* > 0$ be the smallest root and $\xi^* < 1$. Then, as a measure of closeness between the upper bound λ^* and the exact eigenvalue we take $\varepsilon = 1 - \xi^* > 0$. The equality $\lambda^* = \lambda$ takes place if $\varepsilon = 0$, i.e., $\xi^* = 1$. Conversely, if $\lambda^* = \lambda$, then $\xi^* = 1$, which follows from the definition of λ. In fact, the determinant $\Delta(\xi,\lambda^*)$ possesses the properties of the sagittary function $S(x,\lambda)$ of Chapter 6.

If $\xi^* = \xi^*(\lambda^*)$ has been found, then the one-parameter family of functions

$$v(x,\xi^*,\lambda^*) = c[v_1(x,\lambda^*)v_2'(\xi^*,\lambda^*) - v_1'(\xi^*,\lambda^*)v_2(x,\lambda^*)] = cv^*(x,\xi^*,\lambda^*) \tag{7.1.10}$$

consists of solutions of the problems

$$\begin{gathered}(p(x)v'')'' - \lambda^* r(x)v = 0, \quad 0 \le x \le \xi^* \le 1,\\ v(0) = v'(0) = v(\xi^*) = v'(\xi^*) = 0.\end{gathered} \tag{7.1.11}$$

The constant c in (7.1.10) is chosen from some additional conditions, for instance, the condition of normalization. If the value λ^* is a fairly good approximation of λ, i.e., if $1 - \xi \ll 1$, then $v(x)$ can be taken as an approximate solution of the problem. The error $\delta = \lambda^* - \lambda > 0$ of the approximation of λ is related to ε, for $\varepsilon \ll 1$, by the formula

$$\Delta_\xi'(\xi^*,\lambda^*)\varepsilon = \Delta_\lambda'(\xi^*,\lambda^*)\delta + O(\varepsilon^2 + \delta^2); \quad \Delta_\xi'(\xi^*,\lambda^*) < 0, \quad \Delta_\lambda'(\xi^*,\lambda^*) < 0. \tag{7.1.12}$$

Relations (7.1.12) have certain similarity to those which follow from the second Strum theorem about oscillations [34, 36, 49] (see Section 6.3) and reflect the fact that a shorter beam vibrates with a higher frequency. Thus, clamping the beam at a point $x = \xi^* < 1$ increases the fundamental frequency. Conversely, clamping the beam at a point $x = \xi > 1$ reduces the fundamental frequency. An additional confirmation that the inequality $\Delta_\xi' < 0$ holds is the relation $\Delta_\xi'(\xi^*,\lambda^*) = -v^{*\prime\prime}(\xi^*,\xi^*,\lambda^*) < 0$, which follows directly from (7.1.9) and (7.1.10).

Theorem 7.1. *For a sufficiently small abscissa discrepancy $\varepsilon > 0$, the error satisfies the inequality $\delta > 0$. Conversely, if $\varepsilon < 0$, then $\delta < 0$.*

In order to show that $\varepsilon > 0$, we proceed as follows. Suppose that $\lambda^* \ge \lambda$ and $\xi^* > 1$; more precisely, let $\varepsilon < 0$, $|\varepsilon| \ll 1$, $\delta \ll 1$. We introduce a function $U(x)$ such that $U(x) \equiv u(x)$ for $0 \le x \le 1$ and $U(x) \equiv 0$ for $1 \le x \le \xi^*$. This function satisfies equation (7.1.1) and the boundary conditions. Multiplying equation (7.1.1) by v and equation (7.1.11) by $U(x)$, let us subtract one from the other and integrate the difference in x, $0 \le x \le \xi$. Using integration by parts, we obtain

$$(\lambda^* - \lambda)\int_0^1 r(x)u(x)v(x)\,dx = 0, \tag{7.1.13}$$

which is possible only for $\lambda^* = \lambda$, i.e., $\xi^* = 1$, since $r > 0$, $u \ge 0$, $v > 0$ for $0 \le x \le 1$. Thus, we have come to a contradiction with our assumption that $\xi^* > 1$ and proved that the inequality $\delta > 0$ implies that $\varepsilon > 0$.

Thus, we have constructed a solution $v(x,\lambda^*)$ of the boundary value problem (7.1.11) corresponding to the eigenvalue λ^*. This solution can be taken as an approximate solution of the original problem (7.1.1), provided that $1 - \xi^* = \varepsilon \ll 1$. A numerical estimate of the closeness between λ^* and λ can be obtained by constructing a lower bound λ_* for the eigenvalue λ.

7.1.4. Construction of lower bounds and two-sided estimates. Suppose that the parameter ε is sufficiently small. Then, this parameter could be linked to a lower bound λ_*, for instance, by the formula

$$\lambda \geq \lambda_* = [1 - k(\varepsilon)]\lambda^*. \tag{7.1.14}$$

In view of (7.1.12), the inequality (7.1.14) always gives a lower bound, if we take $k(\varepsilon) \gg \varepsilon$, in particular, $k(\varepsilon) = \chi\varepsilon^\gamma$, where $0 < \gamma < 1$, $\chi \sim 1$. From (7.1.12), it also follows that the lower bound λ_* lies in the ε-neighborhood of the values λ, λ^*, i.e. $\gamma = 1$. In some situations, it is possible to find a constant which yields the inequality (7.1.14). In such a case, the expression (7.1.14) can be written in equivalent form as $\lambda_* = \xi^{*\chi}\lambda^*$, $\chi > 1$.

Consider some relations which allow us to obtain the desired lower bound λ_*. Using (7.1.6), we construct the functions $v_{1,2}$ for $\lambda = \lambda_*$ (7.1.14) and find a root ξ_* of equation (7.1.9). If it turns out that $\xi_* \geq 1$, then $\lambda_* \leq \lambda$ and the equality $\xi_* = 1$ yields the exact value. In order to prove this statement, we argue in the same way as when establishing the inequality $\xi^* < 1$. Take the solution $v(x, \xi_*, \lambda_*)$ of problem (7.1.11). Let us multiply equation (7.1.1) by v and equation (7.1.11) by u and integrate their difference in x, $0 \leq x \leq 1$. We get

$$\left(v(pu'')' - v'pu''\right)\Big|_{x=0}^{x=1} = (\lambda - \lambda_*) \int_0^1 r(x)u(x)v(x)\,dx. \tag{7.1.15}$$

Without the loss of generality, we can assume that $u \geq 0$ for $0 \leq x \leq 1$. Since $v > 0$, we have $v' < 0$. Moreover, the following inequalities hold:

$$pu''\Big|_{x=1} > 0, \quad (pu'')'\Big|_0^1 > 0.$$

Therefore, the left-hand side of (7.1.15) is positive for $\xi_* > 1$. Hence, we obtain the desired inequality $\lambda - \lambda_* > 0$. Thus, if $v(1) > 0$, $v'(1) < 0$, then the corresponding λ_* is an upper bound for the first eigenvalue. We have obtained he following two-sided estimate for λ:

$$(\delta < 0, \quad \varepsilon < 0) \quad \lambda_* \leq \lambda \leq \lambda^* \quad (\delta > 0, \quad \varepsilon > 0). \tag{7.1.16}$$

The following remark allows us to use the above reasoning for obtaining a lower bound λ_*.

Let us take $\lambda^*\xi^4$ instead of λ^* in the Cauchy problems (7.1.6). If $v(1) > 0$, $v'(1) < 0$, then $\lambda^*\xi^4$ is a lower bound for the eigenvalue λ_1. The reason for choosing this value is provided by calculations based on the perturbation method.

7.2. Closeness Criterion and Perturbation Theory

7.2.1. Introduction of a small parameter. As a criterion of closeness between λ_1^* and λ_1 we take the inequality

$$1 - \xi = \varepsilon \ll 1, \tag{7.2.1}$$

where ξ is the smallest root of the equation $\Delta(\xi, \lambda^*) = 0$ (7.1.9). In order to facilitate the application of the perturbation method for refining the approximate eigenvalue λ_1^*, we introduce a new argument y, a parameter Λ, and a function U by

$$y = \xi x, \quad 0 \leq y \leq \xi, \quad \Lambda = \lambda\xi^{-4}, \quad U = U(y, \Lambda, \varepsilon) \equiv u\left(y\xi^{-1}, \Lambda\xi^4\right). \tag{7.2.2}$$

Then, the original problem of Sturm–Liouville type (7.1.1) transforms as follows (the primes denote the derivatives in y):

$$\begin{gathered}(p(y)U'')'' - \Lambda r(y)U = -\varepsilon(yp'(y)U'')'' + \varepsilon\Lambda yr'(y)U + \cdots,\\ U(0) = U'(0) = U(\xi) = U'(\xi) = 0.\end{gathered} \tag{7.2.3}$$

7.2.2. An approximate solution of the perturbed problem. We seek a solution of the above problem in the form of expansions in powers of ε. In the first approximation, we have

$$U = U_0(y, \Lambda_0) + \varepsilon U_1 + \cdots, \qquad \Lambda = \Lambda_0 + \varepsilon\Lambda_1 + \cdots, \tag{7.2.4}$$

where $U_0 = v(y, \xi, \lambda^*)$ is the known eigenfunction constructed by the method described in Section 7.1; $\Lambda_0 = \lambda^*$ is the first eigenvalue of the generating problem (7.1.11). The function v is defined by (7.1.10), and λ^* is found with the help of the Rayleigh principle. Note that ξ is the smallest positive root of equation (7.1.9). The constant $c \neq 0$ in (7.1.10) may be arbitrary, say, $c = \pm 1$ or $c = \pm\|v\|^{-1}$, where

$$\|v\|^2 = \int_0^\xi r(x)v^2(x)\,dx.$$

The function $u_1 = U_1(y, \xi, \lambda)$ and the coefficient $\Lambda_1 = \Lambda_1(\xi, \lambda^*)$ are found by solving the boundary value problem

$$\begin{gathered}(pU_1'')'' - \lambda^* rU_1 = \Lambda_1 rU_0 - (yp'U_0'')'' + \lambda^* yr'U_0,\\ U_1(0) = U_1'(0) = U_1(\xi) = U_1'(\xi) = 0.\end{gathered} \tag{7.2.5}$$

The solution U_1 of the nonhomogeneous boundary value problem (7.2.5) exists, provided that the right-hand side is orthogonal to the solution of the homogeneous problem (Fredholm's alternative). Since the homogeneous problem is self-conjugate, this condition has explicit form and can be satisfied by suitably choosing Λ_1, namely,

$$\Lambda_1 = \|U_0\|^{-2}\int_0^\xi [(yp'(y)U_0'')'' - \lambda^* yr'(y)U_0]U_0\,dy. \tag{7.2.6}$$

From (7.2.6), it follows that the value Λ_1 does not depend on the constant c, which is cancelled. Therefore, instead of U_0 we can take $v(y, \xi, \lambda^*)$ (7.1.10). Integrating by parts and taking into account the boundary conditions for U_0, we obtain, instead of (7.2.6), an expression for Λ_1 which is more convenient and does not require that the function $p(y)$ be thrice differentiable. This expression is the following:

$$\Lambda_1 = \|U_0\|^{-2}\int_0^\xi [yp'(y)U_0''^2 - \lambda^* yr'(y)U_0^2]\,dy. \tag{7.2.7}$$

We obtain the refined approximation of the sought eigenvalue λ of problem (7.1.1)

$$\lambda = \xi^4\Lambda = \xi^4\lambda^* + \varepsilon\Lambda_1 + O(\varepsilon^2). \tag{7.2.8}$$

From the representation (7.2.8) for $\varepsilon \ll 1$, we conclude that for $\Lambda_1 > 0$, the value $\lambda_* = \xi^4\lambda^*$ is a lower bound for the eigenvalue λ. A rough sufficient condition for λ_* to be a lower bound is that the integrand in (7.2.7) should be nonnegative, which is the case, for instance, if $p'(y) \geq 0$, $r'(y) \leq 0$. Note that both values λ^* and λ_* have been determined with the error of the order $O(\varepsilon)$. Thus, without calculating the integrals (7.2.6) and (7.2.7), under the above conditions on p and

r, we can construct a lower bound λ_* for the eigenvalue λ. Further, if the test function $\psi(x)$ in (7.1.6), together its first and second derivatives, differs from U_0 and its corresponding derivatives by $O(\varepsilon)$, then we can replace $U_0(y)$ in (7.2.6) and (7.2.7) by the function $\psi(y)$. The function $U_1(y)$ can also be constructed by numerical integration of the corresponding Cauchy problems, just as the function $U_0(y)$.

Using the perturbation theory and the expansions (7.2.4) or successive approximations with respect to powers of the small parameter ε, one can construct approximations of the eigenvalues Λ and eigenfunctions U with any given accuracy, and it is possible to formulate sufficient conditions for the convergence of these approximations. Formulas (7.2.2) allow us to find the desired quantities λ and u. However, this approach is unreasonable and requires much effort.

7.3. The Method of Accelerated Convergence for Fourth-Order Boundary Value Problems

7.3.1. A differential relation between the eigenvalue and the length of the interval. Let us turn to formulas (7.2.6) and (7.2.7), in order to calculate Λ_1. These expressions can be simplified, so that there would be no integrals of the terms in square brackets, as has been done in Chapter 2 (Sections 2.3, 2.4). Indeed, integrating by parts, we eliminate the derivatives p', r' and, taking into account the boundary conditions, obtain

$$\Lambda_1 = ||U_0||^{-2}[4||U_0||^2 - \xi p(\xi) U_0''^2(\xi)]. \tag{7.3.1}$$

Substituting (7.3.1) into (7.2.8), we obtain the following expression for the first approximation of λ:

$$\lambda_{(1)} = \lambda^* - \varepsilon \xi p(\xi) ||U_0||^{-2} U_0''^2(\xi), \quad |\lambda - \lambda_{(1)}| \leq c_\lambda \varepsilon^2, \tag{7.3.2}$$

where c_λ is a constant which can be constructively estimated in terms of p and r.

In (7.3.2), it is required that we know the second derivative of the function U_0, which is calculated in the process of constructing the function v, when integrating equations (7.1.6). Formally, we can take $\xi = 1$ in (7.3.2), without the loss of precision with respect to ε.

The expression (7.3.2) has the following mechanical meaning: a decrease of the beam length (i.e., its clamping at a nearer point $x = \xi^* < 1$) increases the fundamental frequency. Conversely, an increase of the beam length (i.e., its clamping at a farther point $\xi > 1$) reduces the fundamental frequency. This fact has been mentioned above. Formula (7.3.2) also implies the following previously unknown fact:

$$\left.\frac{d\lambda}{d\xi}\right|_{\xi=1} = -\frac{p(1)}{||u||^2} u''^2(1). \tag{7.3.3}$$

7.3.2. Algorithm of the accelerated convergence method. The above relations can be utilized for the creation of a numerical-analytical algorithm which consists of the following operations.

1. On the preliminary stage, we choose a test function $\psi(x)$, $0 \leq x \leq 1$, which possesses the required smoothness and satisfies the boundary conditions. Using this function and (7.1.5), we find an upper bound λ^* for the first eigenvalue, i.e., an approximation of λ. The function ψ should be either convex or concave on the interval and have no intermediate zeroes. Note that one can also use another approximate value $\tilde{\lambda} \simeq \lambda$ obtained for a problem whose the coefficients are "close" to p, r.

2. From (7.1.6), we find the functions $v_{1,2}(x, \lambda^*)$, $v'_{1,2}(x, \lambda^*)$ and with these construct the determinant $\Delta(\xi, \lambda^*)$ (7.1.9). Then, we find the smallest positive root of the determinant, $\xi < 1$,

and calculate the parameter $\varepsilon = 1 - \xi$. The quantity ε indicates to what extent λ is close to λ^* and whether the test function $\psi(x)$ has been chosen properly. Computational experience shows that the value $\varepsilon \sim 0.1$ is acceptable. If the obtained value of ε raises doubts, one can use a two- or three-coordinate approximation by the Rayleigh–Ritz method. As a rule, if the functions $p(x)$ and $r(x)$ are not too "exotic", a one-coordinate approximation is sufficient as the starting point of the calculations by the method of accelerated convergence.

3. Using $\xi \in]0,1[$ and $\varepsilon = 1 - \xi$ found above, we construct the function $v^{(1)}(x, \xi, \lambda^*)$ and calculate the norm $\|v^{(1)}\|$, where $\|v^{(1)}\|^2 = (v^{(1)}, rv^{(1)})$ is the scalar product with the weight $r(x)$, $0 \le x \le \xi$.

4. Now, step 2 of the procedure should be repeated with λ^* replaced by $\lambda_{(1)}$ (the refined value obtained by (7.3.2)). Again, the Cauchy problems are solved and new values of ξ_1 and ε_1 are found. Then, one refines the value $\lambda_{(1)}$ and obtains the next approximation $\lambda_{(2)}$; etc.

Thus, for the calculation of the first eigenvalue, we obtain the following procedure:

$$\lambda_{(j+1)} = \lambda_{(j)} - \varepsilon_j \xi_j p(\xi_j) \frac{U''^2_{(j)}(\xi_j)}{\|U_{(j)}\|^2}. \tag{7.3.4}$$

This algorithm has quadratic convergence, which can be proved by arguments similar to those of Sections 2.10, 2.11.

The above considerations can also be used for finding subsequent eigenvalues, provided that on the preliminary stage, we have somehow constructed (for instance, by the Rayleigh–Ritz method) sufficiently close bounds λ_2^*, λ_3^*, etc.

7.4. Other Types of Boundary Conditions

Observe that the structure of the correction term in (7.3.1)–(7.3.3) is determined by the type of the boundary condition at the right end-point $x = 1$. The boundary conditions at $x = 1$ affect the solution through the function v or U. In previous sections, we have considered in detail the case of boundary conditions (7.1.1) (clamping)

$$u(0) = u'(0) = 0, \quad u(1) = u'(1) = 0. \tag{7.4.1}$$

At the right and the left end-points, other types of self-conjugate boundary conditions may be imposed. For definiteness, consider the case of hinged fixation at the right end-point

$$u(1) = p(1)u''(1) = 0, \tag{7.4.2}$$

while at $x = 0$, the boundary conditions have the form (7.4.1), (7.4.2) or some other (see below). The conditions at $x = 1$ may correspond to a free end-point,

$$p(1)u''(1) = (pu'')'\big|_{x=1} = 0, \tag{7.4.3}$$

and the conditions at $x = 0$ may be of one of the types under consideration. It is also possible to consider the case of the conditions that fix the tangential direction,

$$u'(1) = (pu'')'\big|_{x=1} = 0. \tag{7.4.4}$$

The conditions at the left end-point may be of one of the types (7.4.1)–(7.4.4). According to Chapter 6, one can consider the general conditions of elastic fixation.

All arguments used when studying the case of boundary conditions (7.4.1) remain valid for boundary conditions (7.4.2)–(7.4.4). Boundary conditions at the left end-point are taken into account by a suitable choice of the initial conditions in the Cauchy problems (7.1.6). Then, one has to perform calculations similar to those described above. Omitting the details, we will only give the final formulas allowing us to refine eigenvalue approximations for all types of boundary conditions.

In the case (7.4.2), we have

$$\lambda_{(1)} = \lambda^* + 2\varepsilon p(\xi)\frac{(pU_0'')'U_0'(\xi)}{||U_0||^2}. \tag{7.4.5}$$

Obviously, the coefficient of ε in the correction term (7.4.5) differs from zero and is negative.

In the case of the boundary conditions (7.4.3), the following formula holds:

$$\lambda_{(1)} = \lambda^* - \varepsilon\lambda^* r(\xi)\frac{U_0^2(\xi)}{||U_0||^2}. \tag{7.4.6}$$

Comments similar to those given above can be made regarding the expression (7.4.6). Finally, in the case of condition (7.4.4), we get

$$\lambda_{(1)} = \lambda^* - \varepsilon\lambda^* r(\xi)\frac{U_0^2(\xi)}{||U_0||^2} - \varepsilon p(\xi)\frac{U_0''^2(\xi)}{||U_0||^2}. \tag{7.4.7}$$

In all these formulas, ξ is the root of the corresponding equation $\Delta(\xi, \lambda^*) = 0$, and $\varepsilon = 1 - \xi$.

The above formulas remain valid for subsequent eigenvalues $\lambda = \lambda_n$, with ξ being the nth root of the function $\Delta(x, \lambda^*)$.

Just as in Section 4.2, one can develop a method of accelerated convergence with the refinement formulas (7.4.5)–(7.4.7).

7.5. Procedure of Continuation in a Parameter

The following observation allows us to simplify calculations to a considerable extent. Suppose that the coefficients p and r of equation (7.1.1) also depend on a parameter a, i.e., $p = p(x, a)$; $r = r(x, a)$, where $a_0 \le a \le a_1$. In order to solve the boundary value problems by the method of accelerated convergence, the following simple but very effective trick (referred to as "the procedure of continuation in a parameter"). For $a = a_0$, one calculates the upper bounds λ_1^*, λ_2^*, etc. By the method of accelerated convergence, one finds highly precise approximations of the eigenvalues $\lambda_1 = \lambda_1(a_0)$, $\lambda_2 = \lambda_2(a_0)$, etc. It may happen that for some $a = a_0$, a solution can be constructed analytically or in a relatively simple way.

Let Δa be an increment of the parameter. One finds $\lambda_1 = \lambda_1(a_0 + \Delta a)$, $\lambda_2 = \lambda_2(a_0 + \Delta a), \ldots$ by the method of accelerated convergence, using $\lambda_1(a_0), \lambda_2(a_0), \ldots$ as the initial approximations. Then, taking the highly precise values $\lambda_1 = \lambda_1(a_0 + \Delta a)$, $\lambda_2 = \lambda_2(a_0 + \Delta a), \ldots$ as the initial approximations, one finds $\lambda_1(a_0 + 2\Delta a)$, $\lambda_2(a_0 + 2\Delta a), \ldots$ The process is continued until the desired finite point a_1 is reached. If necessary, the increment of the parameter may be increased or decreased, and this opens up wide possibilities for calculations. Moreover, the parameter a may be introduced artificially. An example of the procedure of continuation in a parameter is given in Section 7.7.

7.6. Model Problems

7.6.1. General remarks about calculations. A computer program has been created for the construction of approximate solutions of the boundary value problem (7.4.1). This program is based on a rapidly convergent algorithm and allows us to construct eigenfunctions $u(x, \lambda)$ and the first eigenvalue λ with high precision for arbitrary coefficients $p(x)$ and $r(x)$. In view of the remark of Section 7.3, this program can also be used for computing subsequent eigenvalues and eigenfunctions.

The first stage of the calculations consists in obtaining an upper bound $\lambda^* = J[\psi]/\Phi[\psi]$ on the basis of (7.1.5). For definiteness, take $\psi = x^2(x-1)^2$ (similar estimates are obtained, if one considers other test functions, for instance, $\psi = \sin^2 \pi x$). In order to find λ^*, it is necessary to calculate integrals of the functions $p(x)\psi''^2(x)$ and $r(x)\psi^2(x)$. As a rule, these integrals are calculated by numerical methods, for instance, the Simpson method or some other.

In order to find refined values of λ, one should use the method of accelerated convergence described in Section 7.3. When constructing the functions $v^*(x, \lambda_{(k)})$ and $\Delta(x, \lambda_{(k)})$ by the fourth-order Runge–Kutta method, one has to integrate the fourth-order system (7.1.2) with various types of initial conditions (at the left end-point $x = 0$); in the case under consideration, the initial conditions are the following:

$$\begin{gathered} u_i' = k_i, \quad k_i' = -\frac{z_i}{p(x)}, \quad z_i' = w_i, \quad w_i' = -\lambda r(x) u_i, \quad i = 1, 2; \\ u_i(0) = k_i(0) = 0, \quad z_2(0) = w_1(0) = 0, \quad z_1(0) = w_2(0) = 1. \end{gathered} \tag{7.6.1}$$

Whenever system (7.6.1) is used, there is no need to impose high regularity assumptions on the functions $r(x)$ and $p(x)$. The error on each step of the integration by the Runge–Kutta method is equal to ch^5, where $c = \text{const}$ depends on the smoothness of $p(x)$ and $r(x)$; h is a controllable integration step. The function v^* is expressed by (7.1.10) in terms of the functions v_1, v_2 and their derivatives as follows:

$$v^*(x, \xi_k, \lambda_{(k)}) = v_1(x, \lambda_{(k)}) v_2'(\xi_k, \lambda_{(k)}) - v_1'(\xi_k, \lambda_{(k)}) v_2(x, \lambda_{(k)}). \tag{7.6.2}$$

Here, ξ_k is the smallest root of the equation $\Delta(x, \lambda_{(k)}) = 0$ (7.1.9). The precision with which v^* and Δ are found depends on the errors of calculations of all solutions of the Cauchy problems and is controlled by the integration step of the Runge–Kutta method. For a fixed integration step, the foot ξ_k lies between some $x_i = ih$ for which Δ is still positive and $x_i = (i+1)h$ for which Δ becomes negative for the first time. To obtain a more precise value of the root ξ_k, linear interpolation should be used.

Keeping in mind the above remarks, consider two model problems, which have been solved by the method of accelerated convergence.

7.6.2. Test examples with analytically integrable equations. In the classical case of a homogeneous beam, we have $p = r \equiv 1$. In the case of boundary conditions (7.4.1), the exact value of λ_1 is known to within four decimal digits as a result of analytical calculations, $\lambda_1 = 500.5639$. After three iterations of the above algorithm with the test function $\psi_1(x) = x^2(x-1)^2$, we obtain $\lambda_{(3)} = 500.563902$. In this case, $\varepsilon_3 = 10^{-10}$. Comparing this value with the analytical result, we see that the relative calculation error is very small, $\Delta\lambda/\lambda_1 < 2 \cdot 10^{-7}$. Here and in what follows, the integration step in the Runge–Kutta method is $h = 10^{-4}$.

The eigenfunction u_1 is represented by curve 1 in Fig. 7.1.

Consider a beam for which $p(x) = (1+x)^2$ and $r(x) = (1+x)^{-2}$. In this case we have an integrable equation (an equation of Euler type). This rigidity distribution $p(x)$ and this linear density $r(x)$ correspond to beams of rectangular or elliptic cross-sections for which $a \sim (1+x)^2$, $b \sim (1+x)^{-4}$, since $p \sim a^3 b$, $r \sim ab$. On the interval $[0, 1]$, the functions p and r undergo substantial variation (four times). For such p and r, it is possible to obtain the following analytical value: $\lambda_1 = 2181.355$. Four iterations based on the method of accelerated convergence with the test function $\psi = x^2(x-1)^2$ and $\lambda_1^* = 2611.677$ yield the following values: first, $\varepsilon = 0.06$; $\lambda_{1(1)} = 2153.216$; then $\varepsilon_1 = -0.005$; $\lambda_{1(2)} = 2181.186$; then $\varepsilon_2 = -3 \cdot 10^{-5}$; $\lambda_{1(3)} =$

2181.3548; finally, $\varepsilon_3 = -3 \cdot 10^{-6}$, $\lambda_{1(4)} = 2181.35506$. The relative error is $\Delta\lambda_1/\lambda_1 < 10^{-7}$. The graph of the eigenfunction $u(x, \lambda)$ in this case is represented by curve 2 in Fig. 7.1.

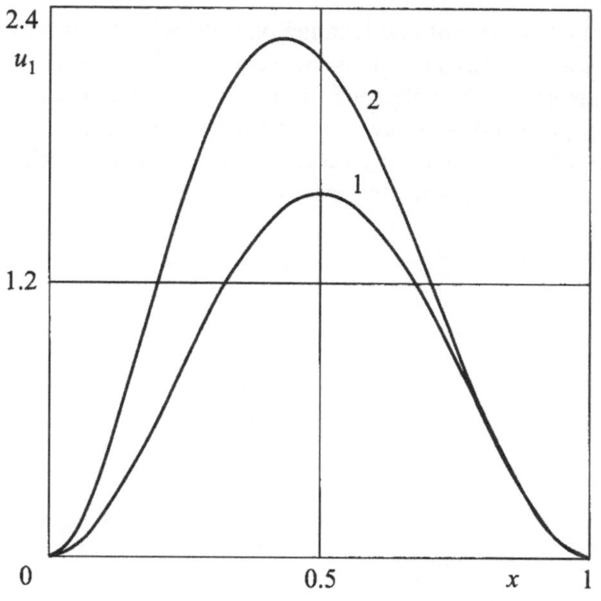

Fig. 7.1

7.6.3. Problem of transverse vibrations of an inhomogeneous beam occurring in applications. We have just considered a beam with variable rigidity and linear density, but its parameters are fairly "exotic" and are unlikely to occur in practice. An important practical problem is that regarding transverse vibrations of conical beams. Fundamental results in that direction were obtained by Kirchhoff; their description can be found in [64]. Kirchhoff integrated equation (7.1.1) with the help of Bessel functions and obtained the first eigenvalue for a conical beam with its right smaller edge free (conditions (7.4.3)) and its left larger edge clamped (conditions (7.4.1)).

Here, we give a solution of the eigenvalue problem for a conical beam with both edges clamped. This case has already been studied in Section 6.5 on the basis of the sagittary function method.

Consider a family of conical beams whose radius linearly depends on the axial coordinate. Without the loss of generality, we can assume that $p(x) = (1 - \alpha x)^4$ and $r(x) = (1 - \alpha x)^2$, where α is a parameter which determines the "sharpness" of the cone, $0 \le \alpha < 1$. In particular, for $\alpha = 0$, we obtain a homogeneous beam considered in Subsection 7.6.2. For the approximation of the eigenvalue λ_1, we used the method of accelerated convergence combined with the method of continuation in a parameter. For $\alpha = 0$, a solution is known: $\lambda_1 = 500.5639$. This value was taken as the initial approximation for $\alpha = 0.1$. Then, the method of accelerated convergence was used for obtaining λ_1 with the relative error 10^{-7}. Then, α was increased up to 0.9 by the increments of length 0.1, upon which the step was decreased to the value $\alpha = 0.99$ and finally, to the value 0.999. For each of these α, the values $\lambda_1(\alpha)$ were calculated by the method of accelerated convergence. Figure 6.4 shows the eigenfunctions $u_1(x, \alpha)$ normalized with the weight $r(x, \alpha)$ for five values of the parameter α: $\alpha_1 = 0$ (curve 1), $\alpha_2 = 0.5$(curve 2), $\alpha_3 = 0.9$ (curve 3), $\alpha_4 = 0.99$ (curve 4), $\alpha_5 = 0.999$ (curve 5). The graph of the function $\lambda_1 = \lambda(\alpha)$ is given in Fig. 6.3. From the graph, we see that the first eigenvalue almost linearly decreases with the increase of α from 0 to 1 and its limit value is equal to $\lambda_1(1) \approx 76.5$. The maximum of the eigenfunction $u_1(x)$ increases and shifts to the right end with the increase of α. It is interesting to observe that the limit value $\lambda_1(1) = 76.5$ coincides with that obtained by Kirchhoff for a beam with its right edge free.

Note that the calculation time of one iteration on a PC with a standard processor amounts to several seconds in the case of $p(x)$ and $r(x)$ being elementary functions, in particular, polynomials.

The above example shows the efficiency of the accelerated convergence method, because it allows us to find λ for $p(x)$ and $r(x)$ having large variation. For instance, if $\alpha = 0.999$, then the parameter $p(x)$ varies 10^{12} times and $r(x)$ varies 10^6 times, while the relative error of the values obtained for the eigenvalue λ_1 does not exceed 10^{-7}.

Chapter 8

Perturbation Method in Eigenvalue Problems for Fourth-Order Equations

In this chapter, we describe the procedure of the perturbation method applied to self-adjoint boundary value problems for fourth-order equations. The coefficients of the equations are assumed nearly constant, and this allows us to introduce a small parameter and construct explicit analytical expressions for the sought quantities. Two schemes of the perturbation method are proposed here and are given mathematical justification: one is based on expansions in powers of the small parameter and the other is based on successive approximations. Approximations of the spectrum and the corresponding orthonormal basis are constructed with given accuracy. These can be used for approximate solution of initial boundary value problems of mathematical physics.

The problem is stated in terms of the theory of elasticity used for the description of the dynamics of quasi-homogeneous elastic beams subjected to various types of boundary conditions. This statement of the problem is important for applications, because it allows one to estimate the effect of perturbations of different physical nature.

8.1. Reduction of the Original Problem to the Standard Perturbed Boundary Value Problem

8.1.1. Statement of the initial boundary value problem; preliminary remarks. For the sake of clarity and definiteness, we consider plane motions of an elastic beam subject to deformations of transverse bending; its longitudinal extensions are neglected. It is assumed that the neutral axis of the undeformed beam is straight and its elastic displacements are small, i.e., motions of the beam can be described in the framework of the linear theory of thin elastic beams [40, 64]. The inertial and the rigidity characteristics are assumed time-independent, and the conditions of motion are such that the dynamical equation for the transverse displacements (the state equation) has the form

$$\rho(x)u_{tt} = -[\sigma(x)u_{xx}]_{xx} + W(x,t), \tag{8.1.1}$$

Here, $u = u(x,t)$ is the transverse displacement of the neutral axis with the Euler coordinate x at the instant t; $W(x,t)$ is the external excitation, a given sufficiently smooth function of $x \in [0,l]$; l is the beam length which is constant. The linear density ρ and the flexural rigidity σ do not depend on t and are sufficiently smooth functions of x whose values vary on sufficiently narrow intervals (see below),

$$0 < \rho_{\min} \le \rho(x) \le \rho_{\max} < \infty, \quad 0 < \sigma_{\min} \le \sigma(x) \le \sigma_{\max} < \infty, \quad 0 \le x \le l. \tag{8.1.2}$$

We consider standard boundary conditions for the function $u(x,t)$ at the points $x = 0$, $x = l$, so that the corresponding boundary value problem is self-adjoint [22, 24, 33, 40]. The nonhomogeneous boundary conditions for $t \in [0,T]$ can be of the following simplest types:

1) Clamping (rigid fixation) of the left end ($x = 0$) or (and) the right end ($x = l$) of the beam,

$$u\big|_{x=0,l} = S_{0,l}(t), \qquad u_x\big|_{x=0,l} = K_{0,l}(t). \tag{8.1.3}$$

2) Free left end ($x = 0$) or (and) right end ($x = l$),

$$-[\sigma(x)u_{xx}]\big|_{x=0,l} = M_{0,l}(t), \qquad -[\sigma(x)u_{xx}]_x\big|_{x=0,l} = P_{0,l}(t). \tag{8.1.4}$$

3) Hinged fixation of the left end ($x = 0$) or (and) the right end ($x = l$),

$$u\big|_{x=0,l} = S_{0,l}(t), \qquad -[\sigma(x)u_{xx}]\big|_{x=0,l} = M_{0,l}(t). \tag{8.1.5}$$

4) Free left end ($x = 0$) or (and) right end ($x = l$) with a fixed direction of the tangential line,

$$u_x\big|_{x=0,l} = K_{0,l}(t), \qquad -[\sigma(x)u_{xx}]_x\big|_{x=0,l} = P_{0,l}(t). \tag{8.1.6}$$

The functions of $t \in [0, T]$ introduced in (8.1.3)–(8.1.6) are assumed sufficiently smooth and have a clear mechanical meaning. Some of these functions characterize kinematic excitations: $S_{0,l}(t)$ is a given displacement, $K_{0,l}(t)$ is a given direction of the tangential line. The others characterize dynamic excitations: $M_{0,l}(t)$ is the moment of external forces applied in the direction orthogonal to the neutral axis; $P_{0,l}(t)$ is the external cutting force orthogonal to that axis. These functions, as well as the distributed external excitations $W(x,t)$ in (8.1.1), may contain perturbations or control variables of kinematic or force character. For simplicity, these functions are assumed given and independent of the unknown function $u(x,t)$ or its derivatives at the points $x = 0$, $x = l$. Note that on the basis of n types of boundary conditions (8.1.3)–(8.1.6) on one or both ends of the beam, we can construct $N = n + \frac{1}{2}n(n-1) = \frac{1}{2}n(n+1)$ types of boundary conditions for the entire beam (in our case, $n = 4$, $N = 10$, and we have 10 types of different boundary value problems). All these problems are self-adjoint, which can be easily shown on the basis of the definition of a self-adjoint problem [22, 24, 33, 45, 46, 54, 66]. If the above functions specifying external actions belong to a certain class, then the solutions will belong to the corresponding class, provided that the equation is supplemented with initial conditions. We consider the standard initial conditions at $t = 0$,

$$u\big|_{t=0} = f^0(x), \qquad u_t\big|_{t=0} = g^0(x). \tag{8.1.7}$$

In control problems, one may also have to specify the so-called terminal conditions at $t = T$,

$$u\big|_{t=T} = f^T(x), \qquad u_t\big|_{t=T} = g^T(x). \tag{8.1.8}$$

The functions $f^{0,T}(x)$, $g^{0,T}(x)$ should be sufficiently smooth, i.e., belong to a certain class of smooth functions, so as to ensure the existence of a solution $u(x,t)$ in the desired class of functions.

Solutions of the above problems are constructed by the method of separation of variables (the Fourier method) in the form of infinite series with the terms $u_n(t,x) = \Theta_n(t)X_n(x)$, $n = 0, \pm 1, \pm 2, \ldots$, where $\{X_n(x)\}$ are eigenfunctions of the corresponding boundary value problems which form an orthonormal basis in L^2 with the weight $\rho(x)$. The boundary value problems for eigenvalues and eigenfunctions corresponding to the boundary conditions (8.1.1), (8.1.3)–(8.1.6) have the form

$$(\sigma X'')'' - \lambda^4 \rho X = 0, \qquad 0 < x < l, \qquad \lambda = \text{const};$$

$$\begin{aligned} &1)\ X|_{x=0,L} = X'|_{x=0,L} = 0, && 2)\ (\sigma X'')|_{x=0,L} = (\sigma X'')'|_{x=0,L} = 0, \\ &3)\ X|_{x=0,L} = (\sigma X'')|_{x=0,L} = 0, && 4)\ X'|_{x=0,L} = (\sigma X'')'|_{x=0,L} = 0, \end{aligned} \tag{8.1.9}$$

where $\sigma = \sigma(x)$ and $\rho = \rho(x)$.

In the case of constant ρ, σ, the solutions of problems (8.1.9) are known [24, 33, 64]: the eigenfunctions $X_n(x)$ are combinations of trigonometric and hyperbolic sines and cosines, and the real eigenvalues λ_n are roots of transcendental characteristic equations. Because of symmetry of the secular equations, it suffices to consider $X_n(x)$, λ_n only for $n = 0, 1, 2, \ldots$, and not $n = -1, -2, \ldots$

8.1.2. Reduction to perturbed boundary value problems. In a more general case, if the beam is nonhomogeneous and the parameters ρ and σ depend on x, the eigenvalues and the eigenfunctions can be found in approximation with the help of constructive algorithms of the perturbation method, provided that $\rho(x)$ and $\sigma(x)$ are close to constants, $\rho(x) \approx \rho_0 = \text{const} > 0$, $\sigma(x) \approx \sigma_0 = \text{const} > 0$. In the perturbation method developed below, this closeness is characterized in terms of a small parameter ε, $0 \le \varepsilon \ll 1$. Taking into account (8.1.2), we can write the identities

$$\begin{gathered} \rho(x) = \rho_0[1 + \varepsilon\delta(x, \varepsilon)], \qquad \sigma(x) = \sigma_0[1 + \varepsilon c(x, \varepsilon)], \\ \varepsilon\delta \equiv \frac{\Delta\rho}{\rho_0}\frac{\rho - \rho_0}{\Delta\rho}, \qquad \varepsilon c \equiv \frac{\Delta\sigma}{\sigma_0}\frac{\sigma - \sigma_0}{\Delta\sigma}, \\ \Delta\rho = \tfrac{1}{2}(\rho_{\max} - \rho_{\min}), \quad \Delta\sigma = \tfrac{1}{2}(\sigma_{\max} - \sigma_{\min}), \\ \rho_0 = \tfrac{1}{2}(\rho_{\max} + \rho_{\min}), \quad \sigma_0 = \tfrac{1}{2}(\sigma_{\max} + \sigma_{\min}). \end{gathered} \tag{8.1.10}$$

Taking, for instance, $\Delta\rho/\rho_0 \sim \varepsilon$, $\Delta\sigma/\sigma_0 \sim \varepsilon$, where ε is a small numerical parameter, $\varepsilon \in [0, \varepsilon_0]$, $0 < \varepsilon_0 \ll 1$, from (8.1.10) we obtain the inequalities $|\delta| \le 1$, $|c| \le 1$ for $x \in [0, l]$. In the limit, as $\varepsilon \to 0$, we obtain self-adjoint boundary value problems for a beam with constant characteristics $\rho = \rho_0$, $\sigma = \sigma_0$. The solutions of these problems can be constructed in the form of quadratures on the basis of known systems of eigenvalues $\{\lambda_n^{(0)}\}$ and eigenfunctions $\{X_n^{(0)}\}$. It can be assumed that $l = \rho_0 = \sigma_0 = 1$ in (8.1.10), if we pass to the variable $x_* = x/l$ and the parameter $\lambda_* = \lambda(\rho_0/\sigma_0)^{1/4}$ and drop the superscript $*$. As a result, we obtain ten self-conjugate boundary value problems with the parameter ε $(0 < \varepsilon \ll 1)$ for the unknown $\lambda(\varepsilon)$ and $X(x, \varepsilon, \lambda)$,

$$\begin{gathered} [(1 + \varepsilon c(x))X'']'' - \lambda^4[1 + \varepsilon\delta(x)]X = 0, \quad 0 < x < l; \\ 1)\ X|_{x=0,L} = X'|_{x=0,L} = 0, \quad 2)\ X''|_{x=0,L} = ((1 + \varepsilon c(x))X'')'|_{x=0,L} = 0, \\ 3)\ X|_{x=0,L} = X''|_{x=0,L} = 0, \quad 4)\ X'|_{x=0,L} = ((1 + \varepsilon c(x))X'')'|_{x=0,L} = 0. \end{gathered} \tag{8.1.11}$$

For $\varepsilon = 0$ (the case of a homogeneous beam), the solutions of the above boundary value problems are known and their properties have been studied in detail. These solutions allow us to construct the solutions $u^{(0)}(x, t)$ of the original Cauchy problems (8.1.1), (8.1.3)–(8.1.7) with the variable t. For $\varepsilon > 0$, the problem is to prove the existence of solutions in the desired classes of regular functions and to construct these solutions by the methods of the perturbation theory [22, 33, 48, 49].

8.1.3. Some features of the standard procedure of the perturbation method. Our problem is to construct approximations for the system of eigenvalues $\{\lambda_n(\varepsilon)\}$ and a system of eigenfunctions $\{X_n(x, \varepsilon)\}$ mutually orthogonal with the weight $(1 + \varepsilon\delta(x))$. These approximations should have given accuracy with respect to ε and be uniform with respect to n, $|n| \to \infty$. Note (see Chapter 3) that if we substitute the expansions

$$\begin{gathered} \lambda_n(\varepsilon) = \lambda_n^{(0)} + \varepsilon\lambda_n^{(1)} + \cdots + \varepsilon^k\lambda_n^{(k)} + \cdots, \\ X_n(x, \varepsilon) = X_n^{(0)}(x) + \varepsilon X_n^{(1)}(x) + \cdots + \varepsilon^k X_n^{(k)}(x) + \cdots \end{gathered} \tag{8.1.12}$$

into (8.1.11), we come to "secular terms" of the form $\varepsilon^p n^q$ (p, q are positive integers). This happens because of the expression $\varepsilon\lambda^4\delta(x)X$, whose smallness can hardly be proved for $\lambda \to \infty$. This fact is a great obstacle (both from the theoretical and the practical standpoints) to the utilization of these approximations as a basis, since with the growth of n ($n \to \infty$), the absolute and the relative errors tend to infinity. The eigenvalue problems considered here are much more difficult to study than the problems for second-order differential equations considered in Chapter 3. Nevertheless, the basic approaches of Chapter 3, which yield a regularized procedure for the construction of $\{X_n(x,\varepsilon)\}$, can also be used in the present case. Below we describe and discuss some relevant algorithms and also consider some questions of justification. From the standpoint of functional analysis, the properties of the approximate basis require further detailed investigation.

8.2. Regularization of the Perturbation Method

8.2.1. Transformation of the independent variable. For the construction of approximate solutions, we propose a method with a perturbed argument y and a parameter ν introduced by nearly identical transformations. For $y = y(x,\varepsilon)$, consider the expression

$$\begin{gathered} y = y(x,\varepsilon) = [x + \varepsilon\varphi(x,\varepsilon)][1 + \varepsilon\varphi(1,\varepsilon)]^{-1} = x + \varepsilon\xi(x,\varepsilon), \\ x = y + \varepsilon\eta(y,\varepsilon), \quad \varphi(x,\varepsilon) = \int_0^x \theta(z,\varepsilon)\,dz, \quad x \in [0,1], \quad y \in [0,1], \\ \theta = \theta(x,\varepsilon) = \frac{1}{\varepsilon}\left[\left(\frac{1+\varepsilon\delta(x)}{1+\varepsilon c(x)}\right)^{1/4} - 1\right] = \frac{1}{4}(\delta(x) - c(x)) + O(\varepsilon), \\ \varphi(0,\varepsilon) = \xi(0,\varepsilon) = \xi(1,\varepsilon) = \eta(0,\varepsilon) = \eta(1,\varepsilon) = 0. \end{gathered} \tag{8.2.1}$$

For $\varepsilon > 0$, formulas (8.2.1) determine a one-to-one correspondence between x and y, so that $y = x$ for $\varepsilon = 0$. Instead of the unknown parameter λ, we introduce the parameter

$$\nu = \lambda(1 + \varepsilon\varphi_1(\varepsilon)), \qquad \varphi_1(\varepsilon) \equiv \varphi(1,\varepsilon). \tag{8.2.2}$$

The unknown function X transforms to

$$X = X(x,\lambda,\varepsilon) \equiv Y(y,\nu,\varepsilon) = Y. \tag{8.2.3}$$

The differential equation (8.1.11) for the unknown function Y becomes

$$\begin{gathered} Y'''' - \nu^4 Y = \varepsilon(AY''' + BY'' + CY'), \qquad 0 < y < 1; \\ A \equiv -2\frac{1+\varepsilon\varphi_1}{1+\varepsilon\theta}\left(\frac{2\theta'}{1+\varepsilon\theta} + \frac{c'}{1+\varepsilon c}\right), \\ B \equiv -\left(\frac{1+\varepsilon\varphi_1}{1+\varepsilon\theta}\right)^2\left(\frac{4\theta' + 3\varepsilon\theta'^2}{1+\varepsilon\theta} + \frac{6\varepsilon c'\theta'}{(1+\varepsilon\theta)(1+\varepsilon c)} + \frac{c''}{1+\varepsilon c}\right), \\ C \equiv -\frac{(1+\varepsilon\varphi_1)^3}{(1+\varepsilon\theta)^4}\left(\theta''' + \varepsilon\frac{2c'\theta'' + c'\theta'}{1+\varepsilon\theta}\right). \end{gathered} \tag{8.2.4}$$

Here, in the expressions of the functions $A = A(y,\varepsilon)$, $B = B(y,\varepsilon)$, $C = C(y,\varepsilon)$, we first differentiate in x, and then make the replacement $x = y + \varepsilon\eta(y,\varepsilon)$. Equation (8.2.4) makes sense if the coefficients $\delta(x)$, $c(x)$ are thrice continuously differentiable functions of $x \in [0,1]$. Note that $A = B = C = 0$, if δ and c are constant for $x \in [0,1]$. The boundary conditions 1)–4) in (8.1.11) are transformed with the help of the relation between x and y and the expressions of X, Y and their derivatives,

$$
\begin{gathered}
X(x,\lambda,\varepsilon) = Y(y,\nu,\varepsilon), \qquad x = y + \varepsilon\eta(y,\varepsilon), \\
y = x + \varepsilon\xi(x,\varepsilon) = (x + \varepsilon\varphi(x,\varepsilon))(1+\varepsilon\varphi(1,\varepsilon))^{-1}, \qquad \lambda = \nu(1+\varepsilon\varphi(1,\varepsilon))^{-1}, \\
X' = Y'(1+\varepsilon\theta)(1+\varepsilon\varphi_1)^{-1}, \quad X'' = Y''(1+\varepsilon\theta)^2(1+\varepsilon\varphi_1)^{-2} + \varepsilon Y'\theta'(1+\varepsilon\varphi_1)^{-1}, \\
X''' = Y'''(1+\varepsilon\theta)^3(1+\varepsilon\varphi_1)^{-3} + 3\varepsilon Y''\theta'(1+\varepsilon\theta)(1+\varepsilon\varphi_1)^{-2} + \varepsilon Y'\theta''(1+\varepsilon\varphi_1)^{-1}.
\end{gathered} \tag{8.2.5}
$$

Here, the prime indicates differentiation of a function in its argument, $X' = dX/dx$, $\theta' = d\theta/dx$, $Y' = dY/dy$, etc. Since $y = 0$ for $x = 0$, and $y = 1$ for $x = 1$, formulas (8.2.5) allow us to pass from the boundary conditions (8.1.11) for X to the corresponding boundary conditions for Y, and we obtain the boundary conditions

$$
\begin{gathered}
Y = 0, \qquad Y' = 0, \qquad Y''(1+\varepsilon\theta)^2(1+\varepsilon\varphi_1)^{-1} + \varepsilon Y'\theta' = 0, \\
(1+\varepsilon c)[Y'''(1+\varepsilon\theta)^3(1+\varepsilon\varphi_1)^{-2} + 3\varepsilon Y''\theta'(1+\varepsilon\theta)(1+\varepsilon\varphi_1)^{-1} + \varepsilon Y'\theta''] \\
+\varepsilon c'[Y''(1+\varepsilon\theta)^2(1+\varepsilon\varphi_1)^{-1} + \varepsilon Y'\theta'] = 0,
\end{gathered} \tag{8.2.6}
$$

where one should take $x = y = 0$ and $x = y = 1$.

Thus, we can replace problem (8.1.11) by the equivalent problem (8.2.4)–(8.2.6) with variable coefficients.

8.2.2. Regular procedure of the perturbation method. In order to solve the eigenvalue problem (8.2.4)–(8.2.6), we construct a general solution of equation (8.2.4). Regular methods of the perturbation theory (based on expansions or successive approximations) yield no secular terms. The perturbed differential equation (8.2.4) is replaced by the corresponding integro-differential equation. Its solution is obtained by the following recurrent procedure of successive approximations applicable for all finite real values of the parameter ν:

$$
\begin{gathered}
Y = Y^{(0)} + \varepsilon L[Y], \quad L = I * D, \quad Y^{(p+1)}(y,\nu,\varepsilon) = Y^{(0)}(y,\nu) + \varepsilon L[Y^{(p)}], \\
Y^{(0)}(y,\nu) = \sum_{i=0}^{3} C_i\Phi_i(y,\nu), \quad \Phi_0(y,\nu) = \frac{\coth\nu y + \cos\nu y}{\coth\nu + \cos\nu}, \quad \Phi_1(y,\nu) = \frac{\sinh\nu y + \sin\nu y}{\sinh\nu + \sin\nu}, \\
\Phi_2(y,\nu) = \frac{\coth\nu y - \cos\nu y}{\coth\nu - \cos\nu}, \quad \Phi_3(y,\nu) = \frac{\sinh\nu y - \sin\nu y}{\sinh\nu - \sin\nu}, \quad y \in [0,1], \quad p = 0,1,2,\ldots
\end{gathered} \tag{8.2.7}
$$

Here, $Y^{(0)}(y,\nu)$ is the so-called generating solution, i.e., a known general solution of the unperturbed equation (8.2.4) for $\varepsilon = 0$; C_i are arbitrary constants. The representation (8.2.7) is convenient for passing to the limit as $\nu \to 0$ and ensuring boundedness of solutions as $\nu \to \infty$. As $\nu \to 0$, the function $Y^{(0)}$ turns into a third-order polynomial of the argument y, since $\Phi_i(y,0) = y^i$. Thus, we have $Y^{(0)}(y,0) = C_0 + C_1 y + C_2 y^2 + C_3 y^3$. For all real ν, the functions Φ_i are uniformly bounded, since $0 \le y \le 1$; moreover, $\Phi_i \to 0$ as $|\nu| \to \infty$. The integro-differential operator L in (8.2.7) is a composition of a third-order differential operator D (in y) and an integral operator I (in y) of Volterra type with difference kernel. The operator D is defined on the set of thrice continuously differentiable functions Y,

$$
F = D[Y] = A(y,\varepsilon)Y''' + B(y,\varepsilon)Y'' + C(y,\varepsilon)Y'. \tag{8.2.8}
$$

The elements $F = F(y,\varepsilon)$ form a set of continuous functions to which the integral operator I is applied,

$$I[F] = \int_0^y G(y-z,\nu)F(z,\varepsilon)dz, \quad G(y,\nu) = \frac{\sinh(\nu y) - \sin(\nu y)}{2\nu^3};$$
$$G(0,\nu) = G'(0,\nu) = G''(0,\nu) = 0, \quad G'''(0,\nu) = 1; \quad G \to \tfrac{1}{6}y^3 \text{ if } (\nu y) \to 0. \tag{8.2.9}$$

The elements $I[F]$ form a set of four times continuously differentiable functions. The difference kernel G possesses the property of fourth-order smoothing. Since Φ_i are analytic functions of y for all real ν, $|\nu| < \infty$, the recurrent scheme (8.2.7) is correctly defined. Note that the application of the operator D to the functions $Y^{(0)}(y,\nu)$, $Y^{(1)}(y,\nu,\varepsilon)$, ..., $Y^{(p)}(y,\nu,\varepsilon)$, ... results in factors of the order ν^3 for $|\nu| \to \infty$, which are annihilated upon the application of the operator I ($\|I\| \sim |\sinh \nu y|\nu^{-3}$). It is important to note that the application of the operator I (8.2.9) according to the procedure (8.2.7) does not result in the exponential growth of the approximations $Y^{(p)}(y,\nu,\varepsilon)$ for $|\nu| \to \infty$, i.e., the functions $Y^{(p)}$ are bounded in $y \in [0,1]$, uniformly in ν, $|\nu| < \infty$, $0 \le \varepsilon \le \varepsilon_0$, where $\varepsilon_0 > 0$ is sufficiently small. Indeed, on the pth step, the leading exponential term with ν in the integrand has the form

$$I_\nu = \left|\sinh \nu(y - z_0) \sinh \nu(z_0 - z_1) \ldots \sinh \nu(z_{q-1} - z_q) \sinh \nu z_q (\cosh \nu)^{-1}\right|,$$
$$1 \ge y \ge z_0 \ge z_1 \ge \ldots \ge z_{q-1} \ge z_q \ge 0, \quad q = 0, 1, \ldots, p, \tag{8.2.10}$$

or is similar to this expression. The coefficients of this term are uniformly bounded functions of the variables z_0, z_1, ..., z_p and the parameters ν, ε. Analysis of (8.2.10) for $\nu \to \infty$ leads us to the following expression:

$$I_\nu \approx Ce^\sigma, \quad \sigma = |\pm \nu(y - z_0) \pm \nu(z_0 - z_1) \pm \ldots \pm \nu(z_{q-1} - z_q) \pm \nu z_q| - |\nu|.$$

The signs $\pm$ of each term in the first expression are independent, and therefore, this expression takes 2^{q+1} values. The maximal value of the first modulus, $|\nu|y$, is attained for $z_k = z_{k-1}$. Therefore, $|\nu|y - |\nu| \le 0$ for $0 \le y \le 1$, which means that the convergence $Y^{(p)} \to 0$ is exponential with respect to $|\nu| \to \infty$ for $y < 1$: $Y^{(p)} \sim \exp(-|\nu|(1-y))$, and it is only for $y = 1$ that $Y = O(1)$ for $\nu \to \infty$.

8.2.3. Justification of the perturbation method. Consider the recurrent procedure (8.2.7). Since the operator L is linear, we have

$$Y^{(p+1)} = Y^{(0)} + \sum_{i=1}^{p} \varepsilon^i L^i[Y^{(0)}], \quad L^{i+1}[Y] \equiv L[L^i[Y]], \quad L^0 = E, \quad L^0[Y] \equiv Y. \tag{8.2.11}$$

For small enough $\varepsilon > 0$, the operator εL is contractive. By the Banach–Schauder–Tikhonov theorem [25], it can be shown that equation (8.2.7) has a unique solution, and this solution is obtained as the limit of the sequence (8.2.11),

$$\lim_{p\to\infty} Y^{(p+1)} = Y^* = Y^{(0)} + \sum_{i=1}^{\infty} \varepsilon^i L^i[Y^{(0)}] = (E - \varepsilon L)^{-1}[Y^{(0)}]. \tag{8.2.12}$$

The operator εL is a contraction for $|\varepsilon| < \varepsilon_0$, $\varepsilon_0 = \|L\|^{-1}$, where $\|L\|$ is the norm of the bounded operator L, which can be expressed in terms of the coefficients A, B, C, and the kernel G. The successive approximations (8.2.11) are uniformly convergent to the sought solution (8.2.12) of equations (8.2.7), (8.2.4).

Suppose that the coefficients θ and c are differentiable four times (see the expressions (8.2.4) of A, B, C). Then, integrating by parts and taking into account the properties of the kernel G (8.2.9), we can transform equation (8.2.7) to an integral equation whose terms, however, are not uniformly bounded in ν. When constructing the desired solution Y in the first approximation in ε (with the error $O(\varepsilon^2)$), it suffices to consider the operator D (8.2.8) for $\varepsilon = 0$, i.e., take

$$D(y,0)[Y^{(0)}] = [-2(3\theta' + c')Y^{(0)'''} - (4\theta'' + C'')Y^{(0)''} - \theta'''Y^{(0)'}]\Big|_{x=y}, \tag{8.2.13}$$

where $Y^{(0)}$ is a known function to within the coefficients C_i. Thus, we obtain the explicit expressions for the sought functions $Y^{(p)}(y,\nu,\varepsilon)$ with any p and the limit function $Y^*(y,\nu,\varepsilon)$,

$$\begin{gathered} Y^{(p)}(y,\nu,\varepsilon) = \sum_{i=0}^{3} C_i \Phi_i^{(p)}(y,\nu), \quad Y^*(y,\nu,\varepsilon) = \sum_{i=0}^{3} C_i \Phi_i^*(y,\nu), \\ \Phi_i^*(y,\nu,\varepsilon) = \lim_{p\to\infty} \Phi_i^{(p)}(y,\nu,\varepsilon), \quad \Phi_i^{(p)}(y,\nu,0) = \Phi_i^*(y,\nu,0) = \Phi_i(y,\nu), \\ Y^{(p)}(y,\nu,0) = Y^*(y,\nu,0) = Y^{(0)}(y,\nu). \end{gathered} \tag{8.2.14}$$

Note that the functions $Y^{(p)}$, Y^* are differentiable in ν and ε; differentiation in y yields a coefficient of the order $O(\nu^k)$, where k is the order of the derivative, since the dependence on y is realized through the product νy.

Since the operators D and L analytically depend on the parameter ε for $|\varepsilon| \le \varepsilon_0$, the sought solution $Y^{(p)}$ (approximated with the error $O(\varepsilon^{p+1})$) or Y^* (the limit as $p \to \infty$) can be respectively represented by a finite sum or a uniformly convergent series in powers of ε,

$$\begin{gathered} Y^{(p)} = Y^{(0)} + \sum_{l=1}^{p} \varepsilon^l Y_l \quad (p \ge 1), \quad Y^* = Y^{(0)} + \sum_{l=1}^{\infty} \varepsilon^l Y_l, \\ Y_1 = L_1 Y^{(0)}, \quad Y_2 = L_2 Y^{(0)} + L_1 Y_1, \quad \ldots, \quad Y_p = L_p Y^{(0)} + L_{p-1} Y_1 + \cdots + L_1 Y_{p-1}, \\ \varepsilon L(y,\nu,\varepsilon) = \sum_{l=1}^{\infty} \varepsilon^l L_l(y,\nu), \quad \varepsilon D(y,\varepsilon) = \sum_{l=1}^{\infty} \varepsilon^l \Big[A_l(y,\varepsilon)\frac{d^3}{dy^3} + B_l(y,\varepsilon)\frac{d^2}{dy^2} + C_l(y,\varepsilon)\frac{d}{dy} \Big]. \end{gathered} \tag{8.2.15}$$

Here, $A_l(y)$, $B_l(y)$, $C_l(y)$ are the coefficients in Taylor's expansions of $A(y,\varepsilon)$, $B(y,\varepsilon)$, $C(y,\varepsilon)$ in powers of ε. Analytical calculations based on (8.2.15) may happen to be much simpler than those based on the recurrent scheme (8.2.7) or (8.2.11), which in most cases is preferable in numerical analysis. Substituting the functions $Y^{(p)}$, Y^* into the boundary conditions (8.2.6) and taking into account (8.1.11), we obtain a transcendental characteristic equation for the unknown parameter ν,

$$\begin{gathered} \Delta^{(p)}(\nu,\varepsilon) \equiv \Delta^{(0)}(\nu) + \varepsilon\Gamma^{(p-1)}(\nu,\varepsilon) = 0, \quad \Gamma^{(-1)} \equiv 0, \\ \Delta^*(\nu,\varepsilon) \equiv \Delta^{(0)}(\nu) + \varepsilon\Gamma^*(\nu,\varepsilon) = 0, \quad \Delta^* = \lim_{p\to\infty} \Delta^{(p)}. \end{gathered} \tag{8.2.16}$$

An approximate and the limit ("exact") solution of equations (8.2.16) are constructed on the basis of the recurrent procedure

$$\begin{gathered} \Delta^{(0)}(\nu_n^{(l)}) = -\varepsilon\Gamma^{(p-1)}(\nu_n^{(l-1)},\varepsilon), \qquad l = 0,1,\ldots,p, \\ \nu_n^{(0)} = \arg\Delta^{(0)}(\nu), \ n = 0,\pm1,\pm2,\ldots, \ |\nu_n^{(p)}(\varepsilon) - \nu_n^{(0)}| \le C\varepsilon, \\ |\nu_n^*(\varepsilon) - \nu_n^{(0)}| \le C\varepsilon, \qquad |\nu_n^*(\varepsilon) - \nu_n^{(p)}| \le C\varepsilon^p. \end{gathered} \tag{8.2.17}$$

Here, $\nu_n^{(0)}$ is a countable set of eigenvalues of the unperturbed problem (for $\varepsilon = 0$) assumed to be known, $\{\nu_n^{(p)}\}$ are approximations of order p, $\{\nu_n^*\}$ are the exact (limit) values.

Similarly to $Y^{(p)}$, Y^*, one can construct series in powers of ε for the characteristic determinants $\Delta^{(p)}$, Δ^* and the eigenvalues $\nu_n^{(p)}$, ν_n^* of the boundary value problem. After finding $\{\nu^{(p)}(\varepsilon)\}$, $\{\nu^*(\varepsilon)\}$ and substituting them into (8.2.9) or (8.2.12), we obtain the desired approximations $\{Y_n^{(p)}\} = \{X_n^{(p)}\}$ or the limit expressions $\{Y_n^*\} = \{X_n^*\}$ $(y = x + \varepsilon\xi)$ of the eigenvalues of the perturbed boundary value problem. The sets of these functions have the properties of an orthogonal basis in the corresponding weighted space,

$$(Y_n^{(p)}, Y_m^{(p)})_\mu = (X_n^{(p)}, X_m^{(p)})_\chi = \|Y_n^{(p)}\|_\mu^2 \delta_{nm} + O(\varepsilon^{p+1}) = \|X_n^{(p)}\|_\chi^2 \delta_{nm} + O(\varepsilon^{p+1}),$$

$$(Y_n^*, Y_m^*)_\mu = (X_n^*, X_m^*)_\chi = \|Y_n^*\|_\mu^2 \delta_{nm} = \|X_{n^2}\|_\chi^2 \delta_{nm},$$

$$\mu = \mu(y,\varepsilon) = 1 + \varepsilon\delta(y + \varepsilon\eta(y,\varepsilon)), \quad \chi = \chi(x,\varepsilon) = 1 + \varepsilon\delta(x), \quad \mu dy = \chi dx, \tag{8.2.18}$$

where $(\cdot,\cdot)_\mu$, $(\cdot,\cdot)_\chi$ are, respectively, the scalar products in L^2 with the weight μ, χ (the integration is, respectively, in x, $y \in [0,1]$); $\|\cdot\|_\mu$, $\|\cdot\|_\chi$ are the respective weighted norms. Note that if both edges of the beam are free (i.e., conditions 2) in (8.1.11) hold on both ends), the zero eigenvalue has double multiplicity ($\lambda = \nu = 0$) and corresponds to two eigenfunctions, which can be orthonormalized with the weights χ, μ, respectively.

8.3. Motion Control Problem

Above, we have constructed systems of eigenvalues $\{\lambda_n(\varepsilon)\}$, $\lambda_n = \nu_n/(1+\varepsilon\varphi_1)$ and orthonormal eigenfunctions $\{X_n(x,\varepsilon)\}$, $X_n(x,\varepsilon) \equiv Y_n(y(x,\varepsilon),\varepsilon)$ (with a given accuracy with respect to the small parameter ε). We seek a solution by the method of separation of variables in the form of a series $u(x,t) = \sum_{n=0}^{\infty} \theta_n(t) X_n(x)$. Then, for the functions $\theta_n(t)$ we obtain the following system of ordinary differential equations:

$$\ddot{\theta}_n + \lambda_n^4 \theta_n = Q_n(t),$$

$$Q_n(t) \equiv W_n(t) + [-((1+\varepsilon c)u'')' X_n + (1+\varepsilon c)(u'' X' - u' X'') + u((1+\varepsilon c)X'')']_{x=0}^{x=1},$$

$$\theta_n = (u, X_n)_\chi, \quad W_n = (W, X_n)_\chi, \quad \theta_n(0) = f_n^0 = (f^0, X_n)_\chi, \quad \dot{\theta}_n(0) = g_n^0 = (g^0, X_n)_\chi. \tag{8.3.1}$$

Here, the dots denote derivatives with respect to t, the dependence of the functions u, X_n, θ_n, etc., on ε is not indicated for the sake of brevity. Taking into account the boundary conditions (8.1.3)–(8.1.6) for $u(x,t)$ and (8.1.11) for $X_n(x)$ with given $S_{0,1}(t)$, $K_{0,1}(t)$, $M_{0,1}(t)$, $P_{0,1}(t)$, we obtain ten types of given right-hand sides $Q_n(t) = Q_n^{j,i}(t) = Q_n^{i,j}(t)$ $(i,j = 1,2,3,4)$ in the countable system (8.3.1). Every Cauchy problem can be easily solved,

$$\theta_n^{i,j} = f_n^0 \cos \lambda_n^2 t + g_n^0 \lambda_n^{-2} \sin(\lambda_n^2 t) + \lambda_n^{-2} \int_0^t \sin\left[\lambda_n^2 (t-\tau)\right] Q_n^{(i,j)}(\tau)\, d\tau. \tag{8.3.2}$$

For the problem of control, i.e., the problem with the conditions (8.1.8) for $t = T$, the control functions $Q_n^{(i,j)}$ from the admissible class should be chosen in such a way that $\theta_n(T) = f_n^T$, $\dot{\theta}_n(T) = g_n^T$. The existence of a solution of the control problem and its actual construction are difficult tasks and require special investigation.

8.4. Finding Eigenvalues and Eigenfunctions in the First Approximation

Consider a beam with ring-shaped or rectangular cross-section. Let r_0 and r_1 be the inner and the outer radii of the ring, respectively; or let a_0 and b_0 be the side lengths of the inner rectangle and a_1, b_1 those of the outer rectangle. If the volume density ρ_V and the Young modulus E are constant, then the linear density and the flexural rigidity of the bean have the form [40, 64]

$$\rho(x) = \rho_V S(x), \qquad \sigma(x) = EI(x), \tag{8.4.1}$$

where S is the area of the cross-section at the point x, and I is the moment of inertia with respect to one of the principal axes. For the ring-shaped cross-section, we have

$$S(x) = \pi[r_1^2(x) - r_0^2(x)], \qquad I(x) = \tfrac{1}{4}\pi[r_1^4(x) - r_0^4(x)]. \tag{8.4.2}$$

For the rectangular cross-section, we obtain

$$S(x) = a_1(x)b_1(x) - a_0(x)b_0(x), \qquad I(x) = \tfrac{1}{12}[a_1(x)b_1^3(x) - a_0(x)b_0^3(x)]. \tag{8.4.3}$$

Let us preserve only linear terms in the expansions of ρ, σ (8.1.10) in powers of the small parameter ε and assume that the cross-section of the beam linearly depends on x, i.e.,

$$\rho(x) = \rho_0(1 + \varepsilon\delta x) + O(\varepsilon^2), \qquad \sigma(x) = \sigma_0(1 + \varepsilon c x) + O(\varepsilon^2).$$

Then, using (8.4.1)–(8.4.3), we obtain the following expressions for the beam with a ring-shaped or rectangular cross-section:

ring-shaped: $\quad r_{0,1}(x) = r_{0,1}^0 + \varepsilon r_{0,1}^1 x, \quad \rho_0 = \pi\rho_V(r_1^{02} - r_0^{02}), \quad \sigma_0 = \tfrac{1}{4}\pi E(r_1^{04} - r_0^{04}),$

$$\delta = \frac{2\pi\rho_V}{\rho_0}(r_1^1 r_1^0 - r_0^1 r_0^0), \quad c = \frac{\pi E}{\sigma_0}(r_1^1 r_1^{03} - r_0^1 r_0^{03});$$

rectangular: $\quad a_{0,1}(x) = a_{0,1}^0 + \varepsilon a_{0,1}^1 x, \quad b_{0,1}(x) = b_{0,1}^0 + \varepsilon b_{0,1}^1 x, \quad \rho_0 = \rho_V(a_1^0 b_1^0 - a_0^0 b_0^0),$

$$\sigma_0 = \tfrac{1}{12}E(a_1^0 b_1^{03} - a_0^0 b_0^{03}), \quad \delta = \frac{\rho_V}{\rho_0}(a_1^0 b_1^1 - a_0^0 b_0^1 + a_1^1 b_1^0 - a_0^1 b_0^0),$$

$$c = \frac{E}{12\sigma_0}(3a_1^0 b_1^{02} b_1^1 - 3a_0^0 b_0^{02} b_0^1 + a_1^1 b_1^{03} - a_0^1 b_0^{03}). \tag{8.4.4}$$

Equation (8.1.11) for eigenfunctions X becomes

$$((1 + \varepsilon c x + O(\varepsilon^2))X'')'' - \lambda^4(1 + \varepsilon\delta x + O(\varepsilon^2))X = 0, \qquad 0 < x < l. \tag{8.4.5}$$

The relation between the variables y and x and that between the parameters ν and λ are simplified to

$$y = x + \varepsilon\varphi_1 x(x-1) + O(\varepsilon^2), \quad x = y + \varepsilon\varphi_1 y(1-y) + O(\varepsilon^2), \quad \nu = \lambda(1 + \varepsilon\varphi_1), \quad \varphi_1 = \tfrac{1}{8}(\delta - c). \tag{8.4.6}$$

In what follows, the equation for Y does not take into account terms of the order ε^2. After the transformations from (8.2.4), we obtain in the first approximation

$$Y'''' - \nu^4 Y = \varepsilon A Y''', \qquad A = -\tfrac{1}{2}(c + 3\delta). \tag{8.4.7}$$

Let us find a solution of the perturbed problem in the linear approximation (with respect to ε) for a weakly inhomogeneous beam with hinged fixation of its left edge and its right edge being free. The boundary conditions for the unperturbed Cauchy problem (8.4.7) read

$$Y\big|_{y=0} = Y''\big|_{y=0,1} = Y'''\big|_{y=1} = 0. \tag{8.4.8}$$

The solution of this problem is known and has the form

$$Y_j^0(y) = \sum_{i=0}^{3} k_i \Phi_i(y, \nu_j), \quad k_0 = k_2 = 0, \quad k_3 = -k_1 \frac{(\sinh\nu - \sin\nu)(\cosh\nu - \cos\nu)}{(\sinh\nu + \sin\nu)(\cosh\nu + \cos\nu)}, \tag{8.4.9}$$

where Φ_i are the functions defined in (8.2.7), and ν_j are found from the known transcendental equation

$$\begin{gathered}\sinh\nu\,\cos\nu = \cosh\nu\,\sin\nu \qquad (\text{or } \tanh\nu = \tan\nu),\\ \nu_0 = 0,\ \ \nu_{\pm1} = \pm3.927,\ \ \nu_{\pm2} = \pm7.069; \qquad \nu_{\pm n} = \pm\tfrac{1}{4}\pi \pm \pi n + O(e^{-2|n|\pi}).\end{gathered} \tag{8.4.10}$$

It can be shown that the first-order approximation of the solution of the perturbed problem (8.4.7) with arbitrary boundary conditions has the form

$$Y(y,\nu) = \sum_{i=0}^{3} k_i^* \Phi_i(y,\nu)\bigl(1 + \tfrac{1}{4}\varepsilon Ay\bigr). \tag{8.4.11}$$

The boundary conditions (8.2.6) for the perturbed problem (hinged fixation) can be written in the form

$$\begin{gathered}Y\big|_{y=0} = 0, \qquad [Y''(1 + 2\varepsilon(\theta - \varphi_1)) + \varepsilon\theta' Y']\big|_{y=0} = 0,\\ [Y'''(1 + 3\varepsilon(\theta - \varphi_1)) + 3\varepsilon\theta' Y'']\big|_{y=1} = 0,\end{gathered} \tag{8.4.12}$$

where $\theta = (\delta - c)/4$ due to (8.2.1). After the transformations (8.4.12), using (8.2.2), we find that

$$\begin{gathered}Y\big|_{y=0} = 0, \quad [Y'' + \tfrac{1}{4}\varepsilon(\delta - c)(Y' - Y'')]\big|_{y=0} = 0,\\ [Y'' + \tfrac{1}{4}\varepsilon(\delta - c)(Y' + Y'')]\big|_{y=1} = 0, \quad [Y''' + \tfrac{3}{4}\varepsilon(\delta - c)(Y'' + \tfrac{1}{2}Y''')]\big|_{y=1} = 0.\end{gathered} \tag{8.4.13}$$

From (8.4.11), we obtain the following relations for the derivatives of Y in y:

$$\begin{aligned} Y' &= \sum_{i=0}^{3} k_i^* \left(\Phi_i'(1 + \tfrac{1}{4}\varepsilon Ay) + \tfrac{1}{4}\varepsilon A\Phi_i\right),\\ Y'' &= \sum_{i=0}^{3} k_i^* \left(\Phi_i''(1 + \tfrac{1}{4}\varepsilon Ay) + \tfrac{1}{2}\varepsilon A\Phi_i'\right),\\ Y''' &= \sum_{i=0}^{3} k_i^* \left(\Phi_i'''(1 + \tfrac{3}{4}\varepsilon Ay) + \tfrac{3}{4}\varepsilon A\Phi_i''\right).\end{aligned} \tag{8.4.14}$$

The first condition in (8.4.13) implies that $k_0^* = 0$. Substituting the values of Y and its derivatives from (8.4.11), (8.4.14) into (8.4.13) and taking into account the values of Φ_i for $y = 0$, $y = 1$, after algebraic transformations, we obtain the expressions of k_2^* and k_3^* through k_1^*,

$$k_2^* = \varepsilon k_1^* \frac{\delta + c}{2\Phi_2'(1,\nu)}, \qquad k_3^* = -k_1^* \frac{\Phi_1''}{\Phi_3''}\left(1 + \frac{\varepsilon(\delta + c)}{2\Phi_2'}\right)\bigg|_{y=1} = 0, \tag{8.4.15}$$

and also the transcendental characteristic equation

$$\left(\Phi_1''' - \frac{\Phi_1''\Phi_3'''}{\Phi_3''} + \varepsilon(\delta + c)\frac{\Phi_2''' - \Phi_1'''}{2\Phi_2'}\right)\bigg|_{y=1} = 0. \tag{8.4.16}$$

Let us insert the derivatives of Φ_i into (8.4.16). After suitable trigonometric transformations, we obtain

$$\begin{gathered}F_0(\nu) + \varepsilon F_1(\nu) = 0;\\ F_0(\nu) = \cosh\nu\,\sin\nu - \sinh\nu\,\cos\nu, \quad F_1(\nu) = \tfrac{1}{2}(\delta + c)\nu^{-1}(1 - \cosh\nu\,\cos\nu).\end{gathered} \tag{8.4.17}$$

Since the solution is sought in the first approximation, ν can be represented in the form

$$\nu = \nu_0 + \varepsilon\nu_1, \tag{8.4.18}$$

where ν_0 are eigenvalues of the unperturbed problem obtained from (8.4.9), and ν_1 is found from (8.2.17),

$$\nu_1 = -\frac{F_1(\nu_0)}{F'_{0,\nu}(\nu_0)}, \qquad F'_{0,\nu}(\nu_0) = 2\sinh\nu_0\,\sin\nu_0. \tag{8.4.19}$$

Substituting ν_1 from (8.4.18), (8.4.19), k_2^*, k_3^* from (8.4.15), and $k_0^* = 0$ into (8.4.11), we find the eigenfunctions Y_l in the first approximation. From (8.2.2), (8.2.3), we find X_l and λ_l,

$$\lambda_l = \nu_l(1 - \varepsilon\varphi_1) = \nu_{0,l} + \varepsilon[\nu_{1,l} - \tfrac{1}{8}\nu_{0,l}(\delta - c)].$$

For instance, for a ring-shaped cross-section with $r_0^0 = r_0^1 = 0$ (solid beam), we have $\delta = 2r_1^1/r_1^0$, $c = 4r_1^1/r_1^0$, $(\delta + c) = 6r_1^1/r_1^0$, $(\delta - c) = -2r_1^1/r_1^0$. This means that in this case, for $l \gg 1$, the eigenvalues λ_l increase ($\nu_l \gg \nu_1 \sim 1$ for $l \gg 1$). For $\varepsilon = 0.1$, $r_1^1/r_1^0 = 1$, and relatively small l, the unperturbed eigenvalues λ_l^0 and the correction coefficients λ_l^1 are the following:

l	1	2	3	4	5
λ_l^0	3.927	7.069	10.21	13.35	16.49
λ_l^1	1.385	1.978	2.699	3.450	4.214

8.5. Exercises

Exercise 1. Solve the boundary value problem (8.1.11) with the boundary conditions:

a) $X(0) = X'(0) = X''(1) = X'''(1) = 0$ (cantilevered beam),

b) $X(0) = X''(0) = X(1) = X''(1) = 0$ (hinged fixation of both ends),

c) $X(0) = X'(0) = X(1) = X''(1) = 0$ (cantilevered beam with hinged support),

d) $X(0) = X'(0) = X(1) = X'(1) = 0$ (both edges clamped).

Exercise 2. Consider the general situation of elastic fixation of the edges:

a) $[\sigma X'' \mp hX']_{x=0,1} = 0$ (elastic fixation with respect to the angle),

b) $[(\sigma X'')' \mp hX]_{x=0,1} = 0$ (elastic fixation with respect to the displacement).

Exercise 3. Extend the perturbation method to a more general problem with $\sigma = \sigma(x, \varepsilon)$, $\rho = \rho(x, \varepsilon)$, assuming that for $\varepsilon = 0$ the exact solution $\{\lambda_n^0\}$, $\{X_n^0(x)\}$ is known.

Chapter 9

Sturm–Liouville Problems for Vector-Valued Functions

The numerical-analytical methods used above for the construction of effective solutions of eigenvalue problems can be directly generalized to the case of multi-dimensional systems, more precisely, to eigenvalue problems for self-conjugate differential operators acting on vector-valued functions of a scalar argument. The differential operators may be of the second, the fourth, or a higher order. In particular, it is interesting to consider eigenvalue problems for coupled subsystems of Sturm–Liouville type. Next, we describe the procedure of the accelerated convergence method for a vector problem which cannot be split into sub-problems of smaller dimensions.

9.1. Setting of the Problem. Preliminary Remarks

9.1.1. Statement of the problem in differential form. Consider a system of second-order differential equations for a vector-valued function $X = X(x)$ with the boundary conditions of the first kind [7]

$$(P(x)X')' + [\lambda R(x) - Q(x)]X = 0, \quad X(0) = X(1) = 0. \tag{9.1.1}$$

Here, $P = P^T$, $R = R^T$, $Q = Q^T$ are symmetric $(N \times N)$ matrices (T indicates the transpose matrix) whose elements are sufficiently smooth functions of x, $0 \le x \le 1$. The matrices P are R positive definite and Q is nonnegative definite. In particular, for diagonal matrices P, R, Q, problem (9.1.1) splits into N scalar Sturm–Liouville problems considered above. Here, we consider problems that cannot be split into sub-problems of a smaller dimension.

Problem (9.1.1) may be regarded as a generalization of the standard Sturm–Liouville problem to the case of vector-valued functions, with P being the rigidity matrix, R the inertia matrix, and the matrix Q characterizing elastic properties of the environment (Winkler's base). The boundary conditions (9.1.1) correspond to rigid fixation of the edges. One can also consider other boundary conditions, such as those of elastic fixation, periodicity, etc. Moreover, the equation may also contain a "gyroscopic" term $G(x)X'$ with a skew-symmetric matrix $G \equiv -G^T$.

Our problem consists in finding values of the scalar parameter $\lambda > 0$ for which problem (9.1.1) has nontrivial solutions $X(x, \lambda)$, called eigenfunctions.

9.1.2. Variational statement of the problem. The above eigenvalue problem for differential equations can be equivalently formulated in variational form. Namely, one seeks the minimum of the functional

$$J[X] = \int_0^1 [X'^T P(x)X' + X^T Q(x)X]\, dx \to \min_X; \qquad X(0) = X(1) = 0, \tag{9.1.2}$$

on the set of continuously differentiable functions $X(x)$ satisfying the boundary conditions from (9.1.2) and the isoperimetric condition

$$\Phi[X] = \|X\|^2 = \int_0^1 X^T R(x) X \, dx = 1, \tag{9.1.3}$$

which plays the role of the condition of normalization of the vector X with the weight $R(x)$.

The absolute minimum of the positive functional J (9.1.2) with the condition $\|X\| = 1$ (9.1.3) is attained on the first vector-valued eigenfunction $X^{(1)}$, and $J[X^{(1)}] = \lambda_1$ is the first eigenvalue of problem (9.1.1). Subsequent eigenfunctions $X^{(m)}(x)$ and eigenvalues λ_m, $m \geq 2$, are defined in terms of minimization of the above functional on more narrow classes of functions satisfying additional orthogonality conditions with the weight $R(x)$:

$$\Psi_k[X] = \int_0^1 X^{(k)T}(x) R(x) X \, dx = 0 \quad (k = 1, 2, \ldots, m-1). \tag{9.1.4}$$

The variational statement (9.1.2)–(9.1.4) can be used in the Rayleigh–Ritz method, the finite element method, or other methods for the construction of upper bounds for lower eigenvalues and the numerical approximation of the corresponding eigenfunctions. Lower bounds for the eigenvalues can be obtained on the basis of the theory of integral equations, the Weinstein–Aronszajn theory, the methods of Fichera [28, 45], etc. It should be observed that these methods require sophisticated analysis and a lot of preliminary work.

Let us use the Rayleigh–Ritz method (in particular, the Rayleigh principle) for finding an upper bound λ_1^* for the first eigenvalue λ_1. Note that for subsequent eigenvalues λ_m a similar approach can be used, and the formulas obtained for λ_m, $m \geq 2$, are analogous those for λ_1.

9.2. Closeness Criterion and Perturbation Theory

9.2.1. Construction of the comparison problem; analysis of its properties. Let us take a known λ^* in equation (9.1.1.) and construct a system of N linearly independent vectors $V = W_i(x, \lambda^*)$ which are solutions of N Cauchy problems

$$(P(x)V')' + [\lambda^* R(x) - Q(x)]V = 0, \quad 0 \leq x \leq \ell, \tag{9.2.1}$$

$$V(0) = 0, \quad V'(0) = E_i, \quad i = 1, 2, \ldots, N.$$

Here, E_i is an N-vector whose ith component is equal to unity and the other components are equal to zero. The solutions $W_i(x, \lambda^*)$ of the Cauchy problems (9.2.1) can be constructed with the desired accuracy, and their construction encounters no principal difficulties. The parameter ℓ is the length of the interval of x and will be determined below. Consider an N-parametric family of functions $V(x, \lambda^*) = W(x, \lambda^*)c$, where W is the matrix formed by the solutions W_i, and c is a constant N-vector. Let us require that $V(x, \lambda^*) = 0$ for some $x = \xi$, where $0 < \xi \leq 1$. The necessary and sufficient condition for the existence of a nontrivial V can be written as the following equation for the unknown x:

$$\det W(x, \lambda^*) = \Delta(x, \lambda^*) = 0, \quad W = (W_1, W_2, \ldots, W_N). \tag{9.2.2}$$

In particular, if $\lambda^* = \lambda_1$ is the exact eigenvalue, then $\xi = 1$. Since λ continuously depends on the length of the interval, it follows that for sufficiently close λ^* and λ_1 in the general situation, there exists a root ξ close to $x = 1$. This is the first real root of equation (9.2.2), which can be found numerically with given accuracy and is assumed known in the sequel.

It will be shown that $\xi < 1$ for $\lambda_1^* - \lambda_1 > 0$ sufficiently small, i.e., an analogue of the corollary from the Sturm's second oscillation theorem holds for the vector equation (9.2.1). Conversely, if $\lambda^* - \lambda_1 < 0$ (λ_* is a lower bound), then $\xi(\lambda_*) > 1$. In this case, the matrix-valued functions $P(x), R(x), Q(x)$ are assumed extended to some interval $1 < x \leq \ell$. Note that the function $\Delta(x, \lambda^*)$ can be constructed together with the integration of the Cauchy problems (9.2.1) for W_i up to the value $x = \xi$ for which $\Delta(\xi, \lambda^*) = 0$. If one considers the nth eigenvalue λ ($n > 1$), then ξ is the nth root of the equation.

To simplify further reasoning, we assume that ξ is a simple root. In general, this can only be claimed under certain additional conditions and requires strict justification (omitted here). Thus, let

$$\Delta'_\xi(\xi, \lambda^*) \neq 0, \quad \xi = \xi(\lambda^*). \tag{9.2.3}$$

Then, $\operatorname{rank} W(\xi, \lambda^*) = N - 1$, which can be proved by *reductio ad absurdum*. Indeed, assuming that all minor determinants of the order $N - 1$ are equal to zero, we see that condition (6.2.3) cannot hold.

Since $\operatorname{rank} W(\xi, \lambda^*) = N - 1$, there exists i such that we can express the components c_k, $k \neq i$, of the vector c through its ith component c_i. We have $c = C(\xi, \lambda^*)c_i$, where C is an N-vector. As a result, we obtain a one-parameter family of solutions of the Cauchy problem (9.2.1) vanishing for $x = \xi$:

$$V(x, \lambda^*) = c_i W(x, \lambda^*) C(\xi, \lambda^*) \equiv c_i V^*(x, \lambda^*), \quad V(\xi, \lambda^*) = 0. \tag{9.2.4}$$

Without the loss of generality, we can take $c_i = 1$ or choose c_i from a condition of normalization of the type (9.1.3). The vector-valued function (9.2.4) is a solution of the boundary value problem (9.1.1) on the interval $0 \leq x \leq \xi(\lambda^*)$, and the corresponding eigenvalue coincides with λ^*.

9.2.2. Introduction of a small parameter. As a measure of closeness between λ^* and λ we take the parameter

$$\varepsilon = 1 - \xi; \quad |\varepsilon| \ll 1, \tag{9.2.5}$$

which is assumed small, and λ^*, $V(x, \lambda^*)$ found above will be regarded as a generating solution of the original perturbed problem corresponding to $\varepsilon = 0$.

Let us transform the boundary value problem (9.1.1) into a perturbed eigenvalue problem, as we have done in previous chapters. Letting $y = \xi x$, $\Lambda = \xi^{-2}\lambda$, we get

$$(P(y)U')' + [\Lambda R(y) - Q(y)]U = H(y, U, \Lambda, \varepsilon), \quad U(0) = U(\xi) = 0. \tag{9.2.6}$$

Here, y is a new argument, ξ is known, and Λ, U are unknown quantities. In (9.2.6) the following notation is used:

$$\begin{aligned} &X(x, \lambda) \equiv U(y, \Lambda, \varepsilon), \\ &H(y, U, \Lambda, \varepsilon) \equiv -\varepsilon[(yP'(y)U')' + y\Lambda R'(y)U - (yQ'(y) + 2Q(y))U] + O(\varepsilon^2), \end{aligned} \tag{9.2.7}$$

and it is assumed that the matrices P, R, Q smoothly depend on x. For $\xi > 1$, these matrices are smoothly extended to a larger interval.

9.2.3. Approximate solution of the problem. In order to find approximate solutions $\Lambda(\varepsilon), U(y, \Lambda, \varepsilon)$ of the perturbed problem (9.2.6), we use the method of expansion in powers of the small parameter ε or the method of successive approximations. The generating solution (for $\varepsilon = 0$) is known,

$$\Lambda = \Lambda(0) = \lambda^*, \quad U^{(0)} = U(y, \Lambda^{(0)}, 0) = V(y, \lambda^*).$$

Consider the first approximation with respect to ε : $\Lambda(\varepsilon) = \lambda^* + \varepsilon\Lambda_1 + \cdots$. Using the Fredholm alternative, we obtain an expression of the coefficient Λ_1 in terms of integrals of quadratic forms of the known vector-valued functions V' and V (9.2.4). We have

$$\Lambda_1 = -\parallel V \parallel^{-2} \int_0^\xi [yV'^T(y,\lambda^*)P'(y)V'(y,\lambda^*) - \lambda^* y V^T(y,\lambda^*)R'(y)V(y,\lambda^*) \\ +V^T(y,\lambda^*)(yQ'(y) + 2Q(y))V(y,\lambda^*)]\,dy, \qquad \parallel V \parallel^2 = \int_0^\xi V^T R V\,dy. \tag{9.2.8}$$

In the general case, Λ_1 is found by numerical integration in (9.2.8). The refined values of Λ and λ with the error $O(\varepsilon^2)$ are equal to

$$\Lambda(\varepsilon) = \lambda^* + \varepsilon\Lambda_1(\lambda^*) + \cdots, \qquad \lambda = \xi^2\Lambda = \xi^2\lambda^* + \varepsilon\Lambda_1 + \cdots = \lambda^* + (\Lambda_1 - 2\lambda^*)\varepsilon + \cdots. \tag{9.2.9}$$

From (9.2.9) for λ, it follows that for $\varepsilon\Lambda_1 < 0$ the value $\xi^2\lambda^* = \lambda_*$ is a lower bound for the sought λ, i.e., $\lambda_* \le \lambda \le \lambda^*$; moreover, $\xi < 1$, $\varepsilon > 0$. Thus, the condition $\varepsilon\Lambda_1 < 0$ reduces to the inequality $\Lambda_1 < 0$, which is always satisfied if the integrand in (9.2.8) is nonnegative.

The sign of ε is determined by a property similar to the corollary from Sturm's second oscillation theorem, namely, $\varepsilon > 0$ if $\lambda_* < \lambda$.

Consider the opposite case: $\Lambda_1 > 0$. This inequality can often be established without integration. In this case, $\xi^2\lambda^*$ is a refined upper bound: $\lambda \le \xi^2\lambda^* \le \lambda^*$. For $\Lambda_1 = 0$, the expression $\xi^2\lambda^*$ determines the sought eigenvalue λ with the error $O(\varepsilon^2)$.

After calculating Λ_1, we find the functions U_1. Then, a refinement procedure can be used for finding Λ_2, U_2, etc., on the basis of the perturbation theory. This approach, however, is inefficient, because the complexity of calculations rapidly increases and the numerical errors caused by roundoff errors accumulate.

9.3. The Method of Accelerated Convergence for the Sturm–Liouville Problem for Vector-Valued Functions

9.3.1. Properties of the first approximation of the solution. To overcome computational difficulties, we will follow the idea of the method of accelerated convergence. For its realization, one has to transform formulas (9.2.8) and (9.2.9) similarly to what has been done in Chapters 2 and 7. Integrating by parts and eliminating the derivatives P', R', Q', one can substantially simplify these expressions. As a result, one obtains the refinement formula

$$\lambda_{(1)} = \lambda^* - \varepsilon \parallel V \parallel^{-2} V'^T(\xi,\lambda^*)P(\xi)V'(\xi,\lambda^*), \tag{9.3.1}$$

which is a vector analogue of the refinement formula obtained in Subsection 2.3.3 in the scalar case.

From (9.3.1), we obtain a differential analogue of the corollary from the Sturm theorem,

$$\lim_{\varepsilon\to 0} \frac{\lambda - \lambda^*(\xi)}{\varepsilon} = -\parallel V \parallel^{-2} V'^T(1,\lambda)P(1)V'(1,\lambda); \\ V(x,\lambda) = X(x,\lambda) \qquad (\xi \to 1 \text{ at } \varepsilon \to 0). \tag{9.3.2}$$

Relation (9.3.2) implies that for a larger interval, the first eigenvalue (frequency) becomes larger, and for a smaller interval, it becomes smaller. This statement agrees with mechanical concepts and the case of a scalar vibration system. Note that the expressions (9.3.1), (9.3.2) do not explicitly contain the matrices R and Q, which are present indirectly, through the matrices

V and ξ. Moreover, when calculating $\lambda_{(1)}$, we can take $\xi = 1$ without the loss of precision in ε (with the error $O(\varepsilon^2)$). Formulas (9.3.1) and (9.3.2) are valid for any eigenvalue λ_m. If other boundary conditions are prescribed at the right end, then new calculations have to be performed to obtain the corresponding refinement formulas. For instance, in the case of the boundary condition $X'(1) = 0$, the refinement formula reads

$$\lambda_{(1)} = \lambda^* - \varepsilon \parallel V \parallel^{-2} V^T(\xi, \lambda^*)[\lambda^* R(\xi) - Q(\xi)]V(\xi, \lambda^*). \tag{9.3.3}$$

Boundary conditions at the left end-point are not explicitly present in the refinement formula and are taken into account implicitly, when integrating the corresponding Cauchy problems.

9.3.2. Algorithm of accelerated convergence for vector problems. Let us describe the recurrent algorithm of the method of accelerated convergence for calculating eigenvalues and eigenfunctions of the vector Sturm–Liouville problem.

On the initial stage of the algorithm, one obtains a sufficiently precise bound λ^* for the sought eigenvalue by the Rayleigh–Ritz or any other method. One constructs the function $V(x, \lambda^*)$ and finds numerically the root $\xi = \xi(\lambda^*)$ of the determinant $\Delta(\xi, \lambda^*)$. By assumption, this root is simple. Then one calculates the small parameter of the problem, $\varepsilon = 1 - \xi$. For practical purposes, it suffices to have the estimate $\varepsilon \sim (0.1 - 0.01)m^{-1}$, where m is the index of the eigenvalue. Above, we have taken $m = 1$, but all formulas are valid for any m, as repeatedly mentioned above.

The next stage consists in refining λ with the help of (9.3.1) or (9.3.3), depending on the boundary conditions at the right end-point $x = 1$. The squared norm is calculated by numerical integration, since the vectors $V(x, \lambda)$ are known after solving the Cauchy problem. Then, the calculations are repeated until the desired precision is obtained. In concise form, the algorithm of accelerated convergence can be represented by the relations

$$\begin{gathered}
\lambda_{(k+1)} = \lambda_{(k)} + \varepsilon_k \mu(\xi_k, \lambda_{(k)}), \quad \mu = - \parallel V_k \parallel^{-2} V_k'^T P V_k' \Big|_{\xi_k}, \quad \varepsilon_k^* = 1 - \xi_k, \\
\xi_k = \xi(\lambda_{(k)}) = \arg \Delta(\xi, \lambda_{(k)}), \quad \Delta = \det W(\xi, \lambda_{(k)}), \\
k = 0, 1, 2, \ldots, \quad \lambda_1^* = \lambda_{(0)}; \quad \xi = \xi(\lambda^*), \quad \varepsilon = 1 - \xi(\lambda^*).
\end{gathered} \tag{9.3.4}$$

The matrix W, the vector V and its norm are found numerically after solving the Cauchy problems (9.2.1) with the known $\lambda_{(k)}$.

Computational experience shows that to ensure the desired precision, only a few iterations are needed, just as in the scalar case. The iteration procedure of the algorithm of accelerated convergence can be numerically realized in a fairly simple way. It reduces to operations of the same type and is based on highly precise integration of the Cauchy problems with various (refined) values $\lambda = \lambda_{(k)}$ and the calculation of the root $\xi_k \to 1$. Upon the construction of the one-parameter family of solutions $cV_{(k)}, cV'_{(k)}$ $(c = 1)$ and the norm $\parallel V_{(k)} \parallel$, a simple expression is used for finding the refined value $\lambda_{(k+1)}$. This algorithm can be tested by examples with analytical solutions, which is done in Section 9.4. The convergence of this algorithm has quadratic character. For small enough $|\varepsilon|$, the following estimate holds ($L = \text{const}$):

$$|\lambda_{(k+1)} - \lambda| \le L\varepsilon_k^2, \quad \varepsilon_k = \varepsilon^{n_k}; \quad n_k = 2^k, \quad k = 0, 1, 2, \ldots; \quad \varepsilon = \varepsilon^{(0)}.$$

As shown by calculations for model problems in Section 9.4, the usual rough upper bound obtained by the Rayleigh-Ritz method yields the values $\varepsilon \sim 0.1 - 0.01$.

One or two iterations of the above algorithm yield highly precise bounds for lower eigenvalues λ_m with the relative error of the order $10^{-4} \div 10^{-6}$. Three to five iterations practically exhaust the precision capacities of modern software.

A characteristic feature of the method of accelerated convergence is that one has to solve Cauchy problems of the same structure. These problems do not become more complex for subsequent iterations, only the parameter λ has to be changed. There is no accumulation of roundoff errors (the convergence of $\lambda_{(k)}$ is controlled in terms of ε_k). The precision of the calculations with the growth of the iteration number $k = 1, 2, \ldots$ grows accordingly. The numerical-analytical approach to the investigation of multi-dimensional systems with nonhomogeneous parameters is closely related to the results obtained for the classical Sturm–Liouville problem.

9.4. Model Problems

Consider some model examples that illustrate the algorithm described in Section 9.3. Problems admitting a fairly complete analytical solution (for instance, problems for systems with constant coefficients, equations of the Euler type, etc.) can be used for debugging or testing computer programs.

9.4.1. A system of Euler type. Consider a system of the Euler type (9.1.1) with $N = 2$ and the (2×2)-matrices P, R, Q of the form

$$P = \operatorname{diag}(1,1); \quad R = (1+x)^{-2}(r_{ij}), \quad r_{11} = 1, \quad r_{12} = r_{21} = 2, \quad r_{22} = 4; \quad Q \equiv 0,$$

where the matrix $R(x)$ is nonnegative. In component notation, the eigenvalue problem for the vector $X = (u_1, u_2)^T$ has the form

$$\begin{aligned} u_1'' + \lambda(1+x)^{-2}(u_1 + 2u_2) &= 0, \\ u_2'' + \lambda(1+x)^{-2}(2u_1 + 4u_2) &= 0, \\ u_1(0) = u_1(1) = u_2(0) = u_2(1) &= 0. \end{aligned} \tag{9.4.1}$$

A solution of system (9.4.4) is sought in the form $X = C(1+x)^p$, where $C = (C_1, C_2)^T$. Substituting this expression into the system and solving the characteristic equation, we obtain

$$p_{1,2} = \tfrac{1}{2} \pm \tfrac{1}{2} i\sqrt{20\lambda - 1}, \qquad i^2 = -1.$$

The vector-valued function X_n satisfying zero boundary conditions at $x = 0$ and $x = 1$ has the form

$$\begin{gathered} X_n(x) = c_n(1,2)^T \sqrt{1+x}\, \sin\!\left(\tfrac{1}{2}\sqrt{20\lambda_n - 1}\, \ln(1+x)\right), \\ \lambda_n = \tfrac{1}{5}\left[(\pi n/\ln 2)^2 + \tfrac{1}{4}\right], \quad n = 1, 2, \ldots, \quad \lambda_1 = 4.1584576. \end{gathered} \tag{9.4.2}$$

Here, n is the mode number (the index of the eigenvalue and that of the corresponding eigenfunction). The constant c_n is found from the conditions of normalization. In particular, $c_1 = \frac{1}{5}(2/\ln 2)^{1/2} = 0.3397287$. Note that the roots $p_{3,4} = 0, 1$ correspond to $X \equiv 0$. Consider the test function $\Psi = (1,2)^T \sin \pi x$, which is fairly distant from the exact eigenfunction $X_1(x)$ (9.4.2). The Rayleigh–Ritz method gives the upper estimate $\lambda_1 \le \lambda_1^* = 4{,}2448420$. Following the procedure described in Section 9.3, we integrate numerically the Cauchy problem, construct the determinant (9.2.8), and find $\xi = \xi(\lambda_1^*) = 0{,}9857031$ and the parameter $\varepsilon = 1 - \xi(\lambda_1^*) = 1.42969 \cdot 10^{-2}$. According to (9.3.1), the refined value satisfies the inequality $\lambda_{1(1)} \le \lambda_1 \le \lambda_1^*$. The second iteration yields $\xi_1 = 1.0001758$ and $\varepsilon_1 = -1.7158 \cdot 10^{-4}$, and we obtain a highly precise upper bound $\lambda_{1(2)} = 4.1585057$. As a result, we obtain the two-sided estimate $\lambda_{1(1)} \le \lambda_1 \le \lambda_{1(2)}$, and the relative error is $\Delta\lambda_1/\lambda_1 = 1.2 \cdot 10^{-5}$. Thus, the method yields very precise results in a more general case of a nonnegative matrix $R(x)$.

9.4.2. A system with periodic coefficients. Let us briefly consider a model problem (9.1.1) for which no analytical solution has been constructed. Let $P = \mathrm{diag}(1,1)$, $R = (r_{ij})$, where $r_{11} = 1$, $r_{12} = r_{21} = \cos^2 \pi x$, $r_{22} = 4$; $Q \equiv 0$. By the Rayleigh–Ritz method with the test function $\Psi = (c_1, c_2)^T \sin \pi x$ we obtain the characteristic equation, which yields upper bounds for λ_1, λ_2 and a relation between C_1, C_2,

$$\lambda_1^* = 2.4547038, \quad C_2 = 12.0827636C_1; \quad \lambda_2^* = 10.0781280, \quad C_2 = -0.0827636C_1. \tag{9.4.3}$$

Then, we obtain a refined value of λ_1 on the basis of (9.4.3). Using the algorithm of accelerated convergence, we find $\xi = 0.9999776$, $\varepsilon = 2.24{\cdot}10^{-5}$. Thus, ε is very small, which shows that λ_1^* is close to the exact value λ_1, and that the test function Ψ is close to the corresponding eigenfunction. The first iteration yields the lower bound $\lambda_{1(1)} = 2.4545938$, and we have $\lambda_{1(1)} \le \lambda_1 \le \lambda_1^*$. Therefore, the relative error is of the order $2.3 \cdot 10^{-5}$.

9.5. Exercises

Exercise 1. Find analytical expressions for all eigenvalues and eigenfunctions for the Sturm–Liouville problem for the following system of equations of Euler type:

$$\begin{gathered} ((1+x)^2 u_1')' + \lambda(u_1 + 2u_2) = 0, \\ ((1+x)^2 u_2')' + \lambda(2u_1 + 4u_2) = 0, \\ u_1(0) = u_1(1) = u_2(0) = u_2(1) = 0. \end{gathered} \tag{9.5.1}$$

Exercise 2. By the method of accelerated convergence find the first two eigenvalues of system (9.5.1). Compare the result with the analytical solution.

Exercise 3. Find analytical expressions for all eigenfunctions and eigenvalues of the Sturm–Liouville problem

$$\begin{gathered} ((1+x)u_1')' + ((1+x)u_2')' + \lambda(u_1 + 2u_2)(1+x)^{-1} = 0, \\ ((1+x)u_1')' + ((4(1+x)u_2')' + \lambda(2u_1 + 4u_2)(1+x)^{-1} = 0, \\ u_1(0) = u_1(1) = u_2(0) = u_2(1) = 0. \end{gathered} \tag{9.5.2}$$

Exercise 4. By the method of accelerated convergence, find the first two eigenvalues of system (9.5.2). Compare the result with the analytical solution.

Exercise 5. Construct an algorithm based on the sagittary function (see Section 6.2) for system (9.1.1).

Chapter 10

Vibrations and Stability of Elastic Systems

In this chapter, the numerical-analytical methods of Chapters 2–4 are applied to the investigation of initial boundary value problems of elasticity. The main focus is on the analysis of the dependence of the first eigenvalue on the essential parameters (parametric synthesis) which determine stability of elastic systems. These results may be interesting for specialists in the theory of elasticity. For the readers with interests in applied mathematics, the problems considered here may serve as examples demonstrating how the general methods can be applied to special cases of equations and boundary conditions.

10.1. Plane Vibrations of a Rotating Heavy Thread and Their Stability

The method of accelerated convergence is applied here to the investigation of plane vibrations of a heavy inhomogeneous thread subjected to additional tension. It is assumed that the planes inhibiting the motion of the thread elements are rotating with constant angular speed about the vertical axis.

10.1.1. Statement of the initial boundary value problem. Its solution by the Fourier method. Our aim is to study small plane transverse vibrations of a rotating heavy inhomogeneous thread subjected to additional tension. We consider the general nonhomogeneous boundary conditions of elastic fixation of the upper and the lower ends of the thread (i.e., boundary conditions of the third kind). The motion of thread is described by the boundary value problem

$$\rho(x)u_{tt} = [W(x)u_x]_x + \rho(x)\omega^2 u + f(x,t), \qquad 0 < x < l; \tag{10.1.1}$$

$$[W(x)u_x - k_0 u]\big|_{x=0} = -h_0(t), \quad [W(x)u_x + k_l u]\big|_{x=l} = h_l(t). \tag{10.1.2}$$

Here, $u = u(x,t)$ denotes transverse displacements of the elements of the string in the plane rotating about the vertical x-axis; the subscripts x and t indicate the corresponding partial derivatives, l is the string length; $\rho(x)$ is its linear density, $W(x)$ is the total tension on the cross-section x; ω is the angular rotation speed of the plane ($\omega = \text{const}$); $f(x,t)$ is the distributed external force; $k_{0,l}$ are coefficients characterizing elastic fixation; $h_{0,l}(t)$ is the external load concentrated at the points $x = 0$, $x = l$. The function $\rho(x)$ is assumed continuous and separated from zero, $0 < \rho_1 \le \rho(x) \le \rho_2 < \infty$; and $f(x,t)$, $h_{0,l}(t)$ are sufficiently smooth functions of their arguments.

It is also assumed that the total string tension $W(x)$ in (10.1.1), (10.1.2) is due to two factors: the weight of the string segment, $m(x)g$, and the additional load W_0 concentrated at the lower end of the string,

$$W(x) = m(x)g + W_0, \quad m(x) = \int_0^x \rho(s)\, ds, \quad W_0 \geq 0, \quad g > 0. \tag{10.1.3}$$

Here, g is gravity (if necessary, its variable character can be taken into account); W_0 is a force concentrated on the lower or (and) upper end of the string ($x = 0$) (this may be the weight of a load acting through a pulley or some other physical force – elastic, electromagnetic, etc.). Stretching of the string will be neglected, since one usually has $W(x) \ll ES(x)$, where E is the Young modulus, $S(x)$ is the area of the cross-section. Coupled transverse and longitudinal string vibrations require a separate study. In order to determine the motion of the string, in addition to (10.1.1)–(10.1.3), one should specify the initial displacements and velocities of the points of the string,

$$u(x,0) = u^0(x), \quad u_t(x,0) = u^1(x), \quad 0 \leq x \leq l. \tag{10.1.4}$$

The functions u^0, u^1 (10.1.4) are assumed sufficiently smooth, so that one can claim the existence of a strong (physical) solution of the initial boundary value problem (10.1.1) – (10.1.4). Our aim is to construct this solution and investigate its properties depending on the parameters of the system.

Note that vibrations of a heavy string, with additional tensions (both longitudinal and transverse) taken into account, are of great interest for theory and applications. In this connection, it is especially important to study free vibrations, which can be done by solving the corresponding eigenvalue problem for an equation with variable coefficients.

Remark. For $\omega = 0$, we obtain a mathematical model of vibrations of a heavy string in an arbitrary plane without rotation (the classical case). In general, small vibrations of a free (i.e., without plane constraints) rotating string have spatial character, because of the Coriolis inertial forces. In the reference frame associated with the rotating system, these forces are described by the following vector relations:

$$\rho(x)\mathbf{u}_{tt} = [W(x)\mathbf{u}_x]_x + \rho(x)\left(\omega^2\mathbf{u} - 2\left[\boldsymbol{\omega} \times \mathbf{u}_t\right]\right) + \mathbf{f}(x,t);$$

$$[W(x)\mathbf{u}_x - k_0\mathbf{u}]\big|_{x=0} = -\mathbf{h}_0(t), \quad [W(x)\mathbf{u}_x + k_l\mathbf{u}]\big|_{x=l} = \mathbf{h}_l(t).$$

By an orthogonal transformation (i.e., passing to a non-rotating coordinate frame), we obtain a boundary value problem which is equivalent to the above one with $\omega = 0$. The vectors $\mathbf{f}$, $\mathbf{h}_{0,l}$, $\mathbf{u}^0$, $\mathbf{u}^1$ should be transformed accordingly. Indeed,

$$\rho(x)\mathbf{U}_{tt} = [W(x)\mathbf{U}_x]_x + \mathbf{F}(x,t,\varphi), \quad \mathbf{u} = \Pi(\varphi)\mathbf{U},$$

$$[W(x)\mathbf{U}_x \mp k_x\mathbf{U}]_{x=0,l} = \mp\mathbf{H}_{0,l}(\varphi), \quad \mathbf{H}_{0,l} = \Pi^{-1}(\varphi)\mathbf{h}_{0,l}, \tag{10.1.5}$$

$$\mathbf{F} = \Pi^{-1}(\varphi)\mathbf{f}, \quad \Pi^{-1}(\varphi) = \Pi(-\varphi) = \Pi^T(\varphi), \quad \varphi = \omega t,$$

where $\Pi(\varphi)$ is a rotation matrix. Thus, we obtain a system of two boundary value problems (10.1.5). Each problem for the components U_1, U_2 of the vector $\mathbf{U}$ can be studied independently, since the components of the vectors $\mathbf{F}$, $\mathbf{H}_{0,l}$, $\mathbf{U}^0$, $\mathbf{U}^1$ are known. The homogeneous boundary value problems and the corresponding Sturm–Liouville problems do not contain the parameter ω, i.e., are equivalent to the case $\omega = 0$ in (10.1.1). Therefore, system (10.1.1), (10.1.2) can be regarded as more meaningful and general than its vector analogue given above. Note that the statements of these two problems are essentially different and their solutions may be qualitatively distinct.

A boundary value problem similar to (10.1.5) may be obtained by the introduction of a complex variable $z = u_1 - iu_2$ and the replacement $z = w\exp(i\varphi)$. The unknown complex-valued function $w(x,t)$ is described by relations of the form (10.1.1) (with $\omega = 0$ in (10.1.1)). Of course, the transformed functions f, $h_{0,l}$ will contain the factor $\exp(-i\varphi)$.

Let us use the standard approach (the Fourier method) based on solving the homogeneous boundary value problem (10.1.1), (10.1.2) (for $f = h_{0,l} \equiv 0$), since this problem admits separation of variables. For the eigenvalues (eigenfrequencies) and eigenfunctions (vibration shapes) we obtain a modified self-conjugate boundary value problem (a Sturm–Liouville problem [6]; see Chapter 2),

$$\begin{gathered}(W(x)X')' + \Lambda\rho(x)X = 0, \qquad \Lambda = \lambda + \omega^2, \\ [W(x)X' - k_0X]_{x=0} = 0, \qquad [W(x)X' + k_lX]_{x=l} = 0.\end{gathered} \tag{10.1.6}$$

Here, Λ is the unknown eigenvalue and $X(x)$ is the unknown eigenfunction of problem (10.1.6), λ is the separation parameter. The problem is to construct a complete countable family of orthonormal functions $\{X_n(x)\}$ (basis) and the family of eigenvalues $\{\Lambda_n\}$. This problem is of principal interest for applications (see below).

Suppose that the said families have been found. Then, for the Fourier coefficients we obtain a countable linear system of independent equations and initial conditions,

$$\begin{gathered}\ddot{\Theta}_n + \lambda_n\Theta_n = f_n(t) + X_n(0)h_0(t) + X_n(l)h_l(t), \qquad f_n = (f, X_n), \\ \Theta_n(0) = (u^0, X_n)_\rho, \quad \dot{\Theta}(0) = (u^1, X_n)_\rho, \quad u(x,t) = \sum_{n=1}^{\infty}\Theta_n(t)X_n(x).\end{gathered} \tag{10.1.7}$$

Here, the dots denote derivatives with respect to t, $(\cdot,\cdot)_\rho$ is the scalar product in the weighted space L^2 with the weight ρ.

The solutions $\Theta_n(t)$ of the Cauchy problem (10.1.7) are easily constructed in terms of simple quadratures. In what follows, we assume that free motions of the string (for $f = h_{0,l} \equiv 0$) have vibrational character, i.e., the smallest eigenvalue is positive, $\lambda_1 > 0$. The case of $\lambda_1 \le 0$ may also be interesting for applications, since it corresponds to the loss of stability of plane vibrations of a rotating string. This state of motion implies that the displacements and the displacement rates of the string elements grow exponentially or linearly with respect to t. In this case, one has to take into account nonlinear factors, dissipation, and the like.

To obtain a highly precise description of the behavior of lower vibration frequencies, in particular, the first frequency is a task of principal importance for theory and applications. The aim of our further investigation is to construct fairly precise approximations of the first eigenvalue λ_1 and to study in detail its dependence on the parameters of system (10.1.1). For this purpose we use the effective numerical-analytical method of accelerated convergence based on the variational approach to the Sturm–Liouville problem (10.1.6) and a differential relation between λ_1 and the string length l [6] (see Chapter 2).

Let us outline the main steps of the method (algorithm) as applied to the Sturm–Liouville problem (10.1.6), using the variational relations from Sections 2.1, 1.4. For definiteness, consider the case $n = 1$, which is of special interest for applications, in particular, for the investigation of stability of plane vibrations of a heavy rotating string subjected to additional tension. For the sake of brevity, the index n is dropped in what follows.

Suppose that we have already constructed an approximate value Λ^0 (for instance, $\Lambda^0 = \Lambda^*$) by the Rayleigh–Ritz or some other method, say, the method of continuation with respect to a parameter (see below). For equation (10.1.6), consider the Cauchy problem

$$\begin{gathered}(W(x)v')' + \Lambda^0\rho(x)v = 0, \qquad v(0) = \alpha_0 l, \quad v'(0) = \beta_0; \\ \alpha_0 = d_0(d_0 + k_0)^{-1}, \quad \beta_0 = k_0(d_0 + k_0)^{-1}, \quad d_0 = W_0 l^{-1}.\end{gathered} \tag{10.1.8}$$

In what follows, we assume that the solution $v(x, \Lambda^0)$, $v'(x, \Lambda^0)$ of the Cauchy problem (10.1.8) is known in the form of a fairly good approximate numerical or analytical solution, or in the form of a computational procedure. This solution automatically satisfies the homogeneous boundary condition (10.1.6) at $x = 0$. The dependence of the functions v, v' on the other parameters (k_0, d_0, W_0, g, and the parameters entering $\rho(x)$) is not indicated explicitly, for the sake of brevity.

Let us calculate the function $E(x, \Lambda^0)$ and require that it should vanish for some minimal value $x = \xi^0 > 0$, i.e., we find the first positive root ξ^0 of the equation

$$E(x, \Lambda^0) = \alpha_l l \nu(x) v'(x, \Lambda^0) + \beta_l v(x, \Lambda^0) = 0,$$
$$\alpha_l = d_l(d_l + k_l)^{-1}, \quad \beta_l = k_l(d_l + k_l)^{-1}, \quad d_l = W(l)l^{-1}, \tag{10.1.9}$$
$$\nu(x) \equiv W(x)W(l)^{-1}, \quad \xi^0 = \xi(\Lambda^0) = \min \arg_x E(x, \Lambda^0) > 0.$$

The relation for the determination of ξ^0 (10.1.9) is constructed in the process of numerical integration of the Cauchy problem (10.1.8). For Λ^0 sufficiently close to Λ, the root ξ^0 exists and is simple. As a measure of relative closeness δ we take a numerical parameter ε defined by

$$\varepsilon = (l - \xi^0)l^{-1}, \quad 0 \leq |\varepsilon| \ll 1, \quad \delta = (\Lambda - \Lambda^0)\Lambda^{-1}, \quad 0 \leq |\delta| \ll 1. \tag{10.1.10}$$

We have $\varepsilon = 0$ if and only if $\delta = 0$. It is assumed that for the given Λ^0, the root ξ^0 and the parameter ε have been found with sufficient precision. In the procedure proposed here, one of the main operations requiring some computational resources is finding the solution of problem (10.1.8)–(10.1.10).

In order to solve the original Sturm–Liouville problem (10.1.6), we use the perturbation methods described in Section 2.3. According to what has been said above, we obtain the refined value $\Lambda^{(1)}$ in the first approximation with respect to ε (with the error of order $O(\varepsilon^2)$),

$$\Lambda^{(1)} = \Lambda^0 + \varepsilon\mu(\xi^0, \Lambda^0), \quad \xi^0 = \xi(\Lambda^0), \quad \mu < 0,$$
$$\mu(\xi^0, \Lambda^0) = -\xi^0 W(\xi^0) v'^2(\xi^0, \Lambda^0) \|v_0\|^{-2} - \Lambda^0 \xi^0 \rho(\xi^0) v^2(\xi^0, \Lambda^0) \|v_{(0)}\|^{-2}. \tag{10.1.11}$$

Here, $\|v_{(0)}\|$ the weighted (with weight $\rho(x)$) norm of $v_{(0)} = v(x, \Lambda^0)$ on the interval $0 \leq x \leq \xi^0$. In view of (10.1.11), the quantity $\mu(l, \Lambda)l^{-1}$ is the derivative of Λ with respect to l. Obviously, this derivative is strictly negative. Therefore, for a larger string length l, the eigenfrequency $(\Lambda - \omega^2)^{1/2}$ increases, and for a smaller l, the eigenfrequency decreases.

The process of refining Λ may be infinitely continued. Let us take the refined value $\Lambda^{(1)}$ as the initial approximation instead of Λ^0 in relations of the type (10.1.8)–(10.1.11). Thus, we obtain a new refined value

$$\Lambda^{(2)} = \Lambda^{(1)} + \varepsilon^{(1)}\mu(\xi^{(1)}, \Lambda^{(1)}),$$

where $\varepsilon^{(1)}$ corresponds to the root $\xi^{(1)} = \xi(\Lambda^{(1)})$, and so on. As a result, we come to a procedure for refining the eigenvalue Λ and the function X. The algorithm of this procedure has quadratic convergence with respect to the original small parameter ε and can be described by the relations (see Section 2.4)

$$\Lambda^{(k+1)} = \Lambda^{(k)} + \varepsilon^{(k)}\mu(\xi^{(k)}, \Lambda^{(k)}), \quad \varepsilon^{(k)} = (l - \xi^{(k)})l^{-1}, \quad k = 0, 1, 2, \ldots,$$
$$\xi^{(k)} = \xi(\Lambda^{(k)}) = \min \arg_x E(x, \Lambda^{(k)}) > 0, \quad \xi^{(0)} = \xi^0 = \xi(\Lambda^0), \quad \varepsilon^{(0)} = \varepsilon,$$
$$E(x, \Lambda^{(k)}) \equiv \alpha_l l \nu(x) v'(x, \Lambda^{(k)}) + \beta_l v(x, \Lambda^{(k)}) = 0, \tag{10.1.12}$$
$$(W(x)v')' + \Lambda^{(k)}\rho(x)v = 0, \quad v(0) = \alpha_0 l, \ v'(0) = \beta_0, \quad v_{(k)} = v(x, \Lambda^{(k)}).$$

The functions $v_{(k)}$, $v'_{(k)}$, E are obtained by numerical integration of the Cauchy problem (10.1.12), i.e., problem (10.1.8) with $\Lambda^0 = \Lambda^{(k)}$. The value $\Lambda^{(k)}$ is assumed to have been calculated on the preceding iteration step. The procedure (10.1.12) yields the following error estimates on the $(k+1)$ th step:

$$\left|\Lambda^{(k+1)}-\Lambda\right| \le C_\Lambda \varepsilon^{(k+1)}, \quad \left|\varepsilon^{(k)}\right| \le d(c\varepsilon)^{n_k}, \quad n_k = 2^k,$$
$$\max_x\Big(\left|X(x,\Lambda)-v(x,\Lambda^{(k)})\right| + l\left|X'(x,\Lambda)-v'(x,\Lambda^{(k)})\right|\Big) \le C_X \varepsilon^{(k)}, \tag{10.1.13}$$
$$0 \le x \le \max\left(l, \xi^{(k)}\right), \quad k = 0,1,2,\dots$$

The constants C_Λ, C_X, c, d can be effectively estimated in terms of the properties of the function $\rho(x)$ and the quantities l, W_0, g, $k_{0,l}$.

Thus, the algorithm (10.1.12) ensures accelerated convergence (quadratic convergence with respect to ε) of the approximations. This convergence is described by the inequalities (10.1.13)). Computational experience testifies to the efficiency of this algorithm: it is easy to implement and immune to errors, and it entails no accumulation of roundoff errors. As a rule, two to four iterations are sufficient for obtaining practically precise values of Λ, $X(x)$, $\|X\|$, X' with a fairly rough choice of the test function $\psi(x)$. The calculation efficiency is greatly increased by combining the accelerated convergence algorithm with the procedure of continuation in the parameters of the system whose values are taken into account when constructing the solution Λ, $X(x)$ (see below). Observe that the precision chosen for the calculations based on (10.1.12) should not be redundant and should merely agree with the index of the iteration step k.

Subsequent eigenvalues Λ_n and eigenfunctions $X_n(x)$, $n \ge 2$, are determined on the basis of the above scheme. The difference is that one calculates the nth root $\xi_n(\Lambda_n^{(k)})$ of the equation $E(x, \Lambda_n^{(k)}) = 0$ (see (10.1.9), (10.1.12)).

The norm $\|v_{(k)}\|$ with the weight $\rho(x)$ in (10.1.11), (10.1.12) is usually calculated by some highly precise method of integration (say, the Simpson method) of numerically defined functions (solutions of the Cauchy problems). This process can be replaced by the calculation of the functions $w = \partial v/\partial\Lambda$, $w' = dw/dx$ by means of joint integration of the Cauchy problem for v, v' (10.1.8) and the Cauchy problem for w, w',

$$[W(x)w']' + \Lambda^{(k)}\rho(x)w = -\rho(x)v, \qquad w(0) = w'(0) = 0;$$
$$\|v_{(k)}\|^2 \equiv \int_0^{\xi^{(k)}} v^2\left(x,\Lambda^{(k)}\right)\rho(x)\,dx \tag{10.1.14}$$
$$= W\left(\xi^{(k)}\right)\left[w\left(\xi^{(k)},\Lambda^{(k)}\right)v'\left(\xi^{(k)},\Lambda^{(k)}\right) - v\left(\xi^{(k)},\Lambda^{(k)}\right)w'\left(\xi^{(k)},\Lambda^{(k)}\right)\right].$$

Instead of integrating system (10.1.12), (10.1.14), it is convenient to integrate the standard systems in the form of Cauchy

$$v' = \frac{y}{W(x)}, \quad y' = -\Lambda^{(k)}\rho(x)v, \quad v(0) = \alpha_0 l, \quad y(0) = \beta_0 W_0,$$
$$w' = \frac{z}{W(x)}, \quad z' = -\Lambda^{(k)}\rho(x)w - \rho(x)v, \quad w(0) = z(0) = 0, \tag{10.1.15}$$

with the corresponding transformations of $E\left(x,\Lambda^{(k)}\right)$ and $\|v_{(k)}\|^2$.

In conclusion, it should be said that in all formulas (10.1.11)–(10.1.15) one can take $\xi^{(k)} = l$ without the loss of precision in $\varepsilon^{(k)}$, except in the expressions for $\varepsilon^{(k)}$; see (10.1.10), (10.1.12).

Next, we consider some examples with linear density $\rho(x)$ and specific boundary conditions, i.e., values of the parameters $\alpha_{0,l}$, $\beta_{0,l}$.

10.1.2. Free vibrations of a rotating heavy homogeneous string subjected to tension.

Assume that the linear density is constant, $\rho = \text{const}$. Then the Cauchy problem (10.1.12) becomes

$$(\sigma(x,\chi)v')' + \gamma v = 0, \quad \gamma = \gamma^{(k)} = \chi\Lambda^{(k)}\Omega^{-2}, \quad v(0) = \alpha_0, \quad v'(0) = \beta_0,$$
$$\sigma(x,\chi) = 1 + \chi(x-1), \quad \chi = mg(W_0+mg)^{-1}, \quad 0 < \chi \le 1, \quad \Omega^2 = gl^{-1}. \tag{10.1.16}$$

Here, the function v and the argument x ($0 \leq x \leq 1$) are normalized by l, $m = \rho l$. Using (10.1.16), for the abscissa $\xi^{(k)}$ (10.1.12) normalized by l we obtain the equation

$$E(x, \gamma^{(k)}, \chi) = \alpha_l \sigma(x, \chi) v'(x, \gamma^{(k)}, \chi) + \beta_l v(x, \gamma^{(k)}, \chi) = 0,$$
$$\xi^{(k)} = \xi(\gamma^{(k)}, \chi), \quad \alpha_l \geq 0, \quad \beta_l \geq 0, \quad \alpha_l + \beta_l = 1. \tag{10.1.17}$$

It is convenient to use expressions (10.1.16), (10.1.17) in the procedure (10.1.12), together with the continuation in the parameter χ, i.e., to seek $\gamma = \gamma(\chi)$, $v = v(x, \chi)$. In the limit case, if the string weight mg is negligibly small relative to the load W_0 (the mass is taken into account only in γ), we obtain a model of a homogeneous string with the conditions of elastic fixation of its ends. The Cauchy problem can be easily solved in terms of trigonometric functions. In order to find $\gamma(0)$, one has to solve the following transcendental equation for $\nu = \sqrt{\gamma(0)}$:

$$\tan \nu = \nu(\alpha_0 \beta_l + \beta_0 \alpha_l)(\nu^2 \alpha_0 \alpha_l - \beta_0 \beta_l)^{-1}, \quad \gamma_n = \nu_n^2, \quad n \geq 1. \tag{10.1.18}$$

The value ν_n depends on two parameters α_k, β_k, which are related by the condition of normalization $\alpha_k + \beta_k = 1$. In particular, for $\alpha_k = 0$ or $\alpha_k = 1$, we obtain the known elementary cases of clamped or free ends. In the general situation, the roots $\nu_n^* = \nu_n(\alpha, \beta)$ of equation (10.1.18) can be found numerically. The eigenvalues $\gamma_n(0)$ and the functions $X_n(x, 0)$ have the form ($n = 1, 2, \ldots$)

$$\gamma_n(0) = \nu_n^2, \quad X_n(x) = \alpha_0 \cos \nu_n x + \beta_0 \nu_n^{-1} \sin \nu_n x. \tag{10.1.19}$$

As mentioned above, the lower eigenvalues and vibration shapes, especially those for $n = 1$, are of main importance for applications.

In what follows, the value $\gamma(0)$ (10.1.19) calculated from equation (10.1.18) is used as the initial bound γ^0 for $\chi = \chi_1 > 0$, where χ_1 is sufficiently small, as shown by numerical experiment. Integrating the Cauchy problem (10.1.16) for $\gamma = \gamma^0$ and finding the root $\xi^{(0)} = \xi(\gamma^0, \chi_1)$ from (10.1.17), we use formulas (10.1.10), (10.1.11), (10.1.14) to obtain the refined value $\gamma^{(1)}(\chi_1)$, and, using the recurrent procedure, we obtain a highly precise approximation $\gamma(\chi_1)$. Then, we implement the process of continuation in the parameter χ: $0 < \chi_1 < \chi_2 < \ldots < \chi_i \leq 1$, with i sufficiently large.

It is known that equation (10.1.16) can be reduced to the Bessel equation of zero order [24, 32, 33, 46, 67]. Indeed, passing from the argument x to θ by a one-to-one transformation and using a standard procedure, we obtain the said equation and its general solution v_0:

$$v'' + \theta^{-1} v' + \eta^2 v = 0, \quad v = v(\eta\theta), \quad \eta^2 = 4\gamma\chi^{-2}, \quad 0 < \chi < 1,$$
$$v_0 = A J_0(z) + B N_0(z), \quad z = \eta\theta, \quad A, B = \text{const}, \tag{10.1.20}$$
$$\theta = \sigma^{1/2}(x, \chi), \quad \theta_1 \leq \theta \leq 1, \quad \theta_1 = (1 - \chi)^{1/2}, \quad 0 < \theta_1 < 1.$$

The prime in (10.1.20) denotes the derivative in θ. Detailed analytical, numerical, and graphical data pertaining to the Bessel functions $J_\nu(z)$ and the Neumann functions $N_\nu(z)$ can be found in [32, 67]. Using the boundary conditions for $\theta = \theta_1$, $\theta = 1$ (see (10.1.16), (10.1.17)), we obtain the characteristic equation for $\eta = \eta(\theta_1, \alpha, \beta)$ related to $\gamma = \gamma(\chi, \alpha, \beta)$ by (10.1.20). This equation also contains the derivatives $J_0'(z) = -J_1(z)$ and $N_0'(z) = -N_1(z)$.

Note that $\theta_1 \to 1$ as $\chi \to 0$ (string model), i.e., the transformation $\chi \to \theta$ (10.1.20) becomes degenerate. This situation has been considered above; see (10.1.18), (10.1.19). For small χ ($0 < \chi \ll 1$), the method of regular perturbations can be applied to problem (10.1.16), (10.1.17), provided that the generating solution is known. An approximate analytical investigation is not very difficult and not very interesting.

Next, we study the general case $0 \le \chi \le 1$. Passing to the limit as $\chi \to 1$ ($\theta_1 \to 0$) results in regular critical points $x = 0$, $\theta = 0$ in equations (10.1.16), (10.1.20), respectively. This situation requires asymptotic analysis, which can be performed on the basis of equation (10.1.20) and the functions J_0, N_0.

A study of solutions of the problem for arbitrary admissible $\alpha_{0,l}$, $\beta_{0,l}$ would be rather lengthy, and therefore, we consider some special cases of boundary conditions corresponding to the limit values $\alpha_{0,l}$, $\beta_{0,l}$. Our main attention will be focused on the first vibration mode.

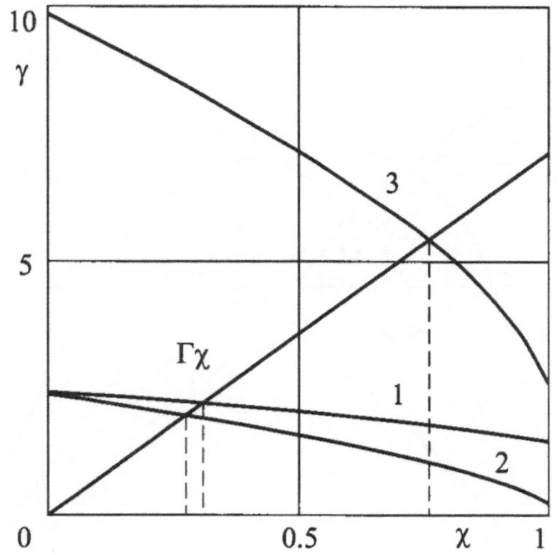

Fig. 10.1

1. Let $\alpha_l = \beta_0 = 0$ ($\alpha_0 = \beta_l = 1$). This is a classical case of vibrations of a heavy inhomogeneous string (chain) subjected to additional tension. Curve 1 in Fig. 10.1 represents the function $\gamma(\chi)$, which admits a better mechanical interpretation than $\eta(\theta_1)$ (see (10.1.16)). The graph shows that $\gamma(0) = \pi^2/4$ and $\gamma(\chi)$ monotonically decreases as $\chi \to 1$; its minimal value is equal to $\gamma(1) = \eta_0^2/4$, where $\eta_0 = \eta(0) \approx 2.4048$ is the first root of the function $J_0(\eta)$ [32]. Asymptotic analysis of the root $\eta(\theta)$ for $0 \le \theta_1 \ll 1$ brings us to the approximate expression $\eta(\theta_1) \approx \eta_0 + O(\theta_1^2)$, which implies that $\gamma(\chi) = \gamma(1) + O((1-\chi))$, i.e., the function γ linearly tends to $\gamma(1)$ as $\chi \to 0$ ($0 \le 1-\chi \ll 1$). As a whole, the decrease of $\gamma(\chi)$ is practically linear (see Fig. 10.1).

The family of normalized eigenfunctions $V(x,\chi) = v\|v\|^{-1}$ is represented by dashed lines in Fig. 10.2. Curves 1 to 3 correspond to the values $\chi = 0.9999$, $\chi = 0.9$, $\chi = 0$. For $\chi = 0$ (curve 3), we have vibrations of a strongly stretched string with its lower end free. The case $\chi \approx 1$ (curve 1) corresponds to vibrations of a heavy string without additional tension, the first vibration shape being described by the function $J_0(\eta_0\theta) = J_0\big(2\sqrt{\gamma(1)x}\big)$. Curve 2 for the intermediate value $\chi = 0,9$ gives the picture of evolution of the family as χ varies from $\chi = 0$ to $\chi = 1$.

2. Consider the "inverse" situation, with $\alpha_0 = \beta_l = 0$ ($\alpha_l = \beta_0 = 1$), i.e., the upper end is free and the lower clamped. The graph of the function $\gamma(\chi)$ is represented in Fig. 10.1 (curve 2). Note that for small χ, curves 1and 2 are close to one another, but for $\chi \to 1$ they diverge and we observe a qualitative difference: $\gamma \to 0$ as $\chi \to 1$. The decrease is rather slow and for $\chi = 0.9999$ we have $\gamma \sim 0.1$. Asymptotic analysis yields the estimate $\eta(\theta_1) = O\left((\ln\theta_1^{-1} - 1/2)^{1/2}\right)$, $\theta_1 \ll 1$. For the sought quantity γ this estimate transforms to $\gamma(\chi) \approx O\big((\ln(1-\chi)^{-1} - 1)^{-1}\big)$, where $0 < 1-\chi \ll 1$.

Figure 10.2 shows normalized eigenfunctions $V(x,\chi)$ (dashed-dotted lines). Curves 4–6 correspond to the values of χ in the case 1. Curve 6 is similar to curve 3 for vibrations of a string with a free end. For $\chi \to 1$ (curve 4), the function $V(x,\chi)$ weakly converges to a function of Heaviside type, which shows that free vibrations of a string with a free end is impossible without additional tension. The functions $\gamma(\chi)$ and $V(x,\chi)$ for $0 < \chi \ll 1$ can be easily studied by the perturbation method.

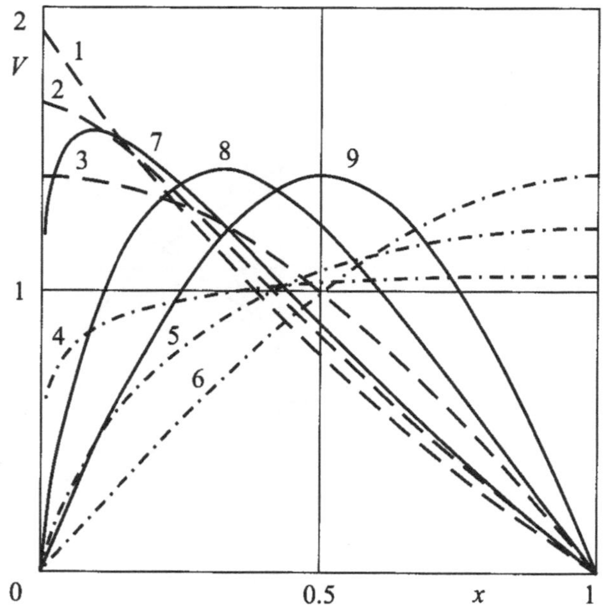

Fig. 10.2

3. Of great interest is the case of vibrations of a heavy string subject to additional tension and having its both ends clamped, i.e., $\alpha_{0,l} = 0$ ($\beta_{0,l} = 1$). The graph of the function $\gamma(\chi)$ is represented in Fig. 10.1 (curve 3). This function has a fairly rapid decrease from the value $\gamma(0) = \pi^2$ to $\gamma(1) = \eta_0^2/4 \approx 1.4$, so that the slope (derivative) becomes unbounded as $\chi \uparrow 1$. Comparison with the case 1 shows that curve 3 remains above curve 1 for $0 \le \chi < 1$. Thus, for nonzero tension, the first eigenfrequency of the "string" is larger than that of a thread with a free end. Asymptotic analysis based on Bessel and Neumann functions yields the approximate expression $\eta(\theta_1) = \eta_0 + O\left(\ln^{-1}(1-\chi)^{-1}\right)$, which implies that curve 3 lies above curve 1 for all $0 \le \chi < 1$.

The normalized eigenfunctions $V(x,\chi)$ are shown in Fig. 10.2 (solid lines). For $\chi = 0$ (curve 9), we obtain the classical case of the first vibration mode of a homogeneous strongly stretched string. For $\chi \uparrow 1$ (curve 7), the eigenfunction converges (in a weak sense) to a discontinuous function which, for $x > 0$, has a shape similar to $J_0(2\sqrt{\gamma(1)\chi})$ (to within a normalizing coefficient); see case 1. Hence, it is clear that vibrations of a "string" without additional tension are impossible.

4. For a heavy stretched thread with its end free (for transverse displacements) ($\beta_{0,l} = 0$, $\alpha_{0,l} = 1$), we easily obtain zero values $\eta(\theta_1) = \gamma(\chi) \equiv 0$ for all admissible $0 \le \theta_1, \chi \le 1$, and the vibration shape is $V(x,\chi) \equiv 1$; see (10.1.16), (10.1.20). The next root of the characteristic equation is different from zero, $\eta(0) = 3.8317$ ($J_1(\eta) = 0$).

Let us study the problem of stability of thread vibrations for the above cases of boundary conditions, 1–4. For this purpose, we use the graphs constructed for the functions $\gamma(\chi)$ (Fig. 10.1).

These values and families of normalized functions $V(x,\chi)$ were constructed by the method of accelerated convergence based on the scheme (10.1.16), (10.1.17) and the procedure of continuation in the parameter χ. It requires a lot of effort to obtain a numerical solution on the basis of Bessel and Neumann functions. Computational experience shows that the rapidly convergent scheme (10.1.12)–(10.1.14) for the construction of eigenvalues and eigenfunctions is very effective even in the above case of "a known exact solution" in terms of special functions (see below).

The inequality $\lambda > 0$ is a sufficient condition for the stability of plane vibrations of a heavy rotating thread subjected to additional tension. The definition of the parameter γ (10.1.16) implies the inequality (condition of stability of vibrations of a given mode)

$$\gamma(\chi) > \Gamma\chi, \quad 0 \le \chi \le 1, \quad \Gamma = \omega^2\Omega^{-2}, \quad \gamma = (\lambda + \omega^2)\chi\Omega^{-2}. \tag{10.1.21}$$

Fix a value of $\Gamma \ge 0$. Then, stable motions of the thread correspond to the values of χ for which the curves $\gamma(\chi)$ are above the straight line $\Gamma\chi$. In cases 1–3, there exist such values of χ, $0 \le \chi < \chi^* \le 1$. A geometrical interpretation of the condition (10.1.20) is given in Fig. 10.1. Moreover, in cases 1, 3 for $0 \le \Gamma < \gamma(1) \approx 1.4$, this inequality holds for all χ, $0 \le \chi \le 1$. As Γ increases, the length of the segment $0 \le \chi \le \chi^*$ decreases, and $\chi^* \to 0$ as $\Gamma \to \infty$, which can be expected. In case 4 ($\gamma(\chi) \equiv 0$), "vibrations" are always unstable: for $\Gamma = 0$ ($\omega = 0$) the instability is secular, and for $\Gamma > 0$ ($\omega \ne 0$) exponential. These conclusion agree with mechanical notions.

10.1.3. Vibrations of an inhomogeneous thread. Let us apply the method of accelerated convergence to the investigation of free vibrations of a heavy inhomogeneous string of fixed total mass m_*. For definiteness, we restrict ourselves to the first eigenvalue and the first eigenfunction and consider two examples.

1. The case of linear variation of mass per unit length. Suppose that the linear density ρ, the mass m, and the tension W have the form

$$\begin{gathered} \rho(x) = \rho_0\left(1 - \kappa(2x - l)l^{-1}\right), \\ m(x) = \rho_0 x\left(1 - \kappa(x - l)l^{-1}\right), \quad -1 < \kappa < 1, \quad m(l) = \rho_0 l = m_*, \\ W(x) = W_0 + P_0 - P_0\left[1 - xl^{-1}\left(1 + \kappa - \kappa x l^{-1}\right)\right], \quad P_0 = m_* g. \end{gathered} \tag{10.1.22}$$

Then, from (10.1.22) it follows that $m(l) = m_*$ and $m(l)$ does not depend on the parameter κ characterizing the admissible range of the linear density: $\rho(x) > 0$ for all $0 \le x \le l$; for $x = l/2$ or $\kappa = 0$, the quantity $\rho \equiv \rho_0$ depends neither on κ nor x. By (10.1.22), we pass to the dimensionless argument x and the parameters γ, κ, χ in the Sturm–Liouville problem (10.1.6) and obtain the equation

$$\begin{gathered} (p(x)X')' + \gamma r(x)X = 0, \quad \gamma = \chi\Lambda\Omega^{-2}, \ \Omega^2 = \tfrac{g}{l}, \\ 0 \le x \le 1, \quad p(x) = 1 - \chi(1 - x(1 + \kappa - \kappa x)), \quad -1 < \kappa < 1, \\ r(x) = 1 - 2\kappa\left(x - \frac{1}{2}\right), \quad \chi = P_0(W_0 + P_0)^{-1}, \quad 0 \le \chi < 1. \end{gathered} \tag{10.1.23}$$

This equation contains two given parameters κ, χ and the unknown parameter γ is to be determined. The parameter χ in (10.1.23) characterizes the ratio of the string weight P_0 to the total tension at the point $x = 1$. Note that $|\kappa| < 1$. For $\kappa \to \pm 1$, the condition that the string is inextensible may be violated. The boundary conditions take the form

$$\left[\alpha_x p(x)X'(x) \mp \beta_x X(x)\right]\big|_{x=0,1} = 0, \qquad \alpha_{0,1} \ge 0, \quad \beta_{0,1} \ge 0, \quad \alpha_{0,1} + \beta_{0,1} = 1, \tag{10.1.24}$$

Consider the boundary conditions of the first kind on one or both ends of the string, just as in cases 1–3 above. Then, the eigenvalues γ of problem (10.1.22)–(10.1.24) depend on two parameters, $\gamma = \gamma(\kappa, \chi)$. The problem considered above corresponds to $\kappa = 0$. We are going to use the method of accelerated convergence and the procedure of continuation in the parameters κ, χ. As the initial precise value of γ, used also as a bound for γ^0, we take the value $\gamma_0 = \gamma(0,0) = \pi^2/4$ (cases 1, 2) and $\gamma^0 = \pi^2$ (case 3). Case 4 (both ends free) is not considered here, since these motions are unstable, as mentioned above. With the help of modern computers, the algorithm of accelerated convergence can be effectively used for the construction of the surface $\gamma = \gamma(\kappa, \chi)$ and the graphical representation of various projections of that surface. However, for the determination of numerical values, it is preferable to describe that surface by a sufficiently dense family of functions, for instance, by its coordinate lines with respect to κ or χ, or level-lines $\gamma = \text{const}$, i.e., cross-sections with respect to γ. The values of γ at intermediate points can be approximated with good precision by interpolation methods.

For the three types of boundary conditions considered here, Figs. 10.3*a*, 10.4*a*, 10.5*a* show level-lines of the surface $\gamma = \gamma_{ka}(\kappa, \chi) = c$, $k = 3, 4, 5$ (the index of each curve corresponds to the value of c) for the parameters κ, χ ranging on the intervals $|\kappa| \leq 0.99$, $0 \leq \chi \leq 0.99$. The case $\kappa = 0$ corresponds to the homogenous thread considered above. Comparing these families of curves, we see that the boundary conditions play a crucial role (conditions of one or both ends clamped). The boundary conditions of example 3 (the classical case) yield strictly positive values $\gamma_{3a}(\kappa, \chi)$ (see Fig. 10.3*a*) which monotonically decrease as $\kappa \to 1$, $\chi \to 1$ and monotonically increase as $\kappa \to -1$, $\chi \to 0$. These conclusions agree with mechanical notions. In particular, it is interesting to observe that for any fixed $0 \leq \chi \leq 1$, the vibration frequency decreases as κ increases from $\kappa = -1$ to $\kappa = 1$. The graphs in Fig. 10.4*a* confirm the great role of the boundary conditions: for $\kappa \to 1$, a considerable part of the thread becomes clamped, which leads to an increase of the frequency. Note that it is quite obvious that γ decreases in χ for a fixed κ. The qualitative difference from the situation in Fig. 10.3*a* consists in that $\gamma_{4a}(\kappa, 1) \equiv 0$, i.e., the vibration "frequency" becomes imaginary ($\lambda_1 = -\omega^2$). This indicates that the motion of the thread is unstable.

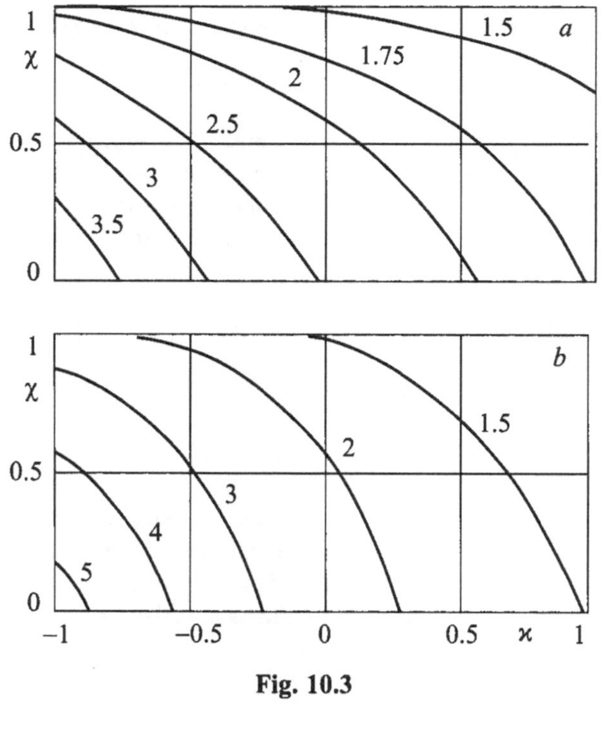

Fig. 10.3

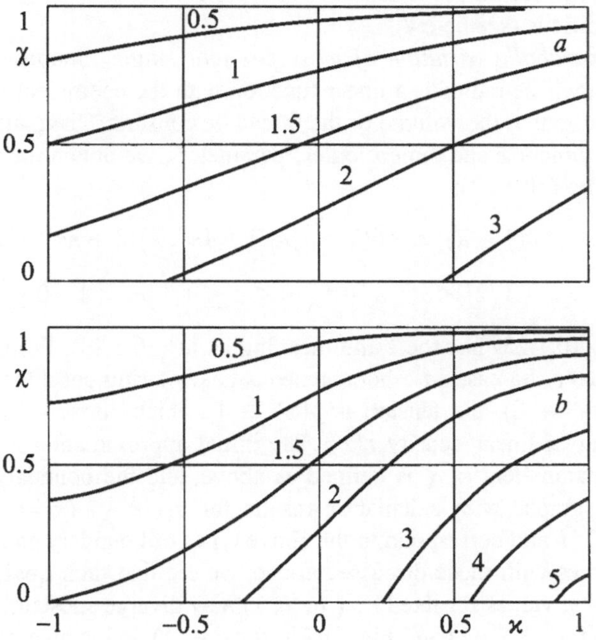

Fig. 10.4

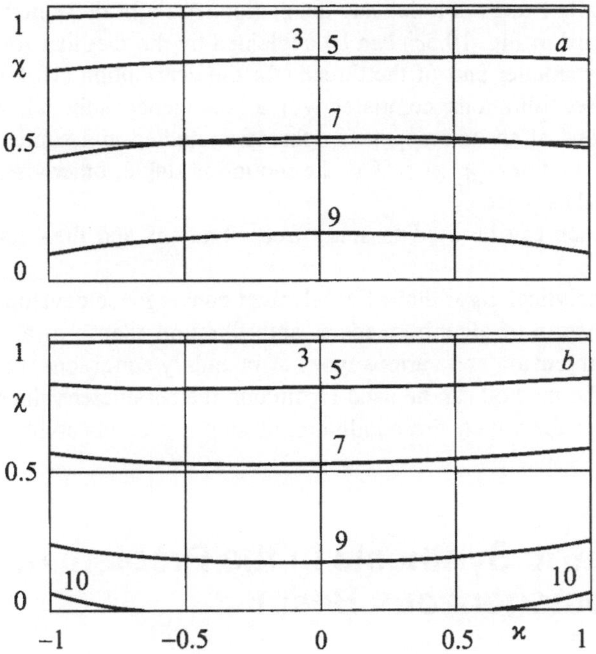

Fig. 10.5

In Fig. 10.5a, similar cross-sections are shown for the boundary conditions of example 3 (both ends clamped). These curves are nearly symmetrical relative to the axis $\varkappa = 0$, which confirms the essential role of the boundary conditions. The effect of mass distribution is insignificant. Note that the values $\gamma_{5a}(\kappa, \chi)$ for all $|\kappa| < 1$, $0 \leq \chi < 1$ are greater than $\gamma_{3,4a}(\kappa, \chi)$ due to the effect of the boundary conditions (see below).

2. *The case of parabolic variation of mass per unit length.* Suppose that the tread has a circular cross-section whose radius is a linear function with the coefficient κ, similarly to $\rho(x)$ in (10.1.22). Let us require that the volume of the thread be constant. Then, after the introduction of the dimensionless argument x and dimensionless parameters, we obtain the following expressions for $p(x)$ and $r(x)$ from (10.1.23):

$$\begin{aligned} p(x) &= 1 - \chi\bigl[1 - \bigl(3(1+\kappa)^2 x - 6(1+\kappa)\kappa x^2 + 4\kappa^2 x^3\bigr)(3+\kappa^2)^{-1}\bigr], \\ r(x) &= 3(1 - 2\kappa(x - 1/2))^2(3+\kappa^2)^{-1}, \quad 0 \leq x \leq 1, \quad |\kappa| < 1, \quad 0 \leq \chi < 1. \end{aligned} \tag{10.1.25}$$

The parameter χ in (10.1.25) has the same meaning as in (10.1.23). From (10.1.25), it follows that for $\kappa = 0$, we have the case of a homogeneous heavy string considered above. In the limit case, $\chi = 0$ ($P_0/W_0 \to 0$), the tension is $p(x) \equiv 1$, which corresponds to the model of an inhomogeneous string of linear density $r(x)$. The initial approximation for γ in the method of continuation in the parameters κ, χ is defined as above, and the boundary conditions have the form (10.1.24). The graphs with calculation results for $\gamma_{kb}(\kappa, \chi)$ (k = 3, 4, 5) are given in Figs. 10.3*b*, 10.4*b*, 10.5*b* and correspond to the above types of boundary conditions; see (10.1.24). Comparing these curves with those discussed above, we see that their qualitative behavior is the same, but the numerical values for fixed κ, χ ($\chi < 1$) may diverge substantially. Somewhat larger values in Fig. 10.3*b* (for $\kappa \to -1$) and Fig. 10.4*b* (for $\kappa \to 1$) and, conversely, somewhat smaller values (for $\kappa \to 1$ and $\kappa \to -1$) are explained by the fact that a relatively larger or smaller part of the thread becomes clamped, i.e., again the effect of boundary conditions is manifest. It is interesting to compare the curves γ_{5a}, $\gamma_{5b} = c$ (see Figs. 10.5*b* and 10.5*a*). Their qualitative behavior is the same, but there is a delicate distinction. A slight deflection upward in Fig. 10.5*b* for $|\kappa| \to 1$ (downward in Fig. 10.5*a*) can be explained by the fact that for the mass distribution (10.1.25), a relatively smaller part of the thread (for the distribution (10.1.23), a relatively larger part) is involved in free vibrations, compared with a homogeneous thread, $\kappa = 0$.

Stability of motions of an inhomogeneous thread is studied in a similar way, on the basis of the inequality (10.1.21). For $\gamma(\kappa, \chi) > \Gamma\chi$, the motion is stable; otherwise, instability of secular or exponential type takes place.

The above approach can be used to study free vibrations and their stability for a thread of arbitrary density.

The numerical-analytical algorithm of accelerated convergence developed here is an efficient tool for the determination of eigenfrequencies and vibration shapes of a heavy inhomogeneous thread with additional tension and various types of boundary conditions. In the case of a rotating plane of vibrations, the method can be used for finding the parameter values (in particular, values of the rotation speed) for which the conditions of stability or instability of these motions are satisfied.

10.2. Parametric Synthesis in the Problem of Instability of an Inhomogeneous Beam

The algorithm of accelerated convergence can be effectively utilized in problems of parametric synthesis, which require highly precise mass calculations. In particular, it can be used in the problem of longitudinal bending of a beam for finding the critical force as a function of the parameters of the problem and choosing optimal values of these parameters.

10.2.1. Setting of the problem of longitudinal bending of an elastic beam. Consider the problem of unstable equilibrium for a rectilinear inhomogeneous beam subjected to a longitudinal force, with the classical boundary conditions of hinged support (Euler's problem). Using the D'Alembert principle, let us write the conditions of equilibrium of any point x of the beam with a concentrated force P applied to at its end, assuming that the deflections of the beam are small [63]. We have

$$\sigma(x)u'' + Pu = 0, \qquad u(0) = u(l) = 0. \tag{10.2.1}$$

Here, $u = u(x)$ is the function that describes the deflection of the beam ($0 \le x \le l$); l is the length of the beam, $\sigma(x) = EI(x)$ is its flexural rigidity, E is the Young modulus of its material, $I(x)$ is the moment of inertia of its cross-section relative to the axis passing through the neutral line and orthogonal to the plane of bending, $I(x) \ge I_0 > 0$, $\sigma(x) \ge \sigma_0 > 0$.

Our aim is to find the smallest value P_* of the compressive load $P_* > 0$ for which $u(x) \not\equiv 0$, i.e., the value for which the unperturbed state $u(x) \equiv 0$ becomes unstable. A mechanical interpretation of instability can be found in [63] and elsewhere.

This problem is equivalent to an eigenvalue problem for a self-conjugate operator with the Dirichlet boundary conditions. Introducing dimensionless variables and parameters in (10.2.1), we can write this Sturm–Liouville problem in standard form,

$$u'' + \lambda r(x)u = 0, \qquad u(0) = u(1) = 0. \tag{10.2.2}$$

Here, $0 < x < 1$; $\lambda = Pl^2/\sigma_0$ is the dimensionless parameter to be determined; σ_0 is a characteristic value of the rigidity $\sigma(x)$, say, its maximal value; the function $r(x) = \sigma_0/\sigma(x)$ describes relative elastic compliance of the beam at the point x. As the unit of length we take l; the variables x, u have been made dimensionless by being divided by l.

It is required to find the smallest value $\lambda_1 > 0$ (the first eigenvalue) for which problem (10.2.2) admits a nontrivial solution $u_1(x) = u(x, \lambda_1)$. The subscript 1 will be omitted for the sake of brevity. The critical force P_* is determined by the value of λ, according to the replacement formula. On the basis of the eigenfunction $u(x, \lambda)$, we determine the equilibrium form after the stability loss under an infinitely small perturbation (of the displacement [63]). It should be mentioned that many problems in theoretical physics, mathematical physics, quantum mechanics, theory of elasticity, hydrodynamics, vibration theory are reduced to the eigenvalue problem (10.2.2) (see Chapters 11-13).

The function $r(x)$ in (10.2.2) may have a very complex structure, and this is a great obstacle to highly precise computation of eigenvalues and eigenfunctions. The linear and the inertial characteristics of the beam cross-section may depend on a vector parameter a,

$$r = r(x, a), \qquad a \in A; \qquad 0 < r^- \le r \le r^+ < \infty, \tag{10.2.3}$$

where A is a set of admissible values of the parameter, and the above inequality may be regarded as a constraint on the compliance function r. Moreover, one often has to find an optimal value of the parameter $a = a^* \in A$ such that

$$\max_a \lambda_1(a) = \lambda_1(a^*), \tag{10.2.4}$$

and additional constraints of geometrical or integral character may be imposed on some function characterizing the cross-section, for instance, its radius ρ or its area S. Usually, constraints are imposed on the linear dimensions and the volume of the beam V,

$$V(a) = \int_0^1 S(x, a)\, dx \le V_0, \qquad S \in \{S\}. \tag{10.2.5}$$

In a more general setting, the problem of maximizing the first eigenvalue λ_1 may involve the compliance function r depending on a "control variable" $h = h(x)$ (a scalar- or vector-valued function), and one has to choose a function $h(x) \in H$ for which some constraints like (10.2.5) hold and the eigenvalue λ_1 attains its maximal value. Thus, one obtains the following variational problem for the unknown $h(x)$, $u_1(x)$, λ_1:

$$\lambda_1[h] \to \sup_h, \qquad h \in H,$$
$$\lambda_1[h] = \min_u \int_0^1 u'^2\,dx, \quad \int_0^1 r(x,h)u^2\,dx = 1, \quad u(0) = u(1) = 0. \tag{10.2.6}$$

An effective numerical solution of problem (10.2.6) is very difficult, and it is common practice to consider parametric optimization of the type (10.2.4), (10.2.5).

10.2.2. Calculation of the critical force for some rigidity distributions. The efficiency of the approach proposed above can be demonstrated by model examples.

Example 1. Consider a rigidity distribution $\sigma(x)$ for which it is possible to construct an exact analytical solution of problem (10.2.2). Suppose that the beam has circular cross-section of radius $\rho(x) = (1 + x^2/l^2)^{1/2}$ which monotonically increases $\sqrt{2}$ times as x varies from $x = 0$ to $x = l$. Then the rigidity function has the form $\sigma(x) = \sigma_0(1 + x^2/l^2)^2$ (it increases 4 times) and the compliance in dimensionless variables is $r(x) = (1 + x^2)^{-2}$ (this increases 4 times). The solution of problem (10.2.2) in this case can be found in explicit analytical form (see Section 2.5),

$$\lambda = 15, \qquad u(x) = 2x(1 - x^2)(1 + x^2)^{-3/2}. \tag{10.2.7}$$

Let us use the procedure based on the Rayleigh principle for finding an approximate solution of this problem. Taking the simplest test function $\psi(x) = \sin \pi x$ (symmetric with respect to the middle point of the interval), we obtain the upper bound $\lambda^* = 15.33728$ for the first eigenvalue λ. Integrating the corresponding Cauchy problem, according to Sections 2.1, 2.2, we find the numerical abscissa value $\xi = 0.98350$ and the parameter $\varepsilon = 1 - \xi = 1.65 \cdot 10^{-2}$. In view of the expressions from Section 2.3, $\lambda^*\xi^2$ is a lower bound: $\lambda \le \lambda_* = \lambda^*\xi^2 = 14.83533$. On the first iteration step, we obtain the refined value $\lambda^{(1)} = 15.00847$ in the first approximation in ε (with the error $0(\varepsilon^2)$). Comparing this with the exact value (10.2.7), we obtain the relative error $|\lambda^{(1)} - \lambda|\lambda^{-1} = 5.7 \cdot 10^{-4} = O(\varepsilon^2)$. Note that our choice of the test function $\psi(x)$ is rather rough, since it does not take into account the non-symmetry properties of the problem (decrease of $r(x)$ or increase of $\sigma(x)$). The procedure of Section 2.3 based on the perturbation method has allowed us to find a highly precise approximation of the first eigenvalue λ (critical force), as well as the instability shape (buckling shape).

For the sake of comparison, consider a beam of constant cross-section $\rho(x) = \rho_0 = 2/\sqrt{3}$ with the same volume as the above beam with variable cross-section. We easily find that $\lambda_0 = \pi^2\rho_0^4 = 16\pi^2/9 \approx 17.6$. Thus, the beam with a constant cross-section is "stiffer", since $P_{0*} > P_*$.

Example 2. Let us briefly consider an edifying problem of the same kind with $\rho(x) = (1 + x)^{1/2}$ (in dimensionless variables). In this case, we come to an equation of Euler type, and for the solution of problem (10.2.2) we obtain the precise expressions (see Section 1.2)

$$\lambda = \frac{1}{4} + \frac{\pi^2}{\ln^2 2} \approx 20.79229, \qquad u(x) = (1 + x^2)^{1/2} \sin\left(\frac{\pi}{\ln 2}\ln(1 + x)\right).$$

The choice of the test function $\psi(x) = \sin(\pi x)$ is even more rough than in the previous case: the value $\lambda^* = 22.22421$ cannot be regarded as a good approximation (the relative error is about 7%). Using formulas from Section 2.1, we find the abscissa $\xi = 0.95459$ and define the parameter $\varepsilon = 4,5 \cdot 10^{-2}$. By formulas from Section 2.2, we obtain the lower bound $\lambda \ge \lambda_* = \lambda^*\xi^2 = 20.25164$. The refined value in the first approximation is $\lambda^{(1)} = 20.80330$, and its relative error is $|\lambda^{(1)} - \lambda|\lambda^{-1} = 5 \cdot 10^{-4} = O(\varepsilon^2)$. Thus, we have obtained a substantial improvement of the critical force bound.

Let us compare these results with those for a beam of constant radius $\rho_0 = (3/2)^{1/2}$ and the same volume. Comparing the critical forces, we see that $\lambda_0 > \lambda$, since $\lambda_0 = 9\pi^2/4 \approx 22.3$. Therefore, the beam of constant cross-section possesses greater rigidity. Therefore, it would be interesting to apply the above method of accelerated convergence to obtain an effective solution of the following optimal control problem: find a distribution $\rho(x)$ for which the critical compressive force P is maximal for a beam of fixed volume (see Example 3).

Example 3. Let us find the critical compressive force for a one-parameter family of beams of circular cross-section. Suppose that in the corresponding Sturm–Liouville problem (10.2.2) the function $r(x,\eta)$ is symmetric with respect to the mid-point $x = 1/2$ and has the form

$$\begin{aligned} r(x,\eta) &= \left[\eta + \gamma(\eta)(x - \tfrac{1}{2})^2\right]^{-4}, \qquad 0 \le x \le 1, \\ \gamma = \gamma(\eta) &= \tfrac{20}{3}\left[-\eta + \tfrac{1}{\sqrt{5}}(9 - 4\eta^2)^{1/2}\right], \quad 0 < \eta \le \tfrac{3}{2}. \end{aligned} \tag{10.2.8}$$

It is easy to show that the volume of the beam with the normalized radius $\rho(x,\eta) = |\eta + \gamma(\eta)(x - \frac{1}{2})^2|$ ($0 < \eta \le 3/2$, according to (10.2.8)) is constant and equal to π. For $\eta > 3/2$, no such beams exist. Further, if $0 < \eta < \eta_* = (15/8)^{1/2} \approx 1.369 < 3/2$, then the function $\rho(x,\eta)$ cannot be equal to zero. For the values $\eta_* \le \eta \le 3/2$, there exist two points $x_{1,2}(\eta)$ equidistant from $x = 1/2$ and such that the beam thickness is equal to zero, $\rho(x_{1,2},\eta) = 0$, i.e., its rigidity vanishes, $\sigma(x_{1,2},\eta) = 0$, and its compliance is infinite, $r(x_{1,2},\eta) = \infty$. From (10.2.8), it also follows that $\gamma(1) = 0$, $\rho(x,1) \equiv 1$, and thus, the beam of constant radius is subjected to symmetric variations in the "class of parabolas". If $\eta = 0$, then at $x = 1/2$ the beam thickness is equal to zero, and its compliance is infinite, $r(1/2, 0) = \infty$. As a geometrical illustration, Fig. 10.6 shows the functions $\rho(x,\eta)$, $0 \le x \le 1/2$, for different values of η, $0 \le \eta \le 3/2$.

For practical purposes, it is important to examine the critical force P_*, i.e., the eigenvalue λ as a function of the parameter η for which $r < \infty$, i.e., for $0 < \eta < \eta_*$. Using the above approach, combined with the method of continuation in the parameter η, we obtain highly precise upper and lower bounds (with relative error $10^{-4} \div 10^{-5}$) for the first eigenvalue $\lambda = \lambda(\eta)$ (proportional to the normalized critical force P, $P_* = \lambda\sigma_0/l^2$) for different values of the parameter η, $0.5 \le \eta \le 1.3$.

From the standpoint of theory and applications, it would be interesting to find an optimal value η^*, i.e., the value realizing the maximum $\lambda(\eta^*)$. This problem, in a more general setting, is considered in numerous investigations. A solution of the optimization problem considered here (i.e., with r of the type (10.2.8)) is represented in Fig. 10.7. The graph gives the relative eigenvalue $\nu(\eta) = \lambda(\eta)/\lambda(1)$. This curve shows that for $\eta = \eta^* \approx 1.175$, the critical force has a sharp maximum which is about 30% greater than the value for the homogeneous beam ($\eta = 1$); this shape is represented in Fig. 10.6.

Our conclusion is that the highly precise method developed in Chapter 2 for the construction of approximations of eigenvalues and eigenfunctions on the basis of the accelerated convergence algorithm and the procedure of continuation in a parameter can be effectively used for solving the problem of longitudinal bending of inhomogeneous beams, in particular, the problem of optimal choice of parameters. One can consider such problems with different types of fixation at the ends and take into account distributed longitudinal loads. The method can also be used for the calculation of subsequent eigenvalues and eigenfunctions in problems of stability of vibrations of inhomogeneous beams. It should be mentioned that highly precise calculation of eigenvalues and eigenfunctions (frequencies and vibration shapes) is important for solving the problems of high precision mechanics arising in tool design, microelectronics, and other branches of modern high-precision technology.

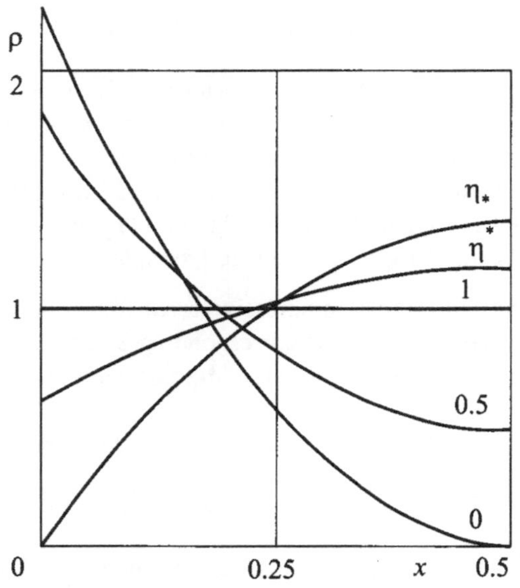

Fig. 10.6

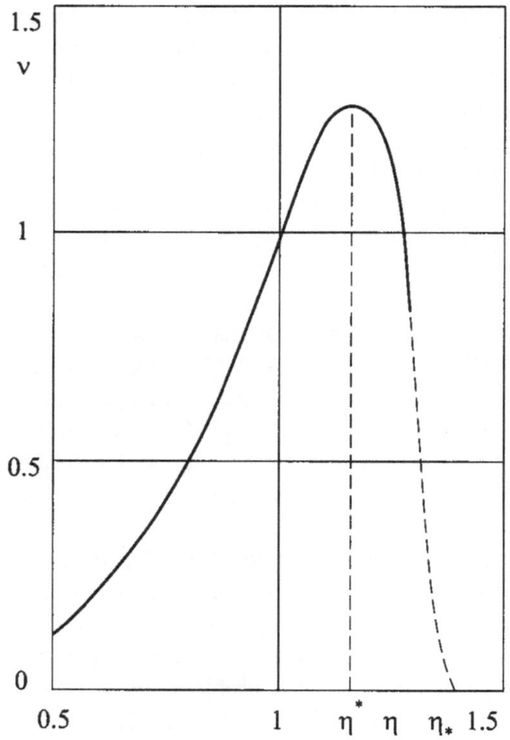

Fig. 10.7

Computational experience shows that this method spares resources, is very fast, stable, and precise, when applied to complex problems with strong variation (by several orders) of stiffness and inertial parameters. Thus, in the model problem of Example 4.3, the compliance coefficient decreases 10^3 times from the ends of the beam to its middle for $\eta = 1.3$. It should be mentioned that satisfactory results can be obtained with the help of a PC.

10.3. The Problem of Lateral Buckling of a Long Beam with Narrow Cross-Section

10.3.1. Statement of the Prandtl problem. Consider the classical problem of lateral buckling of a long cantilever beam with narrow rectangular cross-section [63]. It is assumed that, together with the bending force P concentrated on the edge, great influence is exerted by a distributed mass force, say, the gravity force [63]. The boundary value problem for the twist angle β can be reduced to

$$(\sigma(x)\beta')' + \frac{1}{\alpha(x)}\left[Px + g\int_0^x (x-s)\rho(s)\,ds\right]^2 \beta = 0, \qquad \beta'(0) = \beta(l) = 0. \qquad (10.3.1)$$

Here, $\sigma(x)$ is torsional rigidity; $\alpha(x)$ is flexural rigidity with respect to the deflection orthogonal to the vertical plane; $\rho(x)$ is the linear density; g is the gravity acceleration. The beam is assumed fairly narrow and long, $d^2/hl \ll 1$, where d, h are the characteristic linear dimensions (of the thickness) of the beam in the horizontal and the vertical planes, respectively [63]. The unknown quantities in problem (10.3.1) are the following: the critical value $P = P_*$ for which lateral buckling with twist occurs, i.e., there exists a nontrivial solution $\beta(x, P_*)$; and the corresponding buckling shape. The quantity P_* characterizes the bearing capacity of the beam. The classical results pertain mainly to beams of constant cross-section (σ, α, ρ are constants), and one usually neglects the effect of the (distributed) weight component in (10.3.1). The last assumption for sufficiently long beams may lead to considerable errors — an overestimated value of P_*. Moreover, the beam may become unstable under its own weight.

Consider problem (10.3.1) for a beam of constant cross-section, with its weight taken into account. After the introduction of dimensionless parameters λ, θ and the normalized argument x, we obtain the generalized Sturm–Liouville problem

$$\beta'' + (\lambda x + \theta x^2)^2 \beta = 0, \qquad \beta'(0) = \beta(1) = 0. \qquad (10.3.2)$$

Here, $0 \le x \le 1$, $\lambda = Pl^2(\sigma\alpha)^{-1/2}$, $\theta = \frac{1}{2}\rho g l^3(\sigma\alpha)^{-1/2}$. The boundary conditions in (10.3.2) correspond to the free left end $x = 0$ and the clamped right end $x = 1$. Not that the "compliance coefficient" is equal to zero, $r(x, \lambda, \theta) = 0$ (r is the coefficient of β) at $x = 0$. Moreover, on the left end we have the boundary condition of the second kind.

10.3.2. A numerical-analytical solution. In this case, the scheme of solution described in Chapter 4 is preserved, but the boundary conditions in the Cauchy problems have to be changed, $v(0) = 1$, $v'(0) = 0$. Our aim is to find the first eigenvalue $\lambda_1 = \lambda_1(\theta)$ for a sufficiently wide range of $\theta \ge 0$.

For $\theta = 0$, a solution can be constructed analytically with the help of Bessel functions, and we obtain the known result [63]: $\lambda_1(0) = 4.013$. A more precise value obtained by the method of accelerated convergence is $\lambda_1(0) = 4.012597$. The results of highly precise calculations for $\theta > 0$ are given in Fig. 10.8. The calculations were performed with the relative error $10^{-6} \div 10^{-7}$. The graph of $\lambda_1(\theta)$ shows that as θ increases, the value λ monotonically (almost linearly) decreases to $\lambda_1(\theta_*) = 0$ for $\theta = \theta_* = 6.4269$. As $\theta > \theta_*$ increases further, $\lambda_1(\theta)$ becomes negative, i.e., the beam has negative bearing capacity due to its weight. In order to ensure its stability, the beam has to be "supported" by some load $P^* < P_* \le 0$. But the value of $|P|$ cannot be too large unless lateral buckling arises due to an excessive upward load.

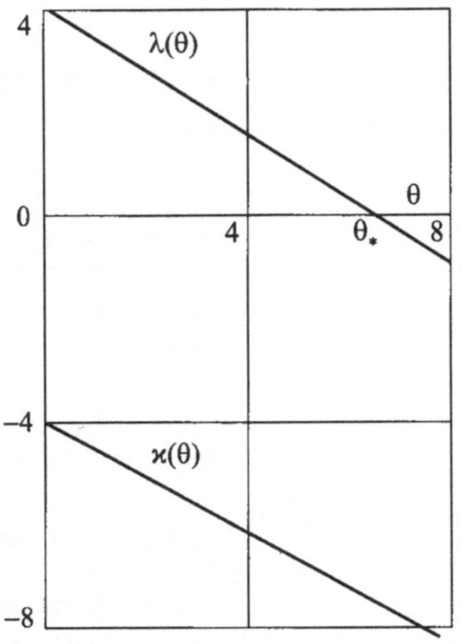

Fig. 10.8

Note that for $\theta = 0$, the problem has also a negative eigenvalue $\kappa(\theta) = -\lambda_1(0)$ with a certain mechanical meaning ($P_* < 0$).

The corresponding values $\kappa(\theta)$, $\theta > 0$, also deserve interest and are shown in the graph. Together with positive values of θ, one can consider its negative values (the acceleration of mass forces is directed upward). Note that the symmetry of the problem (10.3.2) implies the central symmetry $\lambda(\theta) = -\kappa(-\theta)$. By the method of accelerated convergence, problem (10.3.1) can be solved for arbitrary specific functions $\sigma(x)$, $\alpha(x)$, $\rho(x)$, in particular, the problem of optimizing mass distribution in a beam, with natural constraints taken into account.

10.4. Longitudinal Vibrations of an Inhomogeneous Beam with Transverse Inertia

10.4.1. Approaches of Rayleigh and Love. Let us examine longitudinal vibrations of a thin elastic rectilinear beam. The material of the beam is characterized by the elastic modulus E, the density ρ, the Poisson ratio μ. The area of the cross-section of the beam is described by the function $S = S(x)$, and $R = R(x)$ is the radius of gyration of the cross-section.

The problem is to find longitudinal displacements $u = u(x,t)$ of the beam particles. In the framework of traditional concepts, under the above assumptions, the function $u(x,t)$ should satisfy the following equation:

$$\rho S(x) u_{tt} = E[S(x) u_x]_x. \tag{10.4.1}$$

Here, the dots denote derivatives with respect to t. For $S = S_0 = \text{const}$, equation (10.4.1) reduces to

$$\rho u_{tt} = E u_{xx}. \tag{10.4.2}$$

Equations of the form (10.4.1), (10.4.2) commonly occur in many textbooks and monographs. However, it was discovered by Rayleigh [62] that, together with longitudinal vibrations, the material points of a beam undergo transverse displacements in the plane of its cross-section. In this way, the inertia of the beam is increased. Rayleigh called this effect "transverse inertia" and calculated corrections to the eigenvalues due to transverse motions of the beam particles.

With the help of the least-action principle, the effect of transverse inertia was aptly taken into account by Love [40] for a beam of constant cross-section and constant radius of gyration of the cross-section. If we use the least-action principle for the description of vibrations of an inhomogeneous beam and take into account transverse inertia, we obtain the following differential equation for longitudinal displacements of the beam particles:

$$\rho S(x) u_{tt} - \rho\mu^2 [I(x) u_{xtt}]_x = E[S(x) u_x]_x. \tag{10.4.3}$$

Here, $I(x) = R^2(x)S(x)$ is the moment of the beam cross-section. Taking the Poisson ration equal to zero, $\mu = 0$, i.e., neglecting the transverse inertia, we obtain equation (10.4.1), and in the case of constant cross-section, $S = S_0$, we obtain equation (10.4.2). The boundary conditions for the functions $u(x,t)$ describe fixation of the beam at its edges $x = 0$, $x = l$. In particular, one can consider the boundary conditions of the first or the second kind

$$1)\ u(0,t) = u(l,t) = 0, \qquad 2)\ u_x(0,t) = u_x(l,t) = 0. \tag{10.4.4}$$

The differential equation in (10.4.3) is of the second order with respect to x. Time-periodic solutions of equation (10.4.3) may be sought in the form $u(x,t) = U(z)\exp(i\omega t)$, where $z = x/l$. Substituting this $u(x,t)$ into (10.4.3), we obtain the following generalized Sturm–Liouville problem:

$$\begin{gathered}[(S(z) - \lambda\mu^2 l^{-2} I(z))U']' + \lambda S(z) U = 0, \qquad 0 \le z \le 1;\\ 1)\ U(0) = U(1) = 0, \qquad 2)\ U'(0) = U'(1) = 0,\end{gathered} \tag{10.4.5}$$

where the prime denotes differentiation in z, $\lambda = \omega\rho l^2/E$. Using the results of Chapter 4, one can find eigenvalues of problems (10.4.5).

Suppose that the beam has circular-cross section of constant radius a. Then, $S(x) = \pi a^2$, $I(x) = \pi a^4/2$ and equation (10.4.5) becomes

$$\left(1 - \tfrac{1}{2}a^2 l^{-2}\mu^2\lambda\right) U'' + \lambda U = 0. \tag{10.4.6}$$

The solutions of the boundary value problems (10.4.5), (10.4.6) can be found in analytical form, and their eigenvalues are the following:

$$\lambda_n = \pi^2 n^2 \left(1 + \frac{a^2\mu^2\pi^2 n^2}{2l^2}\right)^{-1}, \qquad n = 1, 2, 3, \ldots \tag{10.4.7}$$

Using the relation $\lambda = \omega^2 \rho l^2/E$ (10.4.5), together with (10.4.7), we find the dependence of the eigenfrequencies ω_n on the Poisson ratio μ,

$$\omega_n = \frac{\pi n}{l}\sqrt{\frac{E}{\rho}}\left(1 + \frac{a^2\mu^2\pi^2 n^2}{2l^2}\right)^{-1/2}. \tag{10.4.8}$$

For $a^2\mu^2\pi^2 n^2/(2l^2) \ll 1$, formula (10.4.8) becomes

$$\omega_n \approx \frac{\pi n}{l}\sqrt{\frac{E}{\rho}}\left(1 - \frac{a^2\mu^2\pi^2 n^2}{4l^2}\right). \tag{10.4.9}$$

The value $a^2\mu^2\pi^2n^2/(4l^2)$ was initially found by Rayleigh, who called it the "transverse inertia correction".

10.4.2. Experimental determination of Poisson's ratio on the basis of measurements of longitudinal frequencies by the resonance method. Formula (10.4.9) (and (10.4.8)) can be used for the determination of the Poisson ratio μ of the material. This can be done by analyzing longitudinal vibrations of beams. Experimental observations can be used for finding eigenfrequencies ω_n of some sample beams. With the help of formula (10.4.9), these frequencies can be compared with reference eigenfrequencies in the absence of "transverse inertia".

For the experimental determination of Poisson's ratio the authors have proposed a method based on the analysis of longitudinal vibrations. Next, we give a brief description of that method.

With the help of formula (10.4.9), it is possible to find Poisson's ratios of some materials. The values obtained are acceptable for practical purposes. In the experiments we used rectilinear beams of circular cross-section. We considered beams made of different materials (steel, iron, duralumin, titan, brass, marble).

The dimensions of the beams were measured with great precision. The beams were suspended vertically on thin threads, which ensured the boundary conditions of the second kind (free ends) (10.4.6). By means of an electromagnetic vibrator driven by alternate current of variable frequency, longitudinal vibrations were excited in the beam. Resonance tuning was used for the determination of eigenfrequencies of beams. The ratios of the measured eigenfrequencies ω_n to the reference frequencies $\tilde{\omega}_n = \pi n\sqrt{E/\rho}/l$ were calculated by (10.4.9). The stable relation

$$\frac{\tilde{\omega}_n}{\omega_n} = \left(1 - \frac{\mu^2 a^2}{4l^2}\pi^2 n^2\right)$$

was used for the calculation of the Poisson ratio.

Here, we give some values of the Poisson ratio obtained in our experiments: for steel $\mu = 0.274 \div 0.304$ $(0.28 \div 0.31)$; for brass $\mu = 0.317$ (0.3); for titan $\mu = 0.332$ $(0.33 \div 0.34)$. The values in brackets were borrowed from handbooks.

Thus, the refined statement of the problem of longitudinal vibrations brings us to the generalized Sturm–Liouville problem and gives us wider possibilities for its theoretical and experimental investigation.

10.5. Exercises

Let us formulate some problems in mechanics and vibration theory for which it is possible to use the method of accelerated convergence.

Exercise 1. Find the minimal value of the compressive load P in the Euler problem (Section 10.2) with the flexural rigidity of the form $\sigma(x) = (4 - x^2)^2$.

Exercise 2. Find the first three eigenfrequencies and longitudinal vibration shapes for a beam with constant elastic modulus E and constant density ρ. The cross-sectional area of the beam has the form $S(x) = S_0(1 + \sin \pi x/l)$, where S_0 is the area of the cross-section $x = 0$, l is the length of the beam. In order to apply the formulas of Chapter 2, the equation of vibrations of the beam should be written in dimensionless form. Calculate approximate eigenvalues for two types of boundary conditions: 1) both ends of the beam clamped, 2) both ends are free.

Exercise 3. Find the first three eigenvalues and shapes of transverse vibrations for a string of length l, $\rho = \rho_0(1 + \sin \pi x/l)$, and tension $T = \text{const}$. The string is clamped at both ends.

Chapter 11

Surface and Internal Waves in Heavy Ideal Fluid

Eigenvalue problems play an important role in theoretical and applied hydrodynamics. Such problems have some specific features which make their solution very difficult. Quite often one has to deal with generalized eigenvalue problems. In this chapter, we describe some results of numerical-analytical investigation of two meaningful problems in hydrodynamics by the method of accelerated convergence. The first is the Laplace-Hough problem which describes surface waves on a spherical layer of heavy fluid on a rotating gravitating sphere [21, 57]. The second problem is that of travelling waves in a strongly stratified heavy fluid [26].

11.1. Free Vibrations of the Surface of a Rotating Spherical Layer of Heavy Fluid

11.1.1. Preliminary remarks and statement of the problem. The problem of finding eigenfrequencies and vibration shapes of the surface of a rotating layer of fluid has important applications in geophysics. Its solution is of great interest for the theory of tidal waves in ocean and atmosphere. The main problems for free vibrations of tidal type are formulated in fairly simple terms on the basis of the equations of hydrodynamics (see, for instance, [21, 57]). However, their solution encounters considerable obstacles, even if the influence of continents and other essential factors are neglected. The great difficulty of these problems is confirmed in the cited monographs and in many other publications teeming with extremely cumbersome analytical constructions.

Below, we construct an effective numerical solution of the Sturm–Liouville problem for the Hough equation [57], which is at the basis of the theory of tides. This solution is obtained by the highly precise method of accelerated (quadratic) convergence developed in Chapters 2 and 4. The eigenvalues (frequencies) and the eigenfunctions (vibration shapes) are calculated for a wide range of basic parameters of the problem. These calculations result in small relative errors (of order $10^{-8} \div 10^{-9}$) and may serve as a reference of precise solutions in other calculations.

Let us briefly describe the derivation of the Hough equation, which will be used for the calculation of eigenfrequencies and vibration shapes of the surface of a rotating spherical fluid layer. Consider a solid ball of radius a covered by a homogeneous ideal incompressible fluid forming a thin layer of depth h, $h \ll a$. The fluid particles are attracted to the ball by gravity forces, whose acceleration is equal to g. In view of the assumption $a \gg h$, the gravity acceleration g is constant in the entire fluid layer.

Suppose that the solid ball, together with the fluid layer, is rotating with constant angular speed ω, and $\omega^2 a \ll g$. In this case, the equilibrium surface is an ellipsoid of revolution which differs but slightly from a sphere (by a relative quantity of the order $O(m)$, where $m = \omega^2 a/g \ll 1$).

Therefore, in linear approximation, one can consider small variable deviations ζ of the free surface from the equilibrium ellipsoid. The boundary conditions are imposed on the sphere of radius a, and this entails relatively small errors.

Let us introduce spherical coordinates on the spherical surface with the orthogonal unit vectors $\mathbf{e}_\theta$, $\mathbf{e}_\varphi$, and let the fixed Cartesian axis z be directed along the axis of rotation of the sphere. Denote by v_θ, v_φ the corresponding velocity components of fluid particles, and let ζ denote the small deviation of the free surface from the equilibrium surface in the direction $\mathbf{e}_r$. In the long-wave approximation, the linearized equations of hydrodynamics in these spherical coordinates, with the Coriolis force taken into account, can be written in the form [21, 57]

$$\begin{aligned} &\frac{\partial v_\theta}{\partial t} - 2\omega v_\varphi \cos\theta = -\frac{g}{a}\frac{\partial \zeta}{\partial \theta}, \\ &\frac{\partial v_\varphi}{\partial t} + 2\omega v_\theta \cos\theta = -\frac{g}{a\sin\theta}\frac{\partial \zeta}{\partial \varphi}, \\ &\frac{\partial \zeta}{\partial t} = -\frac{h}{a\sin\theta}\left[\frac{\partial}{\partial\theta}(v_\theta \sin\theta) + \frac{\partial v_\varphi}{\partial \varphi}\right] \end{aligned} \tag{11.1.1}$$

The first two relations of system (11.1.1) are equations of motion and the third is the equation of continuity. The pressure in this approximation is determined by the law of hydrostatics, and the vertical velocity component may be neglected. These variables are not needed in our further investigation. Thus, the system of equations (11.1.1) is closed.

Free vibrations of the spherical fluid layer can be studied in a standard manner. In (11.1.1), we take

$$v_{\theta,\varphi} = V_{\theta,\varphi}(\theta,\varphi)\exp(-i\sigma t), \quad \zeta = Z(\theta,\varphi)\exp(-i\sigma t), \quad i = \sqrt{-1},$$

where $V_{\theta,\varphi}$, Z are unknown functions and σ is an unknown parameter to be determined. Substituting these expressions into system (11.1.1), we obtain the following equation for the function Z:

$$\begin{aligned} 4iamfZ\sin\theta = {} & h\frac{\partial}{\partial\theta}\left[\frac{\sin\theta}{f^2-\cos^2\theta}\left(\frac{\partial Z}{\partial\varphi}\operatorname{cotan}\theta - if\frac{\partial Z}{\partial\theta}\right)\right] \\ & - h\frac{\partial}{\partial\varphi}\left[\frac{1}{f^2-\cos^2\theta}\left(\frac{\partial Z}{\partial\theta}\cos\theta - \frac{if}{\sin\theta}\frac{\partial Z}{\partial\varphi}\right)\right]. \end{aligned} \tag{11.1.2}$$

Here, $f = \sigma/(2\omega)$ is a dimensionless unknown parameter. It is assumed that on both sides of equation (11.1.2) the small quantities are of the same order. The unknown functions V_θ and V_φ are expressed through Z by the formulas

$$\begin{aligned} V_\theta &= \frac{\omega}{2m\left(f^2-\cos^2\theta\right)}\left(\frac{\partial Z}{\partial\varphi}\operatorname{cotan}\theta - if\frac{\partial Z}{\partial\theta}\right), \\ V_\varphi &= \frac{\omega}{2m\left(f^2-\cos^2\theta\right)}\left(\frac{\partial Z}{\partial\theta}\cos\theta + \frac{if}{\sin\theta}\frac{\partial Z}{\partial\varphi}\right). \end{aligned} \tag{11.1.3}$$

Now, we examine equation (11.1.2). For simplicity, we limit ourselves to axially symmetric vibrations satisfying the condition $\partial Z/\partial\varphi \equiv 0$, i.e., consider the null-harmonic with respect to φ.

The latitude θ will be measured from the "north pole" $\theta = 0$ (in contrast to the coordinates adopted in geography), i.e., from the positive direction of the z-axis. The "south pole" corresponds to $\theta = \pi$. It is convenient to introduce a new independent variable, $\mu = \cos\theta$, and represent equation (11.1.2) for eigenfunctions in a form containing no dimensional parameters,

$$\begin{aligned} &\frac{d}{d\mu}\left(\frac{1-\mu^2}{f^2-\mu^2}\frac{dZ}{d\mu}\right) + \beta Z = 0, \quad -1 \le \mu \le 1, \\ &\beta = 4ma/h \sim 1, \quad 0 < \beta_1 \le \beta \le \beta_2 < \infty. \end{aligned} \tag{11.1.4}$$

Equation (11.1.4) is called the Hough equation. This equation has regular critical points $\mu = \pm 1$, and also singular points $\mu = \pm|f|$ for $|f| < 1$.

Equation (11.1.4) can be transformed so that the singularities for $\mu = \pm|f|$ are eliminated. To that end, one introduces a new unknown function $Y(\mu)$ instead of $Z(\mu)$ [27],

$$Y(\mu) = \frac{1-\mu^2}{f^2-\mu^2}\frac{dZ(\mu)}{d\mu}, \qquad Z(\mu) = -\frac{1}{\beta}\frac{dY(\mu)}{d\mu}. \tag{11.1.5}$$

With the help of (11.1.5), we can write equation (11.1.4) in the standard form

$$\frac{d^2Y}{d\mu^2} + \beta\frac{f^2-\mu^2}{1-\mu^2}Y = 0, \qquad -1 \le \mu \le 1. \tag{11.1.6}$$

One can formulate problems of Sturm–Liouville type for equations (11.1.4) or (11.1.6). Consider a spherical layer of fluid bounded by two latitudes θ_1, θ_2:

$$\cos\theta_1 = \mu_1, \quad \cos\theta_2 = \mu_2, \quad \theta_1 > \theta_2, \quad \mu_1 \le \mu \le \mu_2, \quad \mu_1 < \mu_2. \tag{11.1.7}$$

The problem is to find the values of the parameter f^2 for which there exist nontrivial solutions of equation (11.1.4) or (11.1.6) in the layer (11.1.7) with the boundary conditions

$$\frac{dZ(\mu_1)}{d\mu} = \frac{dZ(\mu_2)}{d\mu} = 0, \quad Y(\mu_1) = Y(\mu_2) = 0, \quad -1 < \mu_1 < \mu_2 < 1. \tag{11.1.8}$$

In view of (11.1.3), these conditions ensure that the normal velocity V_θ and the tangential velocity V_φ are equal to zero on the boundaries of the layer. These boundaries are assumed absolutely rigid.

Relations (11.1.4), (11.1.8) specify a generalized Sturm–Liouville problem for the eigenvalues f_n^2 and the functions Z_n ($n = 1, 2, \ldots$) with the boundary conditions of the second kind, since equation (11.1.4) nonlinearly depends on the parameter f_n^2. The modified problem (11.1.6), (11.1.8) is the standard Sturm–Liouville problem with the boundary conditions of the first kind. This problem will be the object of our further analysis, which is based on the numerical-analytical method of accelerated convergence developed by the authors (see Chapters 2 and 4).

Of special interest for our investigation is the case of μ_1 and/or μ_2 attaining their maximal values $\mu_1 = -1$, $\mu_2 = 1$.

Consider an important extreme case of the problem with the fluid completely covering the rotating ball, i.e., the fluid layer covers both the "north pole" ($\mu_2 = 1$) and the "south pole" ($\mu_1 = -1$). It is required to find the values of the parameter f^2 for which there exists a nontrivial solution of equation (11.1.4) or (11.1.6) bounded near the points $\mu = \pm 1$. Numerous calculations for similar problems show that regular critical points cause considerable difficulties in the numerical implementation of the accelerated convergence method for $\mu_1 \to -1$, $\mu_2 \to 1$, since the equations become singular. These difficulties can be overcome, if one takes into account a detailed numerical-analytical investigation of passing to the limit values with the desired precision. Below, we describe calculation results for problem (11.1.6), (11.1.8) for a wide range of μ: $\mu_1 = -0.99$, $\mu_2 = 0.99$.

11.1.2. Solving the eigenvalue problem. In order to construct a highly precise numerical solution of the Sturm–Liouville problem with the boundary conditions of the first kind (11.1.6), (11.1.8), we use the algorithm of accelerated convergence in combination with the continuation in a parameter (see Chapter 2). To that and, normalizing the argument μ, we reduce this problem to the standard form

$$\begin{gathered}X'' + [\lambda r(x,\mu_1,\mu_2) - \beta q(x,\mu_1,\mu_2)]\,X = 0, \qquad X(0) = X(1) = 0;\\ X = X(x), \quad 0 \le x \le 1, \quad x = (\mu - \mu_1)(\mu_2 - \mu_1)^{-1},\\ r(x,\mu_1,\mu_2) = \left[1 - \eta^2(x,\mu_1,\mu_2)\right]^{-1}, \quad X(x) \equiv Y(\mu),\\ q(x,\mu_1,\mu_2) = (\mu_2 - \mu_1)^2\eta^2(x,\mu_1,\mu_2)r(x,\mu_1,\mu_2),\\ \eta(x,\mu_1,\mu_2) = \mu_1 + x(\mu_2 - \mu_1), \quad \lambda = \beta(\mu_2 - \mu_1)^2 f^2,\end{gathered} \tag{11.1.9}$$

where the prime denotes differentiation in x. According to (11.1.7)–(11.1.8), it is required to construct the sets of eigenvalues $\lambda_n(\mu_1, \mu_2)$ and eigenfunctions $X_n(x, \mu_1, \mu_2)$, $n = 1, 2, \dots$ Let us fix the parameter β and take it equal to $\beta = 19.648$, which corresponds to the "average ocean depth" $h = 4.5 \cdot 10^3$ [m], if the Earth is considered as a sphere with a fluid layer. Whenever necessary, this parameter may be varied within a wide range ($\beta_1 \le \beta \le \beta_2$) for which the above assumptions hold. Moreover, this parameter may depend on θ (i.e., on μ or x), say, for variable h.

Using the functions $\lambda_n(\mu_1, \mu_2)$ introduced above and the formula (see (11.1.9))

$$f_n^2 = \frac{\lambda_n(\mu_1,\mu_2)}{\beta(\mu_2 - \mu_1)^2}, \quad n = 1, 2, \dots \tag{11.1.10}$$

we find the relative frequencies f_n of free vibrations, and we have $\sigma_n = 2\omega f_n$; see (11.1.2).

The functions λ_n, f_n^2 turn out symmetrical with respect to the vertical plane $\mu_1 = -\mu_2$, which is obvious from physical considerations, since the "north" and the "south" poles are symmetrical with respect to the equator. It is easy to show that the Sturm–Liouville problem is invariant under the transformation $\mu \to -\mu$ or $\mu_1 \to -\mu_2$, $x \to 1 - x$.

From (11.1.10), it follows that

$$f_n^2 \sim \beta^{-1}(\pi n)^2(1 - \mu_1^2)(\mu_2 - \mu_1)^{-2} \to \infty, \qquad \mu_1 \to \mu_2.$$

This fact, together with the specific features of equation (11.1.6) mentioned above, is an obstacle to the application of numerical methods to the corresponding Sturm–Liouville problem. For this reason, it is preferable to consider the standard problem (11.1.9).

Using the above asymptotic expression of f_n, one can obtain geophysically meaningful approximate formulas for the eigenfrequencies σ_n,

$$\sigma_n \approx \pi n \frac{\sqrt{gh(1 - \mu_*^2)}}{a(\mu_2 - \mu_1)} \approx \pi n \frac{\sqrt{gh}}{a\delta\theta}, \qquad \delta\theta = \theta_1 - \theta_2. \tag{11.1.11}$$

Here, μ_* ia any value of μ from the interval $[\mu_1, \mu_2]$, for instance, the average value, and $\mu_2 < 1$, $\mu_1 > -1$ (far away from the poles). Hence, we conclude that the leading term of the asymptotic expansion of the eigenfrequencies in powers of the parameter $\delta\theta$ characterizing the layer width depends neither on the rotation speed ω nor the latitude θ. Its dependence on the gravity g, the average depth h, the radius a, and the layer width $\delta\theta$ agrees with physical concepts.

For $\mu_2 = -\mu_1$, the functions X_n and Y_n are symmetric with respect to $x = 0.5$ ($\mu = 0$) for odd $n = 1, 3, \dots$ and antisymmetric for even $n = 2, 4, \dots$, while the converse holds for the functions X_n' and Y_n'. For large n ($n \ge 4$), the functions X_n, X_n' are close to Legendre polynomials [32, 67].

Due to (11.1.5), (11.1.9), the original eigenfunctions $Z_n(\mu)$ are related to X_n by

$$Z_n(\mu) = -\frac{1}{\beta(\mu_2 - \mu_1)}\, X_n'\left(\frac{\mu - \mu_1}{\mu_2 - \mu_1}, \mu_1, \mu_2\right). \tag{11.1.12}$$

Just as for the eigenvalues, it is better to carry out calculations for the eigenfunctions on the basis of the standard problem (11.1.9).

The essence of the method of accelerated convergence consists in consecutively refining some known approximate value $\lambda_n^0(\mu_1, \mu_2)$. This process is realized by integrating the Cauchy problem for equation (11.1.9) and finding the abscissa $\xi_n = \xi_n(\mu_1, \mu_2, \lambda_n^0)$ of the nth positive root of the function $X_n(x, \mu_1, \mu_2, \lambda_n^0)$. As the measure of closeness between λ_n^0 and the precise value λ_n, one takes the numerical parameter $\varepsilon_n = 1 - \xi_n$, assuming that $|\varepsilon_n|$ is sufficiently small ($|\varepsilon_n| \ll 1$). For practical calculations, it suffices to have $|\varepsilon_n| \sim (0.1 \div 0.01)n^{-1}$. The approximate value λ_n^0 can be obtained for some μ_1, μ_2 by some variational method (the Rayleigh principle, the Rayleigh–Ritz method, etc.) or the method of continuation in two parameters μ_1, μ_2 (see below).

Thus, suppose that for fixed μ_1, μ_2 (and β) we know an approximate (in the above sense) eigenvalue λ_n^0. Then, the refined value $\lambda_n^{(1)}$ is determined with the error $O(\varepsilon_n^2)$ by the following relations (see [4, 6], and also Sections 2.4, 2.6):

$$\begin{aligned}
&\lambda_n^{(1)}(\mu_1, \mu_2) = \lambda_n^{(0)}(\mu_1, \mu_2) - \varepsilon_n^{(0)} A_n(\mu_1, \mu_2, \lambda_n^{(0)}), \quad -1 < \mu_1 \le \mu_2 < 1, \\
&A_n(\mu_1, \mu_2, \lambda_n^{(0)}) = X_n'^2(\xi_n^{(0)}, \mu_1, \mu_2, \lambda_n^{(0)}) \|X_n\|^{-2}, \\
&\xi_n^{(0)} = \operatorname{Arg}_{nx} X_n(x, \mu_1, \mu_2, \lambda_n^{(0)}), \quad \lambda_n^{(0)} = \lambda_n^0, \quad \varepsilon_n^{(0)} = \varepsilon_n.
\end{aligned} \tag{11.1.13}$$

In (11.1.13), the symbol $\operatorname{Arg}_n X_n$ denotes the nth positive root (in x) of the function X_n, which is found by analytical or numerical integration of the Cauchy problem for equation (11.1.9) with $\lambda = \lambda_n^0$:

$$X_n'' + [\lambda_n^{(0)} r(x, \mu_1, \mu_2) - \beta q(x, \mu_1, \mu_2)] X_n = 0, \qquad X_n(0) = 0, \quad X_n'(0) = 1. \tag{11.1.14}$$

Highly precise integration of the Cauchy problems and the determination of the roots ξ_n are the main points of difficulty (from the computational standpoint) in the algorithm under discussion.

The symbol $\|X_n\|^2$ in (11.1.13) denotes the squared norm of the function X_n in the space L^2 with the weight r on the interval $0 \le x \le \xi_n^{(0)}$. This norm can be expressed by quadratures and, according to [4, 6] and Chapter 2, can be found by joint integration of equations for the unknown function $S_n = \partial X_n / \partial \lambda_n$ with zero initial values at $x = 0$:

$$\begin{aligned}
&S_n'' + [\lambda_n^{(0)} r(x, \mu_1, \mu_2) - \beta q(x, \mu_1, \mu_2)] S_n = -r(x, \mu_1, \mu_2) X_n, \\
&S_n(0) = S_n'(0) = 0, \quad 0 \le x \le \xi_n^{(0)}, \\
&\|X_n\|^2 = \int_0^{\xi_n^{(0)}} r(x, \mu_1, \mu_2) X_n^2 \, dx = X_n'(\xi_n^{(0)}, \lambda_n^{(0)}) \, S_n(\xi_n^{(0)}, \lambda_n^{(0)}).
\end{aligned} \tag{11.1.15}$$

For brevity, the dependence of the functions X_n, S_n on μ_1, μ_2 (β) in (11.1.14), (11.1.15) and below is not indicated. Thus, the refinement procedure for λ_n and X_n for problem (11.1.9) is completely specified by relations (11.1.13)–(11.1.15). This procedure can be continued in a recurrent manner on the basis of the following scheme which endures accelerated (quadratic) convergence in ε_n:

$$\begin{aligned}
&\lambda_n^{(j+1)} = \lambda_n^{(j)} - \varepsilon_n^{(j)} A_n(\lambda_n^{(j)}), \quad j = 0, 1, 2, \ldots, \\
&A_n = X_n'^2(\xi_n^{(j)}, \lambda_n^{(j)}) \|X_n^{(j)}\|^{-2}, \quad \xi_n^{(j)} = \operatorname*{Arg}_{n\,x} X_n(x, \lambda_n^{(j)}), \; 0 \le x \le \xi_n^{(j)}, \\
&X_n'' + (\lambda_n^{(j)} r - \beta q) X_n = 0, \quad X_n(0) = 0, \; X_n'(0) = 1, \\
&S_n'' + (\lambda_n^{(j)} r - \beta q) S_n = -r X_n, \quad S_n(0) = S_n'(0) = 0, \\
&1 - \xi_n^{(j)} = \varepsilon_n^{(j)} \sim (c_n \varepsilon_n)^{\kappa_j}, \quad \kappa_j = 2^j, \quad c_n = \text{const} \sim 1.
\end{aligned} \tag{11.1.16}$$

Numerous calculations show the efficiency of the method of continuation in the parameters μ_1, μ_2, since for $\mu_2 = \mu_1$ $(-1 < \mu_1 < 1)$ the exact solution of the Sturm–Liouville problem (11.1.9) is known,

$$\lambda_n(\mu_1, \mu_1) = (\pi n)^2(1 - \mu_1^2), \quad X_n = \frac{\sqrt{1-\mu_1^2}}{\sqrt{\lambda_n}} \sin \frac{\sqrt{\lambda_n}\, x}{\sqrt{1-\mu_1^2}} = \frac{\sin \pi n x}{\pi n}. \tag{11.1.17}$$

Note that for $\mu_1 = \mu_2$ (infinitely thin layer), the original problem (11.1.6), (11.1.8) is degenerate and has no physical meaning ($f_n^2 = \infty$, see (11.1.9), (11.1.10)). As the initial approximation $\lambda_n(\mu_1, \mu_2)$, for sufficiently small $(\mu_2 - \mu_1)$, we take the values $\lambda_n^0 = \lambda_n(\mu_1, \mu_1)$ obtained by (11.1.17).

The scheme (11.1.16) was used to refine the value λ_n (11.1.17) for $\mu_2 > \mu_1$, with $\mu_2 - \mu_1$ regarded as a sufficiently small $(10^{-2} \div 10^{-4})$ perturbation of μ_2. After several iterations (usually 2 or 3), we obtain highly precise approximations of the sought quantities λ_n, X_n, $\|X_n\|$ with the error of the order $10^{-8} \div 10^{-9}$. The method proposed here is very efficient and has advantages over the other known methods with regard to speed, precision, stability, and use of resources. The calculation results are described below.

11.1.3. Calculation results and their analysis. Here we describe the results of highly precise calculations of the eigenvalues λ_n, the eigenfunctions $U_n = X_n\|X_n\|^{-1}$ (normalized with the weight r) and their derivatives U_n' in x for the standard Sturm–Liouville problem (11.1.9). These numerical results were obtained by the recurrent algorithm (11.1.16) combined with the method of continuation in the parameters μ_1, μ_2. We calculated λ_n, U_n, U_n', $n = 1, 2, \ldots, 6$, for a wide range of μ_1, μ_2 with a small relative error $10^{-8} \div 10^{-9}$. To test the values obtained for $\lambda_n(\mu_1, \mu_2)$, we insert f_n^2 into equation (11.1.4) and integrate the Cauchy problem for this equation up to $\mu = \mu_2$, the initial conditions being $Z_n(\mu_1) = 1$, $Z_n'(\mu_1) = 0$; the discrepancy obtained is of the order $\sim 10^{-8}$. Recall that in view of (11.1.5), (11.1.9), the functions U_n' determine the sought functions $Z_n(\mu)$.

Figures 11.1–11.8 show the eigenvalues $\lambda_{1,2,4}$ and the functions $U_{1,2,4}'$ for μ_1, μ_2 on the interval $-0.99 \le \mu_1 \le \mu_2 \le 0.99$.

The graphs in Figs. 11.1, 11.6 show the surfaces $\lambda_{1,4}(\mu_1, \mu_2)$ in linear perspective. To clarify the picture, the level lines of that surface are drawn to the left of the diagonal with step 2 in Fig. 11.1 and with step 20 in Fig. 11.6.

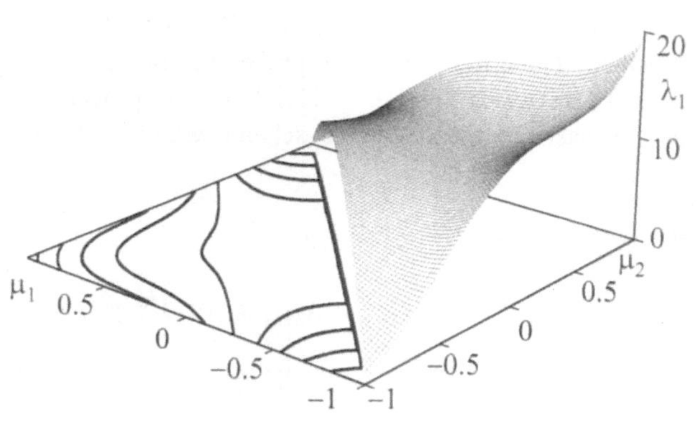

Fig. 11.1

Figures 11.2, 11.4, 11.7 show $\lambda_{1,2,4}(\mu_1)$ as functions of μ_1, $-0.99 \le \mu_1 \le \mu_2$, for a fixed value of μ_2 indicated below the graphs. The surface $\lambda_1(\mu_1, \mu_2)$ (Fig. 11.1) has a fairly complex structure as compared with the surface $\lambda_4(\mu_1, \mu_2)$ (Fig. 11.6); in particular, its level-lines have inflection points (Fig. 11.2). These properties are less pronounced in the function $\lambda_2(\mu_1, \mu_2)$ (Fig. 11.4). Starting from $n = 4$, the surfaces $\lambda_n(\mu_1, \mu_2)$ are convex upwards (Fig. 11.6, 11.7).

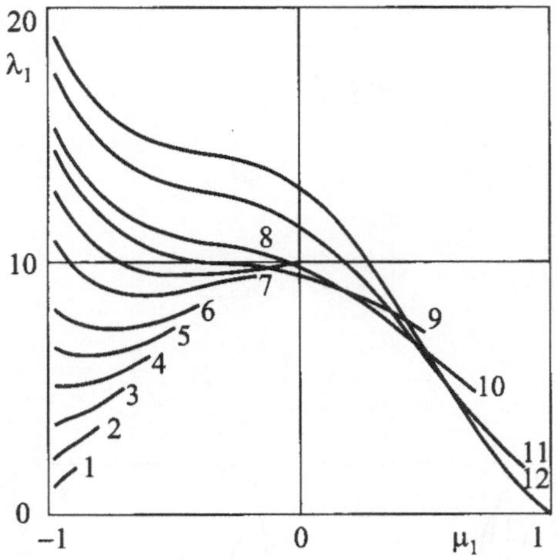

Fig. 11.2

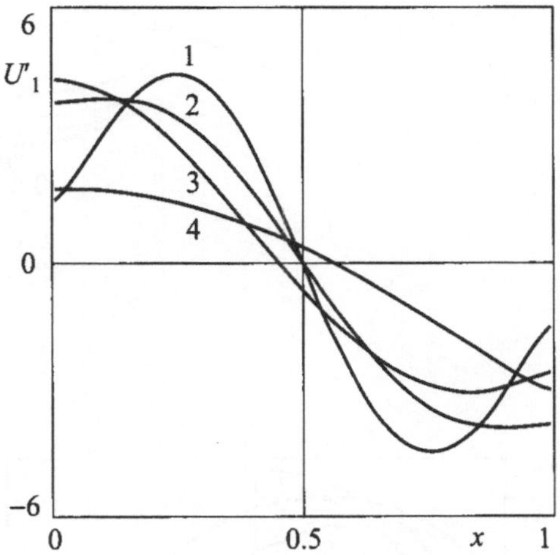

Fig. 11.3

Figures 11.3, 11.5, 11.8 show the functions $U'_{1,2,4}(x, \mu_1, \mu_2)$ corresponding to $\lambda_n(\mu_1, \mu_2)$ constructed for characteristic values of μ_1, μ_2. Curve 1 is constructed for $(\mu_1, \mu_2) = (-0.99; 0.99)$, which corresponds to a region symmetrical "from the north pole to the south pole". Curve 2 is constructed for the values $(-1/\sqrt{2}, 1/\sqrt{2})$ and corresponds to a symmetrical relatively narrow "near-equatorial" region. Nonsymmetric regions are represented by curves 3 and 4. Curve 3 is constructed for the values $(0; 0.99)$ and corresponds to the region "from the equator to the pole". Curve 4 is constructed for the values $(1/\sqrt{2}; 0.99)$ and corresponds to a relatively narrow "polar" region. In conclusion, we perform comparative qualitative analysis of these curves characterizing the vibration shapes Z_n for relative elevations ζ_n.

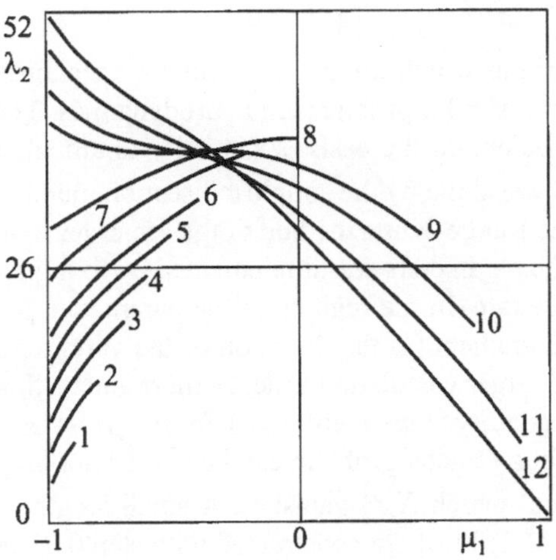

Fig. 11.4

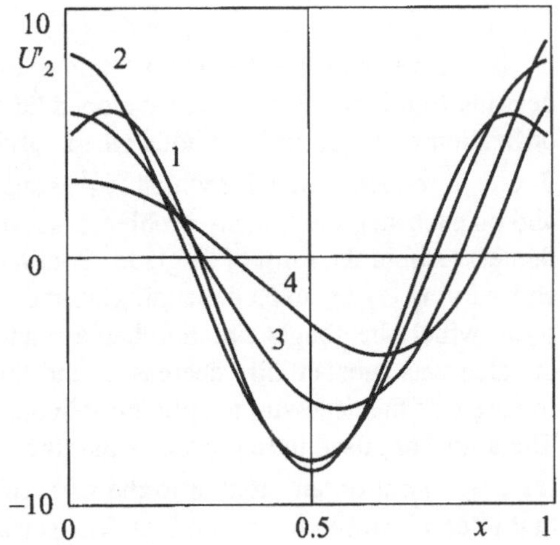

Fig. 11.5

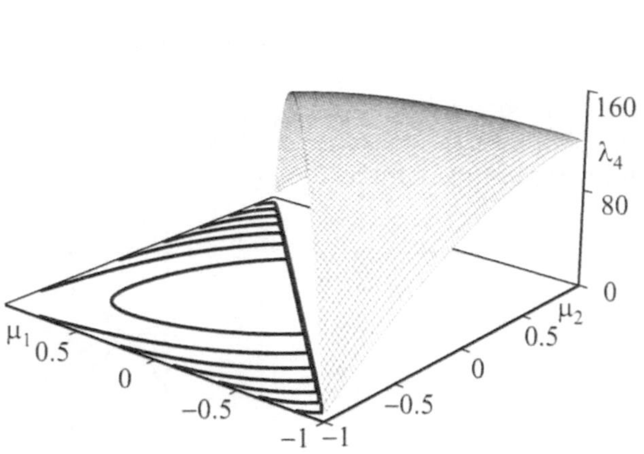

Fig. 11.6

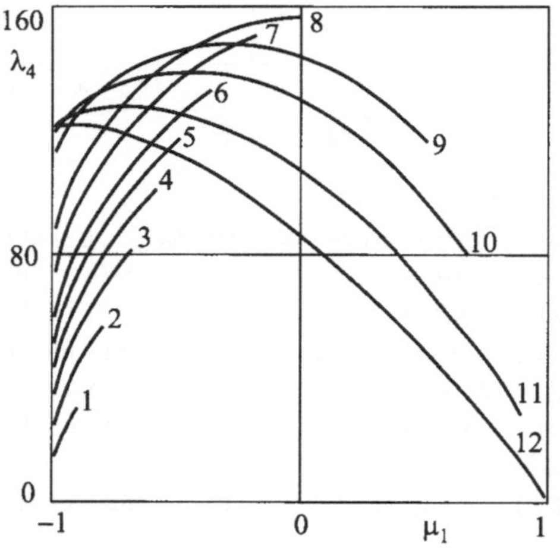

Fig. 11.7

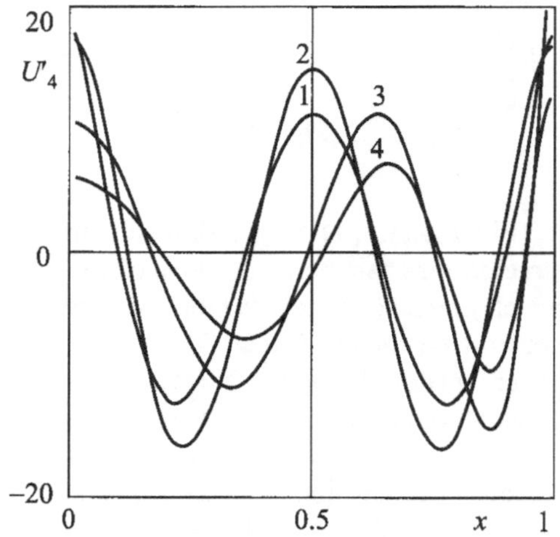

Fig. 11.8

Next, we point out the main features of the algorithm which might be useful for specialists desiring to utilize the method developed here for solving similar problems in hydrodynamics. The utilization of numerical methods when solving the problem on the basis of (11.1.16) is aimed at finding a sequence of x for which $X_n^{(j)}$ vanishes. This was done by the simplest bisection method, and on each step the Cauchy problem was solved by the Runge–Kutta method of the 4th order with the test term in the form of England. Continuation in the parameters was implemented by fixing μ_1 and varying μ_2 by steps depending on the convergence rate. In the regions of the parameters μ_1, μ_2 in which the sought function had a relatively large gradient (in the direction of the variation), the step was substantially decreased, and this entailed larger calculation time. A more interesting feature was the necessity to split the step due to the singularity at the point $x = 1$ for $\mu_1, \mu_2 \to \pm 1$. The algorithm used here presumes that the corresponding Cauchy problem can be solved not only for $x \leq 1$, but on an interval to the right of the point at which $X_n^{(j)}$ vanishes. A small length of that interval caused the said effect. The surfaces (Fig. 11.1, 11.6) are constructed with step 0.01 in both variables, which required about 5000 values of $\lambda_{1,4}(\mu_1, \mu_2)$ to be calculated. Of course, the actual number of the points obtained is at least by an order greater, because calculations have to be performed in a neighborhood of the said singular points.

Thus, in a region fairly distant from the singular points (in our case, the distance is equal to 0.01 in the region of the parameters μ_1 and μ_2) and for relatively small gradients of the sought function in the procedure of continuation in the parameters, the method of accelerated convergence utilizes a fairly simple algorithm, requires nearly the same memory resources as those needed for solving the Cauchy problem of double dimension (compared with the dimension of the original problem), and ensures high precision (of the same order as the solutions of the said Cauchy problem).

In Subsections 11.1.2 and 11.1.3, we give a complete numerical-analytical solution of the Hough problem regarding symmetric wave motions of a rotating spherical layer for different values of the determining parameters θ_1, θ_2, ω, g, h, a. We examine the lower vibration modes, $n = 1, 2, \ldots$, most important for hydrodynamical applications. Our calculations were performed for modes $n = 1, 2, \ldots, 6$, and the graphs were constructed for $n = 1, 2, 4$, which gives a fairly complete picture of the frequencies and shapes of free vibrations. Next, we list some of their basic properties.

Property 1. Our investigation of lower eigenfrequencies σ_n shows that at a fair distance from the poles they are mainly determined by the angular width of the layer, $\delta\theta$, the index of the harmonic n, and the parameter $(gh/a^2)^{1/2}$, provided that $\delta\theta$ is not vary large; and we have $\sigma_n \sim 1/\delta\theta$ (see (11.1.11)). The dependence of the lower eigenfrequencies on the rotation speed ω and the latitudes of the northern θ_2 and the southern θ_1 boundaries is weaker (see Subsection 11.1.2). If the layer is close to the poles ($\theta_2 \to 0$ or $\theta_1 \to \pi$), the increase of $\sigma_n \to \infty$ for $\delta\theta \to 0$ becomes somewhat weaker, however, the tendency remains, which follows from highly precise calculations and the graphs (Figs. 11.1, 11.2, 11.4, 11.6, 11.7). An increase of the layer width or its shift toward the poles lead to a decrease of the eigenfrequencies, which seems natural. More delicate effects characterizing the dependence of the eigenfrequencies on the said parameters are related to the regions of convexity, inflection, and other geometrical properties of the surfaces $\lambda_n(\mu_1, \mu_2)$ in the said graphs.

Property 2. A fairly complete qualitative analysis of the vibration shapes Z_n is hardly possible, since there are too many determining parameters. However, calculations performed for characteristic values of the parameters μ_1, μ_2 allow us to make some qualitative conclusions. Thus, the number of null-points of the functions Z_n characterizing the relative elevation ζ_n correspond to the mode index. For the fundamental mode, $n = 1$, Fig. 11.3 shows that the elevation ζ_1 is not very large near $x^* \approx 1/2$, and therefore, $\mu^* = \cos\theta^* \approx 1/2(\mu_2 + \mu_1)$. In the case of symmetric constraints ($\mu_1 = -\mu_2$, curves 1, 2) this relation is strict, i.e., $\theta^* = \pi/2$ (the elevation ζ_1 is equal to zero on the equator). Extremal values of the elevation ζ_1 occur on the intervals between the equator and the borders. For $\mu_1 \to -1$, $\mu_2 \to 1$, the absolute value of the elevation sharply decreases near $\mu \approx \mu_1$, $\mu \approx \mu_2$. For small $\mu_2 = -\mu_1$ (a narrow layer), largest deviations take place almost on the boundaries (cosine half-wave, curve 2), and this agrees with the above analysis. In the case of essentially asymmetric constraints (curves 3, 4), the elevation extremal values are shifted to the boundaries (become closer to a pole, curve 4).

The second vibration mode (see Fig. 11.5) corresponds to two zero values of relative displacements ζ_2 and three extremal points near $x^* \approx 1/2$, $x^* \approx 0$, $x^* \approx 1$. Relatively narrow symmetric constraints yield a complete cosine wave (curve 2). Small relative displacements take place in the middle of the layer (curve 1) or nearer a pole (curves 3 and 4 correspond to nonsymmetric constraints).

Further increase of the mode index n results in a larger number of zeroes and extremal points of displacements, which grow faster than previously near the poles in the case of nonsymmetric constraints (see Fig. 11.8, curves 3 and 4). Relatively narrow symmetric (and nonsymmetric) constraints result in simple vibration shapes (close to cosine waves).

Conclusion. The method of accelerated convergence allowed us to calculate the sought characteristics with very high precision (the relative error is of the order $10^{-8} \div 10^{-9}$). Basic properties of fluid vibrations, with regard to their dependence on the said parameters, were investigated in detail. In our study, some qualitative effects were detected and described. The results obtained can be used as a reference of exact solutions in calculations for problems of hydrodynamic and atmospheric tides in a more general setting, in particular, problems taking into account tesseral harmonics, variable layer thickness, and other types of boundary conditions [21, 26, 57].

11.2. Internal Waves in Essentially Inhomogeneous Fluids

Wave processes in stratified fluids are of great theoretical and practical interest for marine hydrodynamics. Their properties differ greatly from those of surface waves in homogeneous fluids, since sea water has a complex stratified structure in the vertical direction [26]. For definiteness, we consider a class of horizontal travelling waves for two cases of stratification.

11.2.1. Statement of the problem and some mathematical aspects of its solution. We consider two-dimensional wave motions of a continuously stratified fluid layer which is infinite in the direction of the x-axis and has density $\rho = \rho(y)$. The fluid is in uniform gravity field with acceleration g. Here, (x, y) is a Cartesian reference frame, x is the horizontal coordinate, y is the vertical coordinate, $y = 0$ is the upper boundary. The layer has depth h and its upper and lower boundaries are assumed absolutely rigid. The problem is to find eigenfrequencies and shapes of internal waves propagating in this fluid layer..

For the description of two-dimensional wave motions of the stratified fluid one introduces the flow function $\Psi = \Psi(x, y, t)$ (plane problem). In the Boussinesq approximation, the function Ψ satisfies the equation [26]

$$\frac{\partial^2}{\partial t^2}\left(\frac{\partial^2 \Psi}{\partial x^2} + \frac{\partial^2 \Psi}{\partial y^2}\right) + N^2(y)\frac{\partial^2 \Psi}{\partial x^2} = 0;$$
$$|x| < \infty, \quad -h < y < 0, \quad t \geq t_0; \quad N^2(y) = -g\frac{\rho'(y)}{\rho(y)}. \tag{11.2.1}$$

Here, $N^2(y)$ is the squared Brent–Väisälä frequency depending on the vertical coordinate y, $N^2(y) > 0$. The fluid is assumed stably stratified in the sense that the density $\rho(y)$ increases as y decreases, i.e., $\rho'(y) < 0$.

The flow function Ψ should also satisfy certain boundary and initial conditions. In particular, on the rigid boundaries of the layer, $y = 0$, $y = -h$, the non-penetration conditions should hold,

$$\Psi(x, 0, t) = \Psi(x, -h, t) = 0. \tag{11.2.2}$$

The dependence of Ψ on x, t is chosen in a special manner. We seek a particular solution of equation (11.2.1) with the boundary conditions (11.2.2) in the form of travelling waves

$$\Psi(x, y, t) = \Psi(y) \exp i(\omega t - kx), \tag{11.2.3}$$

where i is the imaginary unit, ω is the frequency, k is the wave number. The function $\Psi(y)$ is unknown and is to be found. Substituting the expression (11.2.3) into equation (11.2.1) and taking into account boundary conditions (11.2.2), we come to the following Sturm–Liouville problem for the unknown ω, Ψ (the wave number k is assumed given):

$$\Psi'' + k^2[\omega^{-2}N^2(y) - 1]\Psi = 0, \qquad \Psi(0) = \Psi(-h) = 0. \tag{11.2.4}$$

Here, the prime denotes the derivative of Ψ in y. For our analysis of the boundary value problem (11.2.4) it is convenient to pass to the dimensionless argument $y^* = -y/h$, the function $\Psi^*(y^*) = \Psi(y)$ (in the sequel, the subscript $*$ is dropped for the sake of brevity), and the parameters μ and p. Thus, we come to a self-conjugate boundary value problem for eigenvalues μ and eigenfunctions Ψ, with a given value of the parameter p, i.e., we have the Sturm–Liouville problem

$$\begin{gathered} \Psi'' + (\mu N^2(y) - p^2)\Psi = 0, \qquad \Psi(0) = \Psi(1) = 0, \\ 0 < y < 1, \quad \mu = p^2\omega^{-2}, \quad p^2 = k^2h^2, \quad 0 < \mu, p^2 < \infty. \end{gathered} \tag{11.2.5}$$

Problem (11.2.5) consists in finding the eigenvalues μ_n (the frequencies ω_n, $\omega_n^2 = \mu_n/p^2$) and the functions $\Psi_n(y) = \Psi(y, \mu_n)$ depending also on the parameter p^2 which may take arbitrary nonnegative values. Of special interest for hydrodynamics are the lower eigenvalues and eigenfunctions, i.e, μ_1, μ_2, $\ldots$ and Ψ_1, Ψ_2, $\ldots$ for relatively small p. For the functions $N^2(y)$ (Brent–Väisälä frequencies), it is extremely difficult to construct highly precise approximations of eigenvalues and eigenfunctions of problem (11.2.5). For these calculations, there exist several methods, some of which are very complex and difficult in implementation (see [21, 28, 45]).

Computational experience shows that the most simple and natural approach (based on the Rayleigh principle or the Rayleigh–Ritz method; see [28, 37]) quite often gives acceptable results, namely, fairly precise upper bounds for eigenvalues and approximations of the corresponding eigenfunctions. Krylov [37] proposed a refinement of the Rayleigh–Ritz method which gives an *a priori* estimate of the error. It was shown that the approximate solutions of the Strum–Liouville problem converge in the corresponding metric to its exact solution, and an estimate was obtained for the rate of this convergence; see Section 1.4.

It should be mentioned that for applications it is important to construct highly precise lower bounds for the eigenvalues μ_1, μ_2, ..., and thereby, obtain acceptable two-sided estimates. With regard to the hydrodynamical problem under consideration, the Rayleigh–Ritz method yields lower bounds for the sought eigenfrequencies ω_1, ω_2, $\ldots$ of internal waves and approximations of the corresponding vibration shapes. For practical purposes, however, it is most interesting to construct highly precise upper bounds for the said higher frequencies (lower vibration modes). Note that the existing methods for obtaining lower bounds of eigenvalues are very cumbersome and ineffective (see, for instance, [28, 45]). These methods are based on approximate solutions of a Fredholm integral equation of the second kind, which is equivalent to the original Sturm–Liouville problem (see [45]).

To obtain a preliminary rough estimate for the eigenvalue μ_1 and a rough approximation of the function $\Psi_1(y)$, the Rayleigh–Ritz approximation with a few test functions is used. In particular, for simple calculations in model examples, we use the Rayleigh principle. This approach allows us to obtain an upper bound for the first eigenvalue μ_1 in terms of simple quadratures. We have (in what follows, the subscript 1 is omitted for the sake of brevity)

$$\begin{gathered} \mu^* = \frac{J[\varphi]}{I[\varphi]}, \qquad \varphi(0) = \varphi(1) = 0, \qquad 0 < \mu \le \mu^*, \\ J[\varphi] = \int_0^1 (\varphi'^2(y) + p^2\varphi^2(y))\,dy, \qquad I[\varphi] = \|\varphi\|_{N^2}^2 = \int_0^1 N^2(y)\varphi^2(y)\,dy. \end{gathered} \tag{11.2.6}$$

Here, $\varphi(y)$ is a continuously differentiable test function which is often chosen fairly roughly on the basis of an intuitive physical idea of the first vibration mode, in particular, $\varphi(y) \neq 0$ for $0 < y < 1$. Acceptable results are obtained if one takes, for instance, $\varphi(y) = \sin \pi y$ or $\varphi(y) = y(1-y)$, provided that $N^2(y)$ is symmetric with respect to the value $y = 1/2$ or $N^2(y)$ has slow variation. A more regular choice of the test function may be connected with the method of continuation in a parameter (see below).

Note that $\mu = \mu^*$ only if $\varphi(y) = C\Psi(y)$, where $C = \text{const}$ and $\Psi(y)$ is the first eigenfunction. The estimate (11.2.6) easily follows from the variational statement of the Sturm–Liouville problem (11.2.5)

$$\min J[\Psi] = \mu, \quad I[\Psi] = 1, \quad \Psi(0) = \Psi(1) = 0, \tag{11.2.7}$$

where the functionals $J[\Psi]$, $I[\Psi]$ are those defined in (11.2.6).

As mentioned above, the estimate (11.2.6) does not allow us to decide whether the values μ and μ^* are close. It seems important to study the following questions.

1. Establish a criterion of closeness and determine whether the upper bound μ^*, $\mu \le \mu^*$, obtained by the Rayleigh–Ritz method or the Rayleigh principle (11.2.6) can be regarded as a good approximation of the eigenvalue μ.
2. Find an effective method for the construction of lower bounds μ_*, $\mu \ge \mu_*$.
3. Find methods for refining the bounds μ^*, μ_*.
4. Develop a constructive method (see Chapter 2) and write an algorithm for obtaining consecutive refinements of the bounds, so that μ could be approximated with any given precision (to obtain the exact value in the limit).
5. Extend these results to subsequent eigenvalues μ_n and eigenfunctions Ψ_n, $n \ge 2$.
6. Investigate the possibilities of effectively solving the Sturm–Liouville problem with boundary conditions of the second or the third kind. Such boundary conditions occur in geophysical problems and correspond to a fluid layer with a free upper boundary [31].

11.2.2. A version of the perturbation method for approximate solution of the Sturm–Liouville Problem. Suppose that an upper bound μ^* (11.2.6) of the first eigenvalue of problem (11.2.5) is known. Let us insert μ^* into equation (11.2.5) and consider the Cauchy problem

$$v'' + (\mu^* N^2(y) - p^2)v = 0; \qquad v(0) = 0, \quad v'(0) = 1. \tag{11.2.8}$$

Assume that the solution of problem (11.2.8) $v = v(y, \mu^*)$ has been constructed in analytical or numerical form, or a procedure for its determination is known. The function $v(y, \mu^*)$ satisfies the conditions

$$v(0, \mu^*) = 0, \quad v(\xi, \mu^*) = 0, \quad 0 < \xi \le 1. \tag{11.2.9}$$

The equality $\xi = 1$ in (11.2.9) takes place only if $\mu^* = \mu$; for $\mu^* > \mu$, we have $\xi < 1$ due to the Sturm's second oscillation theorem. As a measure of closeness between μ^* and the exact eigenvalue (to characterize the small difference $\mu^* - \mu > 0$), we take value $\varepsilon = 1 - \xi \ge 0$, where ξ is a known quantity determined by (11.2.9) and equal to the smallest positive root of the function v. The smallness of the parameter ε can be ensured by the Rayleigh–Ritz method or by the shooting method.

Together with the Cauchy problem (11.2.8), consider a similar problem with $\mu = \eta\mu^*$, where $\eta = \eta(\varepsilon)$ is a numerical parameter such that $0 < \eta < 1$. The solution $v = v(y, \eta\mu^*)$ of the latter problem is also assumed known in analytical or numerical form. By standard arguments, common in mathematical physics, the following statements can be proved.

Statement 1. *If the function $v(y, \eta\mu^*)$ preserves its sign on the interval $0 < y \le 1$, then $\eta\mu^* = \mu_*$ is a lower bound for the first eigenvalue,*

$$0 < \mu_* \le \mu < \mu^*, \qquad \mu_* = \eta\mu^*. \tag{11.2.10}$$

Statement 2. *If $v(y, \eta\mu^*)$ vanishes at a point $y = \xi^*$, $0 < \xi^* < 1$, then $\eta\mu^*$ is a refined upper bound,*

$$\mu < \eta\mu^* < \mu^*. \tag{11.2.11}$$

Note that for eigenfrequencies ω of internal waves, the following estimates corresponding to (11.2.10) and (11.2.11) hold:

$$1)\ \frac{p^2}{\mu^*} \le \omega^2 \le \frac{p^2}{\mu_*}, \qquad \mu_* = \eta\mu^*; \qquad 2)\ 0 < \omega^2 < \frac{p^2}{\eta\mu^*} < \frac{p^2}{\mu^*}. \tag{11.2.12}$$

Below, we describe some constructive approaches to the determination of the parameter $\eta = \eta(\varepsilon)$ on the basis of regular perturbation methods. In what follows, it is assumed that the following strong inequality holds:

$$0 \le \varepsilon \ll 1, \qquad \varepsilon = 1 - \xi, \qquad \xi = \min \arg v. \tag{11.2.13}$$

Let us transform the Sturm–Liouville problem (11.2.5) to a special perturbed problem (see Chapter 2). To that end, we introduce the independent variable $z = \xi y$, the unknown function $\Phi(z, \varepsilon) \equiv \Psi(y)$, and the parameter $\nu = \xi^{-2}\mu$. We obtain

$$\begin{gathered} \Phi'' + (\nu N^2(z) - p^2)\Phi = \varepsilon H(z,\nu)\Phi + O(\varepsilon^2), \qquad \Phi(0) = \Phi(\xi) = 0; \\ 0 < z < \xi, \qquad H(z,\nu) = -\nu z (N^2(z))' + 2p^2, \end{gathered} \tag{11.2.14}$$

where the prime denotes differentiation in z. The eigenvalue ν and the eigenfunction Φ of problem (11.2.14) will be constructed with the help of a regular procedure of expansion in powers of ε, or by the method of successive approximations. Thus, consider the expressions

$$\nu = \nu(\varepsilon) = \mu^* + \varepsilon\nu_1 + \cdots, \qquad \Phi = \Phi(z,\varepsilon) = v(z,\mu^*) + \varepsilon\Phi_1(z) + \cdots \tag{11.2.15}$$

The value $\mu^* = \nu(0)$ and the function $v(z,\mu^*) = \Phi(z,0)$ represent a solution of the corresponding unperturbed (generating) problem (11.2.14) for $\varepsilon = 0$. For the unknown quantities ν_1, Φ_1 we have a nonhomogeneous boundary value problem obtained by substituting the expansions (11.2.15) into (11.2.14) and equating the coefficients of ε,

$$\begin{gathered} \Phi_1'' + (\mu^* N^2 - p^2)\Phi_1 = -\nu_1 N^2 v - \mu^* z {N^2}' v + 2p^2 v, \\ \Phi_1(0) = \Phi_1(\xi) = 0, \qquad N^2 = N^2(z), \qquad 0 \le z \le \xi. \end{gathered} \tag{11.2.16}$$

According to the Fredholm alternative, problem (11.2.16) has a solution if and only if the right-hand side of the equation is orthogonal to the eigenfunction $v(z,\mu^*)$ (since the homogeneous problem is self-conjugate). If the problem has a solution, this orthogonality property is obtained by multiplying equation (11.2.16) by v, integrating the result over the interval $0 \le z \le \xi$, and using equation (11.2.8) for v. In this way, we obtain the desired expression for the coefficient ν_1 in (11.2.15),

$$\nu_1 = -\frac{\mu^*}{\|v\|_{N^2}^2} \int_0^\xi \left[z(N^2(z))' - 2\frac{p^2}{\mu^*} \right] v^2(z,\mu^*)\,dz, \qquad \|v\|_{N^2}^2 = \int_0^\xi N^2(z) v^2(z,\mu^*)\,dz. \tag{11.2.17}$$

Using this expression for ν_1, we obtain the sought function $\Phi_1(z)$ as a solution of problem (11.2.16). A similar procedure can be used for finding subsequent coefficients ν_k, Φ_k, $k \ge 2$, in the expansions (11.2.15). However, this approach to refinement is very cumbersome and inefficient from the standpoint of calculations. In practice, it often suffices to construct only the first approximations for ν, Φ, i.e., for μ, Ψ. Note that without the loss of precision in ε, we can take $\xi = 1$ in (11.2.17). The first approximation of the corresponding eigenfunction is obtained by integrating the Cauchy problem (11.2.8) with the refined value of ν or μ (see below).

Restricting ourselves to the first approximation of ν (11.2.15), we obtain the following expression for the approximation of the eigenvalue μ with the error $O(\varepsilon^2)$:

$$\mu = \xi^2\mu^* + \varepsilon\nu_1\xi^2 + \cdots = \mu^* + \varepsilon(\nu_1 - 2\mu^*) + \cdots. \tag{11.2.18}$$

From (11.2.18), it follows that for small enough $\varepsilon > 0$, Statement 1 holds (i.e., the inequalities (11.2.10)) for $\eta(\varepsilon) = \xi^2$, $\xi = 1 - \varepsilon$, provided that $\nu_1 > 0$ (see (11.2.17)). This condition holds if the integrand in the expression for ν_1 is nonpositive. In particular, if $N'(y) \le 0$, $0 \le y \le 1$, this condition holds and $\mu_* = \xi^2\mu^*$ is a lower bound. This can be claimed *a priori*, without performing the integration in (11.2.17). In a similar way, one establishes Statement 2, i.e., the inequalities (11.2.11) for $\nu_1 < 0$ and $\eta(\varepsilon) = \xi^2$. Thus, the strict inequalities $\nu_1 > 0$, $\nu_1 < 0$ ensure the corresponding estimates 1), 2) in (11.2.12) for the highest eigenfrequency ω ($\omega^2 = p^2/\mu^2$).

Let us apply the method of accelerated convergence for approximate solution of the Sturm–Liouville problem. Using this highly effective numerical–analytical method, we calculate the eigenvalues (eigenfrequencies) and eigenfunctions (vibration shapes) of the Sturm–Liouville problem. These eigenvalues and eigenfunctions determine internal wave motions of the stratified fluid. The corresponding algorithm possesses the property of accelerated (quadratic) convergence, similarly to the Newton's method of tangential lines. The essence of this approach consists in consecutively using the refined value of μ determined by (11.2.18), integrating the Cauchy problem (11.2.8), and finding the abscissa $\xi > 0$ (see Chapter 2).

Formula (11.2.17) requires computational resources, because it is constructed by integrating functions determined in the process of solution (usually numerical) of the Cauchy problem. The quadrature (11.2.17) can be substantially simplified and reduced to canonical form with the help of integration by parts (see Sections 2.3 and 2.4),

$$\mu = \mu^* - \varepsilon\xi v'^2(\xi, \mu^*)\|v\|_{N^2}^{-2} + O(\varepsilon^2) \equiv \mu_{(1)} + O(\varepsilon^2). \tag{11.2.19}$$

The squared norm $\|v\|_{N^2}^2$ of the function v with the weight N^2 is defined in (11.2.17). One can take $\xi = 1$ in (11.2.19) without any loss in precision with respect to ε.

Inserting the refined value $\mu_{(1)}$ into equation (11.2.8) instead of μ^*, let us integrate the Cauchy problem. Relations (11.2.9) will contain $\mu_{(1)}$ and the unknown ξ equal to $\xi_{(1)} = 1 - \varepsilon_{(1)}$, where $\varepsilon_{(1)} = O(\varepsilon^2)$, since $v'(1, \mu^*) \ne 0$. Let us calculate the refined value $\mu = \mu_{(2)} + \varepsilon_{(1)}...$, where ε and ξ are taken equal to $\varepsilon_{(1)}$ and $\xi_{(1)}$, respectively; and without the loss of precision with respect to $\varepsilon_{(1)}$ in (11.2.19), we can take $\xi_{(1)} = 1$. From the estimate $\varepsilon_{(1)} = O(\varepsilon^2)$, it follows that $\mu - \mu_{(2)} = O(\varepsilon_{(1)}^2)$, i.e., the error is of the order $O(\varepsilon^4)$.

The process of consecutively refining the value μ and the function $v(y, \mu)$ may be infinitely continued, so that on the kth step, the error of the approximations is of the order $O(\varepsilon_{(k)}) = O(\varepsilon^{\theta_k})$, where $\theta_k = 2^k$, $k = 1, 2, \ldots$ The formal algorithm of the recurrent process of calculation of the eigenvalues and eigenfunctions is described by the following relations (see (11.2.19), (11.2.13), (11.2.8), (11.2.9)):

$$\begin{aligned} &\mu_{(k+1)} = \mu_{(k)} - \varepsilon_{(k)}\xi_{(k)}v'^2(\xi_{(k)}, \mu_{(k)})\|v_{(k)}\|_{N^2}^{-2}, \qquad k = 0, 1, 2, \ldots, \\ &v'' + (\mu_{(k+1)}N^2(y) - p^2)v = 0, \qquad v(0) = 0, \quad v'(0) = 1, \\ &\varepsilon_{(k+1)} = 1 - \xi_{(k+1)}, \qquad \xi_{(k+1)} = \min \arg_y v(y, \mu_{(k+1)}) > 0. \end{aligned} \tag{11.2.20}$$

The scheme (11.2.20) is resource sparing, for it requires only the integration of the Cauchy problem (11.2.19) for $\mu = \mu_{(k)}$ and the calculation of the smallest positive root $\xi_{(k)}(\mu_{(k)})$ of the function $v(y, \mu_{(k)})$ nearest to $\xi = 1$. The squared norm $\|v_{(k)}\|_{N^2}^2$ can be calculated by some efficient numerical method (for instance, the Simpson method) with a sufficiently small step in y. Note that without the loss of precision with respect to $\varepsilon_{(k)}$, one can take $\xi_{(k)} = 1$ in formula (11.2.20) for $\mu_{(k+1)}$. The calculation of the squared norm can be replaced by the integration of the Cauchy problem for the function $w(y, \mu_{(k)}) = \partial v/\partial\mu$,

$$\|v_{(k)}\|^2_{N^2} = \int_0^{\xi_{(k)}} v^2(y, \mu_{(k)}) N^2(y)\, dy = v'(\xi_{(k)}, \mu_{(k)}) w(\xi_{(k)}, \mu_{(k)}),$$

$$w'' + (\mu_{(k)} N^2(y) - p^2) w = -N^2(y) v(y, \mu_{(k)}), \qquad w = \frac{\partial v}{\partial \mu}, \tag{11.2.21}$$

$$w(0) = w'(0) = 0, \qquad v'^2(\xi, \mu)\|v\|^{-2}_{N^2} = v'(\xi, \mu) \left(\frac{\partial v}{\partial \mu}\right)^{-1}.$$

Computational practice shows that for most applications it suffices to consider a single iteration in ε (11.2.19), provided that the test function has been chosen properly (for $\varepsilon \lesssim 0.1$). Two or three iterations of the process (11.2.20), (11.2.21) yield highly precise approximations (the error being of the order $10^{-4} \div 10^{-8}$). Four or more iterations yield practically precise values attainable for modern computers (errors of the order 10^{-16} or even less). The simplicity of implementation of this recurrent procedure allows one to obtain acceptable results on programmable micro computers.

Note that the approach developed here is effective also in calculations of subsequent eigenvalues μ_n and eigenfunctions $\Psi_n(y)$, $n \geq 2$. A method of quadratic convergence can also be developed for boundary value problems with boundary conditions of the second or the third kinds, as well as mixed boundary conditions.

11.2.3. Calculations for some specific stratified fluids. Next, we describe calculations for two cases of stratified fluid. These examples demonstrate the efficiency of our algorithm for the calculation of the first mode of internal waves.

1. Suppose that the Brent–Väisälä frequency in (11.2.1) is modelled by the function $N^2(y) = N_0^2 \exp(2\beta y)$, where $-h \leq y \leq 0$. Introducing the dimensionless variable $y^* = -y/h$ and dropping the superscript $*$, we obtain the following Sturm–Liouville problem corresponding to (11.2.5)

$$\begin{gathered} \Psi'' + (\mu \exp(-2\gamma y) - p^2)\Psi = 0, \qquad \Psi(0) = \Psi(1) = 0; \\ \gamma = \beta h, \qquad \mu = p^2 N_0^2 \omega^{-2}, \qquad 0 \leq y^* \leq 1. \end{gathered} \tag{11.2.22}$$

This problem can be solved analytically in terms of cylindrical functions [32, 65, 67]. This, however, requires lengthy calculations of the first eigenvalue μ (and the frequency ω). Using one iteration from (11.2.18), we obtain the following estimates for the frequency ω for $\gamma = \ln 2 = 0.6931$ (and give its exact tabular values) and the corresponding values of p:

$$\begin{aligned} &0.0788 < \tfrac{\omega}{N_0} = 0.0795 < 0.0796, \quad p = \tfrac{\gamma}{2}, \quad \varepsilon = 0.0148, \\ &0.1547 < \tfrac{\omega}{N_0} = 0.1562 < 0.1564, \quad p = \gamma, \quad \varepsilon = 0.0160, \\ &0.2899 < \tfrac{\omega}{N_0} = 0.2942 < 0.2946, \quad p = 2\gamma, \quad \varepsilon = 0.0566. \end{aligned} \tag{11.2.23}$$

Thus, the estimates obtained are in agreement with tabular values of cylindrical functions. The left-hand inequalities (the upper estimates for μ and the lower estimates for ω) follow from the Rayleigh principle (with the test function $\varphi(y) = \sin \pi y$). The right-hand inequalities (the lower estimates for μ and the upper estimates for ω) were obtained on the basis of (11.2.18) by the algorithm proposed here. Thus, we have established two-sided estimates (11.2.10), (11.2.12), since $\nu_1 > 0$, because $N'(y) < 0$, $0 \leq y \leq 1$. From the two-sided estimates (11.2.23) it follows that the calculation error does not exceed 1%.

2. In contrast to the above case of monotone dependence of the Brent–Väisälä frequency on the depth, consider the case of $N^2(y)$ with an extremal point, say, a minimum. We are going to study problem (11.2.5) of the form

$$\Psi'' + (\mu(1 - \alpha \sin \pi y)^{-1} - p^2)\Psi = 0, \qquad \Psi(0) = \Psi(1) = 0, \tag{11.2.24}$$

where $0 \leq \alpha < 1$, $\mu = p^2 N_0^2/\omega^2$. An approximate analytical solution of problem (11.2.24) can be constructed for $\alpha \ll 1$. Figure 11.9 shows the dispersion curves $\chi = \omega(p,\alpha)/N_0$ for $\alpha = 0, 0.5$, and 0.8 (curves 1, 2, and 3, respectively) as $p = kh$ varies from 0 to 5. The relative calculation error does not exceed 0.5%. It should be observed that to ensure this precision for $0 \leq p \leq 2$, two iterations were needed, and three iterations for $2 < p \leq 5$.

Using the relation between ω, p, and μ, we obtain an approximate expression for ω for $p \to 0$ (the case of long waves): $\chi = \omega/N_0 = p\mu^{-1/2}\ (0,\alpha) + qp^3 + \cdots$. The constant $q < 0$ can be effectively calculated by the perturbation theory methods, by expanding μ and Ψ in powers of p^2.

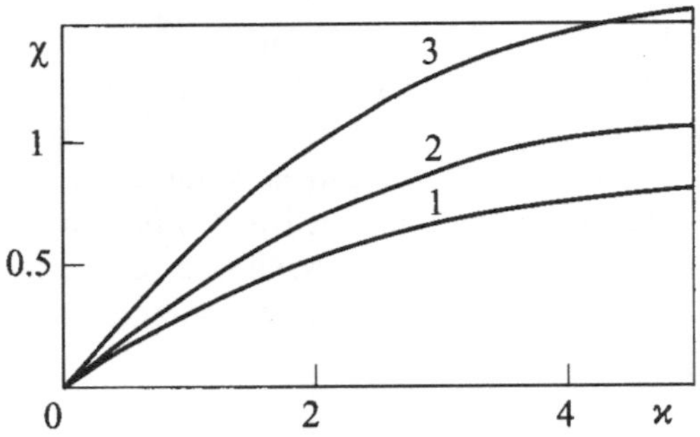

Fig. 11.9

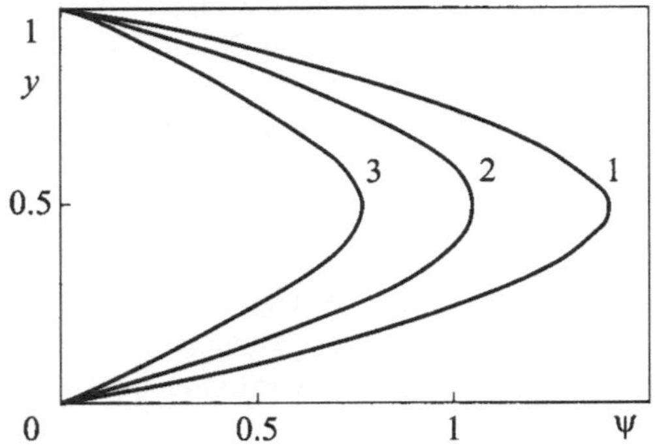

Fig. 11.10

Figure 11.10 shows normalized eigenfunctions $\Psi(y,\alpha,p)$, which are symmetric with respect to the point $y = 1/2$, for $\alpha = 0$, 0.5, and 0.8 (curves, 1, 2, and 3, respectively) for $p = 0.6$, 2.0. As found in the process of calculations, the eigenfunctions undergo substantial variation with respect to the parameter α. The dependence of these functions on p, for $p \sim 1$, is much weaker. In particular, for $\alpha = 0$, there is no such dependence. For $\alpha > 0$, the dependence on p, for $0 \leq p \leq 3$, is insignificant, and for this reason it is not reflected in the graphs. It was found, however, that with the growth of p ($p \gg 1$) for $\alpha > 0$ fixed, the maximum of the function $\Psi(y,\alpha,p)$ in y increases. Observe that to reduce calculations, we have used the method of continuation in the parameters α, p.

Conclusion. The method developed in Chapter 2 for solving the Sturm–Liouville problem can be effectively used for finding eigenfrequencies and shapes of internal waves in a layer consisting of a stably stratified fluid whose density is an arbitrary function of the depth. This conclusion is confirmed by examples with calculations of eigenfrequencies and shapes of internal waves in fluids whose density is described by given functions of the depth. We have established (in dimensionless variables) the dependence of the first eigenfrequency and the first vibration shape on the wave number and the parameter characterizing the maximum of the Brent–Väisälä frequency.

11.3. Exercises

Exercise 1. Formulate and perform a preliminary investigation of the Laplace–Hough problem taking into account tesseral harmonics (see Section 11.1) [21, 57].

Exercise 2. Describe the dependence of the first eigenfrequency on the wave number if the squared Brent–Väisälä frequency is approximated by the expression $N^2(y) = N_0^2 \cosh \gamma y$ (see Section 11.2).

Chapter 12

Parametric Vibrations of One-Dimensional Systems

There is a huge number of papers and monographs dedicated to the investigation of parametric vibrations and parametric instability [13, 16, 19, 22, 33, 42, 43, 49, 64, 68]. Almost all these works are based on the classical perturbation theory presuming smallness of the modulation depth, and this is an obstacle to obtaining a complete picture of the regions of parametric instability. An exception is provided by the fundamental monographs [43, 61] which describe detailed investigations of the Mathieu equation and the Meissner equation for large modulation depths. There is also a method of infinite determinants [61, 67, 68] which allows one to calculate the boundaries of parametric instability regions (parametric resonances) for arbitrary modulation depths. But this method is fairly cumbersome and its numerical realization requires a lot of preliminary analytical work. Eventually, this method reduces to an algebraic eigenvalue problem. The precision of the results obtained by that method can hardly be estimated. There is also the Floquet theory which, in principle, gives the possibility of reducing systems with periodic coefficients to those with constant coefficients. However, calculations based on the Floquet theory yield tangible results only in the framework of the perturbation theory.

The results described below were obtained by the methods of Chapters 2 and 4. These methods allow us to construct periodic solutions and determine the boundaries of parametric instability regions with any given precision.

12.1. Parametric Vibrations of Systems of Hill's Type

12.1.1. Setting of the problem. We are going to find the boundaries of parametric instability regions for the modified Mathieu equation (see Section 2.9)

$$\ddot{u} + [\lambda - q(t,e)]u = 0, \qquad u = u(t,\lambda,e),$$
$$|\lambda| < \infty, \quad q(t,e) \equiv \frac{e\cos(2\pi t)}{1+e(\cos 2\pi t)}, \quad |e| < 1. \tag{12.1.1}$$

Note that for $|e| \ll 1$, equation (12.1.1) differs from the classical Mathieu equation [33, 43, 61] by terms of the order e^2.

In order to find the function $\lambda(e)$ which determines the boundaries of parametric instability [19, 42, 43], one has to construct periodic solutions of equation (12.1.1), i.e., solutions satisfying the following periodicity conditions for $t = \pm 1$:

$$u(-1) = u(1), \quad u'(-1) = u'(1). \tag{12.1.2}$$

The dependence of the functions on λ, e is not indicated for brevity.

Since the function $q(t,e)$ is even with respect to t, $q(-t,e) \equiv q(t,e)$, the conditions of periodicity (12.1.2) are equivalent to the conditions of the first and the second kind on the half-interval $0 \le t \le 1$. The conditions of the first kind

$$u(0) = u(1) = 0 \tag{12.1.3}$$

correspond to odd functions of t, and the conditions of the second kind

$$u'(0) = u'(1) = 0 \tag{12.1.4}$$

correspond to even functions of t.

The problem of the determination of the boundaries of parametric instability regions is formulated as follows: find the values of the parameter λ for which there exist nontrivial solutions of equation (12.1.1) with the boundary conditions (12.1.3) (odd functions) or (12.1.4) (even function). Thus, the problem of the determination of the boundaries of parametric instability regions in the case of even q reduces to two Sturm–Liouville problem (12.1.1), (12.1.3) and (12.1.1), (12.1.4). These problems will be solved by the method of accelerated convergence (Chapter 2). Naturally, the sought parameter λ is a function of the parameter e, which is called the depth (or the coefficient) of modulation. From (12.1.1), it obviously follows that $0 \le |e| < 1$.

12.1.2. Perturbation method. Before describing our calculation results for the boundaries of parametric instability regions in the case of arbitrary modulation depth e, we give some formulas obtained by the perturbation theory for $|e| \ll 1$. The eigenvalues λ_1^s and λ_2^s in the case of boundary conditions (12.1.3) can be represented by the asymptotic formulas

$$\begin{aligned} \lambda_1^s &= \pi^2 - \frac{e}{2} - \left(\frac{1}{2} + \frac{1}{32\pi^2}\right) e^2 + O(e^3), \\ \lambda_2^s &= 4\pi^2 - \left(\frac{1}{4} + \frac{1}{48\pi^2}\right) e^2 + O(e^4). \end{aligned} \tag{12.1.5}$$

Note that for any $n = 1, 2, \ldots$, we have $\lambda_n^s = (\pi n)^2 + O(e^n)$.

The eigenvalues (12.1.5) correspond to odd eigenfunctions

$$\begin{aligned} u_1^s &= \sin(\pi t) - \frac{e}{16\pi^2} \sin(3\pi t) + O(e^2), \\ u_2^s &= \sin(2\pi t) - \frac{e}{24\pi^2} \sin(4\pi t) + O(e^2). \end{aligned} \tag{12.1.6}$$

The eigenvalues $\lambda_n(e)$ for $n = 0, 1, 2$ in the case of boundary conditions (12.1.4) are approximated by the expressions

$$\begin{aligned} \lambda_0^c &= -\frac{1}{2}\left(1 + \frac{1}{4\pi^2}\right) e^2 + O(e^3), \\ \lambda_1^c &= \pi^2 + \frac{e}{2} - \left(\frac{1}{2} + \frac{1}{32\pi^2}\right) e^2 + O(e^3), \\ \lambda_2^c &= 4\pi^2 - \left(\frac{3}{4} - \frac{5}{48\pi^2}\right) e^2 + O(e^3). \end{aligned} \tag{12.1.7}$$

Just as for λ_n^s, we have $\lambda_n^c(e) = (\pi n)^2 + O(e^n)$. The eigenvalues (12.1.7) correspond to even eigenfunctions

$$\begin{aligned} u_0^c &= 1 - \frac{e}{4\pi^2} \cos(2\pi t) + O(e^3), \\ u_1^c &= \cos(\pi t) - \frac{e}{16\pi^2} \cos(3\pi t) + O(e^2), \\ u_2^c &= \cos(2\pi t) + \frac{1}{8\pi^2}\left(1 - \frac{1}{3}\cos(4\pi t)\right) e + O(e^2). \end{aligned} \tag{12.1.8}$$

Formulas (12.1.5)–(12.1.8) will be used in the sequel for comparing our highly precise results obtained by the method of accelerated convergence. Upon solving boundary value problems (12.1.1), (12.1.3) and (12.1.1), (12.1.4) by the method of accelerated convergence, we have obtained the eigenvalues listed in Table 12.1. The relative calculation error for $\lambda_n^{c,s}$ does not exceed 10^{-6}. Table 12.2 allows one to compare the results obtained by the formulas (12.1.5), (12.1.7) of the perturbation theory with those obtained by the method of accelerated convergence.

Table 12.1

	Odd Solutions				Even Solutions		
e	λ_1^s	λ_2^s	λ_3^s	λ_0	λ_1^c	λ_2^c	λ_3^c
0	π^2	$4\pi^2$	$9\pi^2$	0	π^2	$4\pi^2$	$9\pi^2$
0.1	9.81382	39.47591	88.82140	$-5.167 \cdot 10^{-3}$	9.91492	39.47085	88.82171
0.2	9.74579	39.46821	88.80504	−0.02113	9.95198	39.44793	88.80717
0.3	9.65995	39.45484	88.77484	−0.04963	9.98201	39.40653	88.78224
0.4	9.54980	39.43481	88.72619	−0.09384	10.00582	39.34208	88.74544
0.5	9.40357	39.40623	88.65102	−0.15969	10.02367	39.24482	88.69416
0.7	8.88837	39.30961	88.33677	−0.25909	10.03524	39.09659	88.62277
0.9	7.05716	39.07410	87.01250	−0.71267	10.03384	38.42080	88.36675
0.95	5.09459	38.94050	85.52097	−2.54667	9.97888	35.75942	87.82713

Table 12.2

	Perturbation theory	Method of accelerated convergence
$e = 0.1$	$\lambda_1^s = 9.81457$	$\lambda_1^s = 9.81382$
$e = 0.2$	$\lambda_1^s = 9.74948$	$\lambda_1^s = 9.74579$
$e = 0.1$	$\lambda_1^c = 9.91457$	$\lambda_1^c = 9.91492$
$e = 0.2$	$\lambda_1^c = 9.94948$	$\lambda_1^c = 9.95198$

In a similar way, one can compare the other eigenvalues obtained here. The error of the formulas obtained by the perturbation theory rapidly increases with the growth of the modulation depth e. Comparison with the results obtained by the accelerated convergence method allows us to claim that for small modulation depths, $e \leq 0.2$, the formulas obtained by the perturbation method yield acceptable results.

12.1.3. Numerical-analytical analysis. Note that for small e, equation (12.1.1) in the linear approximation in e coincides with the canonical Mathieu equation. However, if subsequent powers of e are taken into account, the boundaries of instability regions diverge greatly, even for relatively small e. However, the construction of these boundaries by the perturbation method is of little interest. Figure 2.2 gives a graphical representation of the dependence of the eigenvalues on the parameter e. These results were obtained for e varying from $e = 0$ to $e = 0.99$. Thus, the coefficient of the equation is varied by two orders. The graphs of the eigenfunctions $\lambda_n^{s,c}(e)$ issuing from the point $(\pi n)^2$ for $e = 0$ represent the boundaries of parametric instability regions. A characteristic feature of these diagrams is that the only curve $\lambda_0^c(e)$ corresponding to the zero eigenvalue for $e = 0$ becomes asymptotically close to $\lambda_0^c = -\infty$ as $e \to 1$; and we always have the negative value $\lambda_0^c < 0$. Next, we describe the function $\lambda_1^s(e)$. This eigenvalue corresponds to an odd eigenfunction. We have $\lambda_1^s \to \infty$ as $e \to 1$, and these values lie to the left and above the curve $\lambda_0^c(e)$. The behavior of the eigenvalue $\lambda_1^c(e)$ is quite different. Numerical analysis shows that $\lambda_1^c \to \pi^2$ as $e \to 1$. This result was obtained by fairly lengthy analytical calculations connected with the construction of a generalized solution. There is a region of instability between $\lambda_1^s(e)$ and $\lambda_1^c(e)$.

For $n = 2$, the curve $\lambda_2^c(e)$ is on the left and the curve $\lambda_2^s(e)$ is on the right. The behavior of these curves for $e \to 1$ cannot be accurately predicted on the basis of the data obtained in our calculations. Again, for $n = 3$, the curve $\lambda_3^s(e)$ is on the left and the curve $\lambda_3^c(e)$ is on the right. The general rule claims that for odd $n = 2k+1$, we always have $\lambda_{2k+1}^s(e)$ on the left and $\lambda_{2k+1}^c(e)$ on the right. For even $n = 2k$, the curve $\lambda_{2k}^c(e)$ is on the left and the curve $\lambda_{2k}^s(e)$ is on the right. The curves $\lambda_1^s(e)$ and $\lambda_1^c(e)$ form a finite angle for $e = 0$, and the curves $\lambda_2^c(e)$ and $\lambda_2^s(e)$ have second-order contact. Their contact order increases with the growth of n. This behavior of the boundaries of stability and instability regions for the modified Mathieu equation has an essential qualitative difference from the well-known Ince–Strutt diagram for the canonical Mathieu equation [43, 61]. The boundaries of parametric stability and instability regions for equation (12.1.1) are shown in Fig. 2.2 (curves $\lambda_0^c(e)$, $\lambda_n^{c,s}(e)$, $n = 1, 2, 3$).

12.1.4. Vibrations of crankshafts. Investigations of vibrations of elastic crankshafts with concentrated loads yield the following equation with a periodic coefficient [58]:

$$\ddot{u} + \frac{\lambda}{1 - e\cos(2\pi t)} u = 0. \tag{12.1.9}$$

Just as for the modified Mathieu equation (12.1.1), one seeks the values of the parameter λ for which there exist nontrivial solutions satisfying the boundary conditions (12.1.3), (12.1.4). Naturally, λ is a function of the modulation depth e, and we have $0 \le |e| < 1$.

Before we describe the calculation results obtained for the boundaries $\lambda_n^{c,s}(e)$ of parametric instability regions for arbitrary modulation depths e, we give some formulas obtained by the classical perturbation method in the case of small modulation depths.

The eigenvalues $\lambda_1^s(e)$ and $\lambda_2^s(e)$ in the case of boundary conditions (12.1.3) have the form

$$\begin{aligned} \lambda_1^s &= \pi^2\left(1 + \tfrac{1}{2}e - \tfrac{9}{32}e^2 + \cdots\right), \\ \lambda_2^s &= 4\pi^2\left(1 - \tfrac{1}{3}e^2 - \tfrac{1}{8}e^4 + \cdots\right). \end{aligned} \tag{12.1.10}$$

As above, $\lambda_n^s(e) = (\pi n)^2 + O(e^n)$. The eigenvalues (12.1.10) correspond to odd eigenfunctions

$$\begin{aligned} u_1^s &= \sin(\pi t) + \tfrac{1}{16}e\sin(3\pi t) + O(e^2), \\ u_2^s &= \sin(2\pi t) + \tfrac{1}{6}e\sin(4\pi t) + O(e^2). \end{aligned} \tag{12.1.11}$$

The eigenvalues λ_0^c, λ_1^c, λ_2^c in the case of boundary conditions (12.1.4) have the form

$$\begin{aligned} &\lambda_0^c = 0, \quad \lambda_1^c = \pi^2\left(1 - \tfrac{1}{2}e - \tfrac{9}{32}e^2 + \cdots\right), \\ &\lambda_2^c = 4\pi^2\left(1 - \tfrac{1}{3}e^2 - \tfrac{1}{8}e^4 + \cdots\right). \end{aligned} \tag{12.1.12}$$

The eigenvalues λ_0^c, λ_1^c, λ_2^c correspond to even eigenfunctions

$$\begin{aligned} &u_0^c = 1, \quad u_1^c = \cos(\pi t) + \tfrac{1}{16}e\cos(3\pi t) + (e^2), \\ &u_2^c = \cos(2\pi t) - \tfrac{1}{2}e\left[1 - \tfrac{1}{3}\cos(4\pi t)\right] + O(e^2). \end{aligned} \tag{12.1.13}$$

Formulas (12.1.10) and (12.1.12) show that λ_2^s is very close to λ_2^c for small $|e|$,

$$\lambda_1^s - \lambda_2^c = O(e^6). \tag{12.1.14}$$

Some comments should be made concerning the relation (12.1.14). It has been shown that $\lambda_{2k}^s \equiv \lambda_{2k}^c$ for $|e| \ll 1$. This property follows from the expression of the Hill's determinant. The authors have shown by analytical means, as well as by calculations based on the method of accelerated convergence, that $\lambda_2^s \equiv \lambda_2^c$ for all $|e| < 1$. Therefore, all even eigenvalues $\lambda_{2k}^s = \lambda_{2k}^c$ are degenerate. The eigenfunctions u_{2k}^c and u_{2k}^s are linearly independent and satisfy the orthogonality condition

$$\int_0^1 \frac{u_{2k}^c u_{2k}^s}{1 - e\cos 2\pi t}\, dt \equiv 0, \quad |e| < 1.$$

Thus, no further orthogonalization is needed.

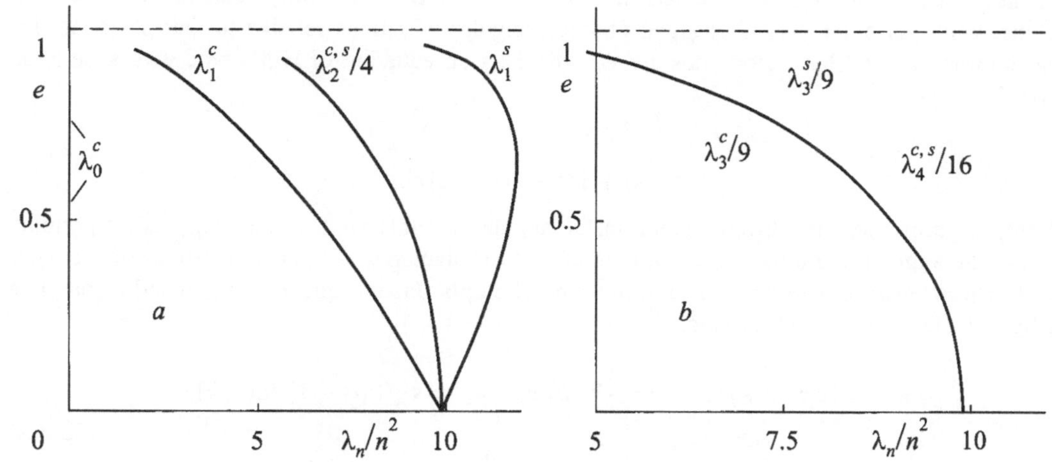

Fig. 12.1

Solving the boundary value problems (12.1.9), (12.1.3) and (12.1.9), (12.1.4) by the method of accelerated convergence, we obtain the eigenvalues listed in Table 12.3. The relative calculation error for $\lambda^{s,c}$ does not exceed 10^{-6}. The boundaries $\lambda_n^{c,s}(e)$ of parametric instability regions constructed on the basis of Table 12.3 are shown in Fig. 12.1 *a, b*. Table 12.3 and Fig. 12.1 *a, b* show that the regions corresponding to even $\lambda_{2k}^{c,s}(e)$ degenerate into curves. Thus, the curves for problem (12.1.9), (12.1.2) are also quite different from the Ince–Strutt diagram for the canonical Mathieu equation.

Table 12.3

	Odd Solutions				Even Solutions			
e	λ_1^s	λ_2^s	λ_3^s	λ_4^s	λ_1	λ_2^c	λ_3^c	λ_4^c
0	π^2	$4\pi^2$	$9\pi^2$	$16\pi^2$	π^2	$4\pi^2$	$9\pi^2$	$16\pi^2$
0.1	10.3347	39.3465	88.5071	157.335	9.3485	39.3465	88.5062	157.335
0.2	10.7428	38.3946	87.5372	155.575	8.7720	38.9464	87.5343	155.575
0.3	11.0890	38.2645	85.8865	152.576	8.1381	38.2645	85.8785	152.576
0.4	11.3667	37.2757	83.4982	148.225	7.4434	37.2757	83.4723	148.225
0.5	11.5652	35.9379	80.2661	142.329	6.6814	35.9379	80.2130	142.329
0.6	11.6663	34.1802	76.0182	134.597	5.8399	34.1802	75.9248	134.597
0.7	11.6383	31.8786	70.4657	124.451	4.8995	31.8786	70.2987	124.451
0.8	11.4149	28.7821	63.0112	110.790	3.8199	28.7821	62.7034	110.790
0.9	10.8342	24.2200	52.0278	90.628	2.5085	24.2200	51.4741	90.628

12.2. Stability of Plane Vibrations and Rotations of a Satellite on a Circular Orbit

12.2.1. Setting of the problem. Our aim is to study stability of perturbed plane nonlinear vibrations and rotations of an axially symmetric satellite (solid body) moving on a circular orbit in a central gravity field. The satellite equatorial axis remains nearly orthogonal to the orbit plane, while its polar axis is engaged in small angular vibrations relative to that plane. In linear approximation, these vibrations are described by an equation of Hill's type with a periodic coefficient [13],

$$q'' + f(\psi, p_\psi, \alpha, k)\omega^{-2}(\alpha, k)q = 0,$$
$$f \equiv (p_\psi + 1)^2 - 3(\alpha - 1)\sin^2\psi. \tag{12.2.1}$$

Here, the prime denotes differentiation in the angular variable w. The function f is 2π-periodic in w. The angular coordinate ψ and the impulse p_ψ characterize the plane vibrations of the polar axis. These variables can be expressed in terms of elliptic Jacobi functions sn, cn and a complete elliptic integral $\mathbf{K}$ of the first kind,

$$\psi = \arcsin\{k\mathrm{sn}(v, k)\}, \quad v = \frac{2}{\pi}\mathbf{K}(k)w, \quad p_\psi = k\sqrt{3(\alpha - 1)}\,\mathrm{cn}(v, k),$$
$$\omega = \frac{\pi\sqrt{3(\alpha - 1)}}{2\mathbf{K}(k)}, \quad k^2 = \frac{2h_0}{3(\alpha - 1)}, \quad 0 \le k < 1. \tag{12.2.2}$$

The parameter α $(1 \le \alpha \le 2)$ is equal to the ratio of the polar and the equatorial moments of inertia. The equation of small vibrations (12.2.1) and the expressions (12.2.2) take place in the case of an oblate ellipsoid. In the case of a prolate ellipsoid $(0 \le \alpha \le 1)$, the replacement $\psi \to \psi + \frac{\pi}{2}$ in (12.2.2) is used. We restrict ourselves to the first case. In what follows, the modulus k of the elliptic functions is expressed through the constant energy h_0 of satellite vibrations, which should be sufficiently small, $2h_0 < 3(\alpha - 1)$. For $2h_0 > 3(\alpha - 1)$, the body is engaged in rotations which can be studied in a similar manner. In view of (12.2.2), the modulus k determines the amplitude ψ_0 of plane vibrations, $\psi_0 = \arcsin k$, $0 \le \psi_0 < \frac{\pi}{2}$.

Thus, equation (12.2.1) contains two independent parameters, α and h_0. For applications, it is important to construct periodic solutions, i.e., to find the values of thé parameters for which there exist such solutions. From the standpoint of mechanics, the parameters α and ψ_0 admit a clearer interpretation. It is required to find the function $\alpha(\psi_0)$ (or $\psi_0(\alpha)$) for which there exist periodic solutions.

In order to apply the method of accelerated convergence, it is convenient to introduce the parameters $\lambda = (3(\alpha - 1))^{1/2}$ and k such that $\lambda \ge 3^{-1/2} \approx 0.577$, $0 \le k < 1$, and also the dimensionless time $t = \frac{w}{2\pi}$, with respect to which the coefficient of equation (12.2.1) is 1-periodic. After these transformations, we come to the following generalized Sturm–Liouville problem with periodic boundary conditions:

$$\ddot{q} + 16\mathbf{K}^2(k)[(\lambda + k\cos\Phi)^2 - k^2\sin^2\Phi]q = 0,$$
$$q(t+1) \equiv q(t), \quad \dot{q}(t+1) \equiv \dot{q}(t), \tag{12.2.3}$$
$$\dot{\Phi} = 4\mathbf{K}(k)\sqrt{1 - k^2\sin^2\Phi}, \quad \Phi(0) = 0, \quad 3^{-1/2} \le \lambda < \infty, \quad 0 \le k < 1.$$

For small k, the terms of the order $O(k^2)$ can be neglected and the first equation in (12.2.3) reduces to the classical Mathieu equation $(\lambda^2 \to \lambda, \lambda k \to \gamma)$. Independent numerical integration of the equation for Φ makes the calculation of the coefficient of q defined in terms of Jacobi elliptic functions much easier. In our further calculations, various analytical representations of elliptic functions are of little use. It is preferable to calculate the complete elliptic integral of the first kind $\mathbf{K}(k)$ for $0 \le k^2 \le 0.99$ with the help of quadrature on the basis of its definition, and for $0.99 \le k^2 < 1$, it is convenient to use the asymptotic expression [32]

$$K(k) = \kappa + \tfrac{1}{4}(\kappa - 1)k'^2 + \tfrac{9}{64}(\kappa - \tfrac{7}{6})k'^4 + \tfrac{25}{256}(\kappa - \tfrac{37}{30})k'^6 + O(k'^8),$$
$$\kappa = \ln(4/k'); \quad k'^2 = 1 - k^2. \tag{12.2.4}$$

The representation (12.2.4) leads to a small error of the order $k'^8 \approx 10^{-8}$ for $k'^2 \le 10^{-2}$. Note that equation (12.2.3) for $k^2 \to 1$ has a singularity, since $\mathbf{K}(k) \to \infty$. This case requires a separate asymptotic analysis (see below).

Since the coefficient $r(t, \lambda, k)$ of q in (12.2.3) is a symmetric function of t, the generalized Sturm–Liouville problem with periodic boundary conditions reduces to two generalized Sturm–Liouville problems:

1) problem of the first kind with an odd function q^-;

2) problem of the second kind with an even function q^+.

The boundary conditions read as follows:

$$\begin{aligned} &1)\ q^-(0) = q^-(1) = 0, \quad \lambda = \lambda_n^-(k), \quad \lambda_n^-(0) = \tfrac{n}{2}, \\ &\quad q_n^-(t,k) = q^-(t, \lambda_n(k), k), \quad q_n^-(t,0) = (\pi n)^{-1}\sin(\pi n t); \\ &2)\ \dot q^+(0) = \dot q^-(1) = 0, \quad \lambda = \lambda_n^+(k), \quad \lambda_n^+(0) = \tfrac{n}{2}, \\ &\quad q_n^+(t,k) = q^+(t, \lambda_n(k), k), \quad q_n^+(t,0) = \cos(\pi n t), \end{aligned} \tag{12.2.5}$$

where $n = 1, 2, \ldots$

12.2.2. Results of numerical-analytical investigation. For $k > 0$ $(0 \le k^2 < 0.999)$, we use the procedure of continuation in the parameter k and the method of accelerated convergence (see Chapter 4). The corresponding highly precise calculations of $\lambda_n^\mp(k)$ are represented in Fig. 12.2 for $n = 1, 2, \ldots, 6$. It should be observed that $\lambda_1^+(k) \le \frac{1}{2}$. This leads us to physically unrealistic values $\lambda_1^+(\alpha_1^+ > 2)$. The branch of $\lambda_1^-(k)$ for $0 \le k \le 0.15$ also yields physically unrealistic values $\lambda_1^-(\alpha_1^- > 2)$. However, for $k > 0.15$, we have $\lambda_1^+(k) > 0.577$, i.e. $\alpha_1^- < 2$. All other branches of the resonance curves $\lambda_n^\mp$, $n \ge 2$, can be physically realized $(1 \le \alpha_n^\mp < 2)$ and for $k = 0$ their contact order depends on n.

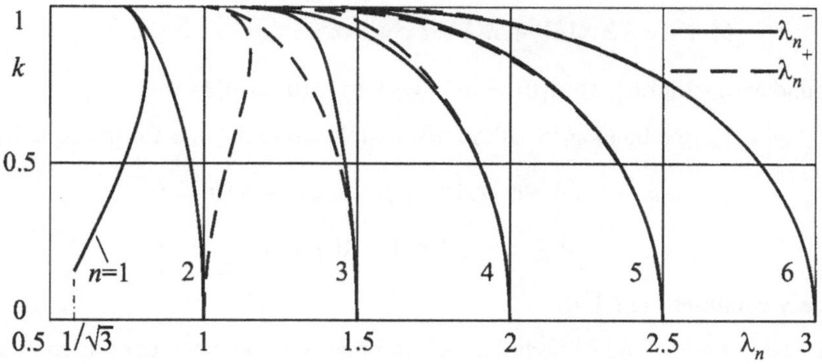

Fig. 12.2

For theoretical and practical purposes, it is interesting to describe the asymptotic behavior of $\lambda_n^\mp(k)$ as $k \to 1$. The structure of the coefficient $r(t, \lambda, k)$ for $k \to 1$ and the properties of the elliptic functions, cn $\to 0$, sn $\to 1$, $0 < t < \frac{1}{2}$, $\frac{1}{2} < t < 1$, together with (12.2.4), imply that $\lambda_n^{\pm}(k) \to 1$. In contrast to classical curves (for instance, the Ince–Strutt or the Haupt diagrams [43, 61]), the even and the odd curves alternate by pairs.

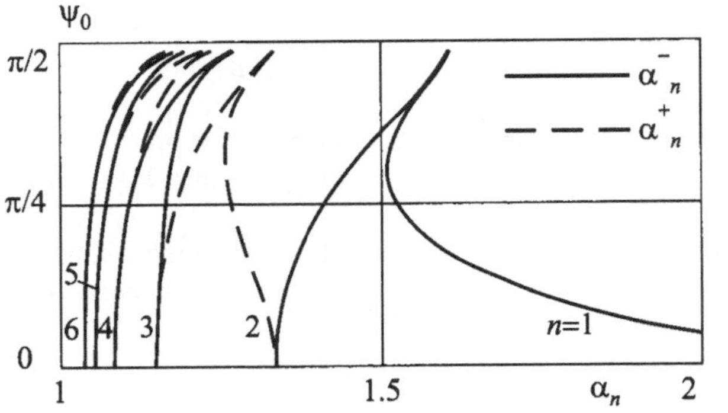

Fig. 12.3

Figure 12.3 shows the resonance curves $\alpha_n^{\pm}(\psi_0)$ for the parameters α, ψ_0 with a more clear mechanical meaning. These curves are easily obtained with the help of the formulas $\alpha_n = 1 + (3\lambda_n^2)^{-1}$, $\psi_0 = \arcsin k$. It would be useful to represent both sets of resonance curves (Figs. 12.2 and 12.3). It is interesting to note that the resonance curve $\alpha_1^-(\psi_0)$ for $\psi \geq 0.16$ cuts out a considerable part of the region of admissible values of the parameters α, ψ_0. The above analysis implies that $\alpha_1^{\pm}(\psi_0) \to \frac{4}{3}$ as $\psi_0 \to \frac{\pi}{2}$. The curves α_n^-, α_n^+ are also alternating by pairs.

A similar approach can be used for the construction of resonance curves $\lambda(k)$ or $\alpha(\psi_0)$ in the case of a rotating satellite $(2h_0 > 2(\alpha - 1))$. The above calculation results demonstrate the high efficiency of the method of accelerated convergence for solving complex generalized periodic problems that cannot be approached by other methods.

12.3. Exercises

Exercise 1. Determine the boundaries of the first and the second zones of parametric instability for the equation

$$((1 + e\cos 2\pi t)\dot{u})^{\cdot} + \lambda(1 + e\cos 2\pi t)u = 0, \qquad 0 \leq e < 1,$$

with the boundary conditions: a) $u(0) = u(1) = 0$, b) $\dot{u}(0) = \dot{u}(1) = 0$.

Exercise 2. Construct the boundaries of the first and the second zones for the equation

$$\ddot{u} + \lambda(1 + e\cos(2\pi t) + \mu\cos(4\pi t + \alpha))u = 0,$$

$$0 \leq e^2 + \mu^2 < 1, \quad 0 \leq \alpha < \pi$$

with boundary conditions (12.1.2).

Exercise 3. Using the method of perturbations and that of accelerated convergence, construct the boundary of the first zone for the equation

$$\ddot{u} + \lambda(1 + e\cos 2\pi t + \mu\cos 2\pi\nu t)u = 0, \quad 0 \leq e^2 + \mu^2 < 1,$$

with boundary conditions (12.1.2). Consider the cases: 1) ν is a rational number; 2) ν is an irrational number.

Chapter 13

Vibrations of a Distributed Inhomogeneous System in a Rectangular Domain

In this chapter, we consider vibrations of an inhomogeneous system described by hyperbolic equations with the Dirichlet boundary conditions. It is assumed that the corresponding eigenvalue problem in a rectangular domain admits partial separation of the variables and the separation parameters enter both equations. Thus, we obtain a system with two parameters whose eigenvalues are to be determined jointly. For this purpose, we develop a method of accelerated convergence, which allows us to calculate approximations of any given order for eigenvalues and eigenfunctions of the problem with variable coefficients.

Our exposition is concerned with a model of a rectangular membrane. We consider the case of variable surface density and membrane tension, and thus, the inertial and the elastic characteristics of the membrane may be nonhomogeneous. However, the main attention is given to the investigation of a rectangular membrane with essentially variable surface density. An effective algorithm of accelerated convergence is developed for solving the corresponding self-conjugate boundary value problem for eigenvalues and eigenfunctions, which allows for partial separation of spatial variables. The results are generalized for the case of nonhomogeneous surface tension and the presence of an external elastic medium. An algorithm is proposed for taking into account perturbations of general character. The efficiency of the method of accelerated convergence is demonstrated by examples of membranes with specific mass distribution.

13.1. Vibrations of an Inhomogeneous Membrane

13.1.1. Statement of the initial boundary value problem. Consider small vibrations of an inhomogeneous rectangular membrane subject to uniform surface tension [9, 22, 24, 46, 62, 64, 65]. The equation of vibrations which takes into account distributed surface forces has the form

$$\begin{gathered}\rho(x,y)\frac{\partial^2 w}{\partial t^2} = \sigma\left(\frac{\partial^2 w}{\partial x^2} + \frac{\partial^2 w}{\partial y^2}\right) + f(x,y,t),\\ 0 < x < a, \quad 0 < y < b, \quad t \geq t_0.\end{gathered} \tag{13.1.1}$$

Here, $w = w(x,y,t)$ is the orthogonal displacement of the membrane element with the Euler coordinates x, y at the instant t; $\rho(x,y)$ is the variable surface density of the membrane $(0 < \rho^- \leq \rho \leq \rho^+ < \infty)$; σ is isotropic surface tension, $\sigma = \text{const}$; $f(x,y,t)$ is the external force density per unit area. On the boundary of the rectangle, the nonzero Dirichlet conditions are imposed

$$w\big|_{x=0,a} = h^{0,a}(y,t), \qquad w\big|_{y=0,b} = g^{0,b}(x,t). \tag{13.1.2}$$

The functions $h^{0,a}$, $g^{0,b}$ in (13.1.2), which determine displacements of boundary elements, are assumed sufficiently smooth, in particular, at the angular points (end-points). One usually considers membranes with clamped boundary, $h^{0,a} = g^{0,b} \equiv 0$.

In order to determine the motion of system (13.1.1), (13.1.2), one has to prescribe the displacements and the displacement rates at an initial instant,

$$w(x, y, t_0) = w^0(x, y), \qquad \frac{\partial w}{\partial t}(x, y, t_0) = w^1(x, y). \tag{13.1.3}$$

The initial boundary value problem (13.1.1)–(13.1.3) is solved by the standard method of separation of variables, $w(x, y, t) \sim u(x, y)\theta(t)$. For the unknown function u one obtains a self-conjugate eigenvalue problem

$$\begin{gathered} \frac{\partial^2 u}{\partial x^2} + \frac{\partial^2 u}{\partial y^2} = -\lambda \rho(x, y) u, \\ u\big|_{x=0,a} = u\big|_{y=0,b} = 0. \end{gathered} \tag{13.1.4}$$

One seeks the countable set of eigenvalues $\{\lambda_n\}$ and the corresponding orthonormal system of eigenfunctions $\{u_n(x, y)\}$ $(n = 1, 2, ...)$ forming a basis in L^2 with the weight ρ.

13.1.2. Separation of variables. A solution of the original problem (13.1.1)–(13.1.3) is constructed in the form of a series in terms of the said basis consisting of eigenfunctions. This procedure amounts to elementary operations and integration,

$$\begin{gathered} w(x, y, t) = \sum_{n=1}^{\infty} u_n(x, y)\theta_n(t), \qquad \nu_n^2 = \lambda_n \sigma, \\ \theta_n(t) = w_n^0 \cos(\nu_n t) + \frac{w_n^1}{\nu_n} \sin(\nu_n t) + \frac{1}{\nu_n} \int_{t_0}^{t} \sin[\nu_n(t - \tau)] F_n(\tau) d\tau, \quad \dot{\theta}_n = \frac{d\theta_n}{dt}, \end{gathered} \tag{13.1.5}$$

where the dots denote differentiation in t, the constants w_n^0, w_n^1 and the functions $F_n(t)$ are defined by the standard relations

$$\begin{gathered} w_n^0 = (w^0, u_n)_\rho, \quad w_n^1 = (w^1, u_n)_\rho, \quad f_n(t) = (f, u_n)_1, \\ F_n(t) = f_n(t) - \sigma \int_0^b \left[h^a(y, t) u'_{nx}(a, y) - h^0(y, t) u'_{nx}(0, y)\right] dy \\ -\sigma \int_0^a \left[g^b(x, t) u'_{ny}(x, b) - g^0(x, t) u'_{ny}(x, 0)\right] dx. \end{gathered} \tag{13.1.6}$$

Here, ν_n are frequencies of free vibrations of the inhomogeneous membrane and the functions $u_n(x, y)$ characterize the corresponding vibration shapes. By $(\cdot, \cdot)_\rho$ and $(\cdot, \cdot)_1$ in (13.1.6) we denote the scalar products in L^2 with the weights $\rho(x, y)$ and 1, respectively; the primes with the subscripts x, y denote the corresponding partial derivatives. If the given functions in problem (13.1.1)–(13.1.3) are sufficiently smooth, then the series (13.1.5) is convergent and represents the strong or the classical solution of the problem [24, 65].

Thus, the above standard approach to the construction of an effective solution of specific problems brings us to the necessity of constructing systems of eigenvalues and eigenfunctions. Theoretical questions pertaining to eigenvalue problems of the type (13.1.4) have been studied in great detail. There exist computational procedures for their solution based on the Rayleigh–Ritz method and the method of finite elements. However, there is a need of efficient methods and algorithms for highly precise approximation of lower eigenvalues and eigenfunctions, for instance, in order to test the precision of the said procedures. A highly precise resource-sparing method is also needed for numerous calculations aimed at studying the dependence of eigenvalues on the parameters of the system (parametric synthesis). We propose to use the numerical-analytical method of accelerated convergence for finding lower eigenvalues and eigenfunctions of problem (13.1.4), which is of particular interest for applications.

13.2. Scheme of the Construction of a Solution of the Membrane Eigenvalue Problem

13.2.1. Separation of spatial variables. An effective solution of problem (13.1.4) with arbitrary $\rho(x,y)$ is hardly possible. If $\rho(x,y)$ has an additive structure, it is possible to separate the variables x, y, and we obtain two coupled eigenvalue problems,

$$\begin{aligned} &\rho_0(x,y) = \varphi(x) + \psi(y), \qquad u(x,y) = X(x)Y(y); \\ &X'' + (\lambda\varphi(x) - \mu)X = 0, \quad X(0) = X(a) = 0; \\ &Y'' + (\lambda\psi(y) + \mu)Y = 0, \quad Y(0) = Y(b) = 0. \end{aligned} \tag{13.2.1}$$

Here λ, μ and X, Y are the parameters and the functions to be determined. In (13.2.1), it is assumed that the term $\varphi(x)$ in $\rho_0(x,y)$ is positive and dominating, $\varphi(x) > \psi(y)$ for $(x,y) \in \Pi$ (this condition can always be ensured by adding a sufficiently large constant to $\varphi(x)$ and subtracting that constant from $\psi(y)$). We do not assume that $\psi(y)$ is positive; in fact, it is preferable that $\psi(y) \leq 0$ (see below). If for some reason it is natural to assume that $\psi(y) > 0$, then (by adding and subtracting a suitable constant) it can be ensured that $\psi(y) > \varphi(x)$ and μ in equations (13.2.1) should be replaced by $-\mu$.

13.2.2. Structural properties of eigenvalues and eigenfunctions. After separating the variables as above, we can consider two families of boundary value problems: one for the unknown quantities λ, X and the parameter of the family μ, the other for μ, Y with the parameter of the family λ. On the basis of the highly efficient method of accelerated convergence described in Chapter 2, we propose a procedure for solving problem (13.2.1). This procedure consists of consecutively constructing eigenvalues and eigenfunctions of the said two families of problems,

$$\begin{aligned} &\lambda = \lambda_n(\mu), \quad X_n(x,\mu) = X(x,\lambda_n(\mu),\mu) \quad (n = 1,2,\ldots), \\ &\mu = \mu_m(\lambda), \quad Y_m(y,\lambda) = Y(y,\mu_m(\lambda),\lambda) \quad (m = 1,2,\ldots). \end{aligned} \tag{13.2.2}$$

Here, the independent indices n, m are assumed fixed. One usually considers lower vibration modes, for instance, $n, m = 1, 2$. After the functions $\lambda_n(\mu)$, $\mu_m(\lambda)$ (13.2.2) have been found, the value λ_{nm} is determined as the root of the equation

$$\lambda = \lambda_n(\mu_m(\lambda)), \quad \lambda = \lambda_{nm}; \qquad \mu_{nm} = \mu_m(\lambda_{nm}). \tag{13.2.3}$$

The eigenfunction $u_{nm}(x,y)$ corresponding to λ_{nm} (13.2.3) is calculated with the help (13.2.1), (13.2.2):

$$u_{nm}(x,y) = X(x,\lambda_{nm},\mu_{nm})Y(y,\mu_{nm},\lambda_{nm}) \equiv X_{nm}(x)Y_{nm}(y). \tag{13.2.4}$$

By construction, the functions $u_{nm}(x,y)$ (13.2.4) are orthogonal with the weight $\rho(x,y)$ in the rectangle $0 \leq x \leq a$, $0 \leq y \leq b$. Note that in contrast to the one-dimensional case (string), the eigenvalues λ_{nm}, (13.2.3) may be multiply degenerate.

Let us estimate the range S_{nm} of the parameters λ, μ for which it is required to construct the solutions of problems (13.2.1) in the form (13.2.2). Taking into account the above assumptions, we obtain the desired estimates and the representation of S_{nm}:

$$\Lambda_n = \{\lambda, \mu : \ \lambda_n^+(\mu) \le \lambda \le \lambda_n^-(\mu), \ |\mu| < \infty\}, \quad \lambda_n^\pm(\mu) = \frac{1}{\varphi^\pm}\left(\mu + \left(\frac{\pi n}{a}\right)^2\right),$$

$$M_m = \{\lambda, \mu : \ \lambda \ge 0, \ \mu_m^\pm(\lambda) \le \mu \le \mu^-(\lambda)\}, \quad \mu_m^\pm(\lambda) = \frac{\pi m}{b^2} - \lambda\psi^\pm, \qquad (13.2.5)$$

$$S_{nm} = \Lambda_n \cap M_m, \qquad (\lambda, \mu) \in S_{nm}.$$

Here, $\varphi^\pm > 0$ are the maximal and the minimal values of the function $\varphi(x)$, $0 \le x \le a$; $\psi^\pm$ are the maximal and the minimal values of the function $\psi(y)$, $0 \le y \le b$. The characteristic rectangular form of the region S_{nm} (13.2.5) is represented in Fig. 13.1. In particular, for $\psi(y) = \text{const}$ or $\varphi(x) = \text{const}$, this region degenerates into a linear segment (points A, B and D, C or D, A and C, B, respectively, coincide). If one of the functions has a relatively small variation, then S_{nm} is a narrow strip. Thus, the curves $\lambda_n(\mu)$, $\mu_m(\lambda)$ (13.2.2) are inside the region S_{nm}, if $(\mu, \lambda) \in S_{nm}$, and have a single common point (13.2.3), (13.2.4). The typical behavior of these curves and the corresponding graphical solution of equation (13.2.3) are also shown in Fig. 13.1.

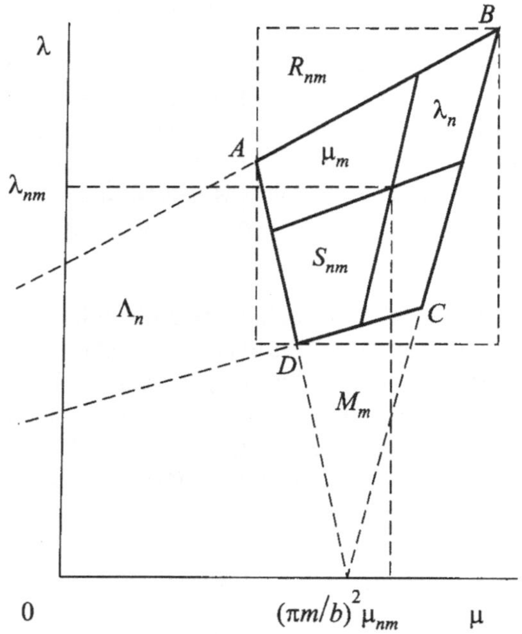

Fig. 13.1

Thus, we come to the problem of constructing two one-parameter families of solutions (13.2.2) for $(\lambda, \mu) \in S_{nm}$, (13.2.5). To facilitate calculations, the region of λ, μ under consideration can be approximated by a rectangle R_{nm} whose sides are parallel to the coordinate axes (see Fig. 13.1). This rectangle is determined by the extremal values of the abscissas and the ordinates of the vertices A, B, C, D of the rectangle S_{nm}.

13.3. Method of Accelerated Convergence

13.3.1. Two-parameter scheme of solution. Let us describe a recurrent procedure for solving problems (13.2.1) on the basis of the method of accelerated convergence and the variational approach of Chapter 2. These problems can be stated as the following variational problems:

$$I_\mu[X] = \int_0^a (X'^2 + \mu X^2)\,dx \to \min_X, \quad \Phi[X] = \int_0^a \varphi(x) X^2 dx = 1, \quad X(0) = X(a) = 0;$$

$$J_\lambda[Y] = \int_0^b (Y'^2 - \lambda\psi(y) Y^2)\,dy \to \min_Y, \quad \Psi[Y] = \int_0^b Y^2 dy = 1, \quad Y(0) = Y(b) = 0. \tag{13.3.1}$$

The absolute minima of the functionals $I\mu$, J_λ (13.3.1) are equal to the first eigenvalues $\lambda_1(\mu)$, $\mu_1(\lambda)$ (13.2.2) of problems (13.2.1) in the domain S_{11} (13.2.5). The functions on which these minima are attained are the first eigenfunctions of the corresponding problems, $X_1(x,\mu)$, $Y_1(y,\lambda)$ (13.2.2). Subsequent eigenvalues $\lambda_n(\mu)$, $\mu_m(\lambda)$ and eigenfunctions$X_n(x,\mu)$, $Y_m(y,\lambda)$ are constructed in a recurrent manner on narrower classes of admissible functions satisfying the additional orthogonality conditions

$$\begin{aligned} \Phi_{\mu i}[X] &= \int_0^a \varphi(x) X_i(x,\mu) X\,dx = 0 \qquad (i = 1, 2, \ldots, n-1), \\ \Psi_{\lambda j}[Y] &= \int_0^b Y_j(y,\lambda) Y\,dy = 0 \qquad (j = 1, 2, \ldots, m-1). \end{aligned} \tag{13.3.2}$$

Isoperimetric variational problems of the type (13.3.1), (13.3.2) are often solved by approximate numerical-analytical methods (the Rayleigh principle, the Rayleigh–Ritz method, the finite element method, and others). The convergence of the Rayleigh–Ritz method was proved in the works of Krylov [37]. This approach, in particular, the Rayleigh principle, is convenient for effectively constructing initial approximations which represent upper bounds for the eigenvalues $\lambda_n(\mu)$ and $\mu_m(\lambda)$. In order to construct highly precise approximations for λ_n, X_n and μ_m, Y_m with a view to solving equation (13.2.3) and constructing the functions (13.2.4), it is convenient to use the method of accelerated convergence of Chapter 2.

For definiteness, let us consider problems (13.3.1) and calculate upper bounds $\lambda_1^*(\mu)$, $\mu_1^*(\lambda)$ on the basis of the Rayleigh principle,

$$\begin{aligned} &0 < \lambda_1(\mu) \le \lambda_1^*(\mu) = \frac{I_\mu[p]}{\Phi[p]}, \qquad p = p(x,\mu), \qquad p|_{x=0,a} = 0, \\ &\mu_1(\lambda) \le \mu_1^*(\lambda) = \frac{J_\lambda[q]}{\Psi[q]}, \qquad q = q(y,\lambda), \qquad q|_{y=0,b} = 0. \end{aligned} \tag{13.3.3}$$

Here, p, q are continuously differentiable test functions satisfying the boundary conditions (13.3.3) and chosen on the basis of intuitive physical ideas regarding the first vibration shape. For our further calculations, it suffices to obtain the estimates (13.3.3) for some fixed μ and λ in the domain S_{11}, and then use the procedure of continuation in parameters. The precision of these estimates can be verified by the Rayleigh–Ritz method [22–24, 28, 37, 46, 62, 64], for instance, by using approximations with two test functions.

13.3.2. Introduction of small parameters. We take another, more regular approach to the solution of these problems. This approach is based on the introduction of small parameters in problems (13.2.1). Taking known bounds $\lambda_1^0(\mu)$, $\mu_1^0(\lambda)$ (μ, λ are fixed), in particular, the upper bounds $\lambda_1^*(\mu)$, $\mu_1^*(\lambda)$ calculated in (13.3.3), we propose to solve two Cauchy problems,

$$
\begin{gathered}
X'' + (\lambda_1^0(\mu)\varphi(x) - \mu)X = 0, \\
X(0) = 0, \quad X'(0) = 1, \qquad X = U^0(x,\mu); \\
Y'' + (\lambda\psi(y) + \mu_1^0(\lambda))Y = 0, \\
Y(0) = 0, \quad Y'(0) = 1, \qquad Y = V^0(y,\lambda).
\end{gathered} \tag{13.3.4}
$$

The functions U_1^0, V_1^0 are assumed known in analytical or numerical form, or in the form of a computational procedure; these functions are constructed independently of one another. Subscripts $n = m = 1$ are omitted.

Next, one finds the smallest (first) roots α, β of the functions U^0, V^0, and determines their relative distances ε, δ from the values $x = a$ and $y = b$, respectively,

$$
\begin{gathered}
\alpha(\mu) = \arg_x U^0(x,\mu), \quad \varepsilon(\mu) = 1 - \xi(\mu), \quad \xi = \tfrac{\alpha}{a}; \\
\beta(\lambda) = \arg_y V^0(y,\lambda), \quad \delta(\lambda) = 1 - \eta(\lambda), \quad \eta = \tfrac{\beta}{b}.
\end{gathered} \tag{13.3.5}
$$

Note that $\varepsilon = 0$, if and only if $\lambda_1^0 = \lambda_1$. Similarly, $\delta = 0$, if and only if $\mu_1^0 = \mu$. If $\varepsilon < 0$ or (and) $\delta < 0$, then the functions $\varphi(x)$ or (and) $\psi(y)$ can be smoothly extended to the intervals $a < x \le \alpha(\mu)$ and $b < y \le \beta(\lambda)$, respectively. Thus, the numerical parameters ε and δ also characterize closeness of $\lambda_1^0(\mu)$ to $\lambda_1(\mu)$ and $\mu_1^0(\lambda)$ to $\mu_1(\lambda)$, respectively, in the region $(\lambda,\mu) \in S_{11}$. Moreover, for small enough ε, δ, we have the following two-sided estimates:

$$
d_\lambda|\varepsilon| \le |\lambda_1^0 - \lambda_1| \le c_\lambda|\varepsilon|, \quad d_\mu|\delta| \le |\mu_1^0 - \mu_1| \le c_\mu|\delta|, \quad d_{\lambda,\mu} < c_{\lambda,\mu},
$$

where $d_{\lambda,\mu}$, $c_{\lambda,\mu}$ are positive constants.

13.3.3. A parallel scheme of the algorithm of accelerated convergence. Now, in order to obtain a highly precise approximation of the solution of the Sturm–Liouville problem (13.2.1) for sufficiently small $|\varepsilon|$, $|\delta|$, one can use the recurrent procedure of the method of accelerated convergence of Chapter 2. This procedure consists in consecutively refining the values $\lambda_1(\mu)$, $\mu_1(\lambda)$ and the functions X, Y, on the basis of solutions of the Cauchy problems (13.3.4) and determining the smallest roots $\alpha_1^{(k)}$, $\beta_1^{(l)}$ (13.3.5) of the functions $U^{(k)}$, $V^{(l)}$, according to the scheme $(k, l = 0, 1, 2, \ldots)$

$$
\begin{gathered}
\lambda_1^{(k+1)}(\mu) = \lambda_1^{(k)}(\mu) - \varepsilon^{(k)}(\mu)\alpha^{(k)}(\mu)\frac{(U^{(k)\prime}(\alpha^{(k)}(\mu),\mu))^2}{\varPhi[U^{(k)}]}, \quad \lambda_1^{(0)} = \lambda_1^0, \\
\varepsilon^{(k)}(\mu) = 1 - \xi^{(k)}(\mu), \quad \varepsilon^{(0)} = \varepsilon, \quad \xi^{(k)} = \frac{\alpha^{(k)}}{a}, \quad \alpha^{(k)}(\mu) = \arg_x U^{(k)}(x,\mu); \\
\mu_1^{(l+1)} = \mu_1^{(l)}(\lambda) - \delta^{(l)}(\lambda)\beta^{(l)}(\lambda)\frac{(V^{(l)\prime}(\beta^{(l)}(\lambda),\lambda))^2}{\varPsi[V^{(l)}]}, \quad \mu_1^{(0)} = \mu_1^{(0)}, \\
\delta^{(l)}(\lambda) = 1 - \eta^{(l)}(\lambda), \quad \delta^{(0)} = \delta, \quad \eta^{(l)} = \frac{\beta^{(l)}}{b}, \quad \beta^{(l)}(\lambda) = \arg_y V^{(l)}(y,\lambda).
\end{gathered} \tag{13.3.6}
$$

The functions $U^{(k)}(x,\mu)$, $V^{(l)}(y,\lambda)$ and their squared norms, i.e., the functionals $\varPhi$, $\varPsi$, are obtained after the integration of the Cauchy problems with the help of the relations

$$X'' + (\lambda_1^{(k)}(\mu)\varphi(x) - \mu)X = 0, \qquad X(0) = 0, \quad X'(0) = 1; \qquad X \equiv U^{(k)}(x, \mu),$$

$$\Phi[U^{(k)}] \equiv \int_0^{\alpha^{(k)}} \varphi(x) U^{(k)2}(x, \mu)\, dx = U^{(k)\prime} P^{(k)} \Big|_{x=\alpha^{(k)}};$$

$$Y'' + (\mu_1^{(l)}(\lambda) + \lambda\psi(y))Y = 0, \qquad Y(0) = 0, \quad Y'(0) = 1; \qquad Y \equiv V^{(l)}(y, \lambda), \tag{13.3.7}$$

$$\Psi[V^{(l)}] \equiv \int_0^{\beta^{(l)}} V^{(l)2}(y, \lambda)\, dy = V^{(l)\prime} Q^{(l)} \Big|_{y=\beta^{(l)}}.$$

The first expressions of the functionals Φ, Ψ in (13.3.7) are squared norms and are written in the traditional form of quadratures. Their second expressions are obtained by joint integration of the Cauchy problems for additional unknown functions $P = \partial X/\partial\lambda$, $Q = \partial Y/\partial\mu$:

$$P'' + (\lambda_1^{(k)}(\mu)\varphi(x) - \mu)P = -\varphi(x)X, \quad P(0) = P'(0) = 0; \quad P \equiv P^{(k)}(x, \mu);$$

$$Q'' + (\mu_1^{(l)}(\lambda) + \lambda\psi(y))Q = -Y, \qquad Q(0) = Q'(0) = 0; \quad Q \equiv Q^{(l)}(y, \lambda). \tag{13.3.8}$$

Thus, the operations of highly precise numerical integration of the Cauchy problems (13.3.8) are replaced by the calculation of quadratures (13.3.7).

The rate of convergence of iterations (13.3.6)–(13.3.8) is quadratic with respect to ε, δ,

$$|\lambda_1^{(k)} - \lambda_1| \le C_\lambda |\varepsilon|^{n_k}, \quad |\mu_1^{(l)} - \mu_1| \le C_\mu |\delta|^{n_l}, \quad n_k = 2^k, \qquad (\lambda, \mu) \in S_{11};$$

$$|X^{(k)} - X| + |X^{(k)\prime} - X'|a \le C_X |\varepsilon|^{n_k}, \quad |Y^{(l)} - Y| + |Y^{(l)\prime} - Y'|b \le C_Y |\delta|^{n_l}. \tag{13.3.9}$$

Computational experience (see Section 13.5 and Chapter 2) shows that for $|\varepsilon|$, $|\delta| \approx 0.1$, a few iterations (usually from 2 to 4) yield highly precise results quite acceptable in practical problems (relative calculation errors are of the order $10^{-4} \div 10^{-6}$).

13.3.4. Numerical-graphical solution of the problem. The next step in the construction of an approximation of the first eigenvalue λ_{11} consists of the construction of the curves $\lambda_1(\mu)$, $\mu_1(\lambda)$. For this purpose, it is proposed to use the procedure of continuation in parameters. For sufficiently close values $\mu + \Delta\mu$, $\lambda + \Delta\lambda$ in S_{11}, the highly precise approximations $\lambda_1(\mu)$, $\mu_1(\lambda)$ obtained above are taken as the initial bounds, just as previously we have taken as such $\lambda_1^0(\mu)$, $\mu_1^0(\lambda)$. Then we use the iteration procedure (13.3.6)–(13.3.8) of the method of accelerated convergence. The steps $\Delta\mu$, $\Delta\lambda$ may be varied and are chosen on the basis of numerical experiment from the conditions ensuring convergence of the algorithm. Thus, the procedure of highly precise solution of the Sturm–Liouville problem (13.2.1) in the region S_{11} (13.2.5) reduces to the integration of the Cauchy problems (13.3.4), the determination of the first roots of the functions $U^{(k)}$, $V^{(k)}$, and the refinement of the values $\lambda_1^{(k)}$, $\mu_1^{(l)}$ according to (13.3.6). After the functions $\lambda_1(\mu)$, $\mu_1(\lambda)$, $(\mu, \lambda) \in S_{11}$, have been constructed, the method of continuation in parameters is used for finding the solution λ_{11} of equation (13.2.3) for $n = m = 1$. This value is used for finding the functions $X_{11}(x)$, $Y_{11}(y)$ and the lowest vibration shape $u_{11}(x, y)$ (13.2.4).

Subsequent eigenvalues λ_{nm} and eigenfunctions $u_{nm}(x, y)$, $n + m \ge 3$, can be constructed in a similar manner, since almost no properties specific of the first eigenvalue have been used in the algorithm. The only difference is that for the bounds $\lambda_n^0(\mu)$, $\mu_m^0(\lambda)$ obtained, for instance, by the Rayleigh–Ritz method, and for subsequent approximations, one has to seek the n-th root of the function $U_n(x, \mu)$ and the m-th root of the function $V_m(y, \lambda)$ (as in (13.3.5), where one seeks the first positive roots $\alpha^{(0)}(\mu)$ and $\beta^{(0)}(\lambda)$).

13.3.5. Iterative refinement procedure. Consider an iterative procedure that would allow us to reduce the amount of calculations needed for obtaining approximations of the eigenvalue

λ_{nm}. For definiteness, suppose that we have obtained an upper bound λ^*_{nm} by the Rayleigh–Ritz method. This λ^*_{nm} should be inserted into equation (13.3.7) for Y instead of arbitrary $\lambda \in S_{nm}$ and by the method of accelerated convergence, a sufficiently precise value $\mu^{(1)}_m(\lambda^*_{nm})$ should be found. Then, $\mu^{(1)}_m(\lambda^*_{nm})$ is inserted into equation (13.3.7) for X and by the method of accelerated convergence, the next approximation $\lambda^{(1)}_{nm} = \lambda^{(1)}_n(\mu^{(1)}_m(\lambda^*_{nm}))$ is found. In a similar recurrent manner, one finds $\mu^{(2)}_m(\lambda^{(1)}_{nm})$, $\lambda^{(2)}_{nm} = \lambda^{(2)}_n(\mu^{(2)}_m)$, etc. Under certain conditions (see below), this procedure is convergent and the convergence rate is that of a geometrical progression, namely, $O(d^i)$, where i is the iteration index, $0 < d < 1$, d is the geometric ratio to be found.

The conditions of convergence of the above process and the conditions of its divergence can be given a local geometrical interpretation. These conditions are connected with the orientation of the bisectrix of the angle at the crossing point of the curves $\lambda_n(\mu)$ and $\mu_m(\lambda)$, i.e., the angle between their tangential lines measured from the vertical direction (see Fig. 13.2). If the angle ζ between these lines is negative, then there is convergence ($d < 1$). If the angle ζ is positive, then the process diverges ($d > 1$). Zero angle represents a critical case ($d \approx 1$). Convergence of the process can be ensured by the domination of $\varphi(x)$ over $\psi(y)$ (see Section 13.2 below, formula (13.2.1)). It is assumed that $\lambda_n(\mu)$ is an increasing function of μ, and $\mu_m(\lambda)$ is a decreasing function of λ and its decrease rate is not very fast. This assumption is confirmed by numerical experiment (see Subsection 13.5.2). As the initial approximation λ^0_{nm} one can take a lower bound determined, for instance, by solving the original problem (13.1.4) with $\rho = \rho^0(x, y) \geq \rho_0(x, y)$ $(0 \leq x \leq a,\ 0 \leq y \leq b)$.

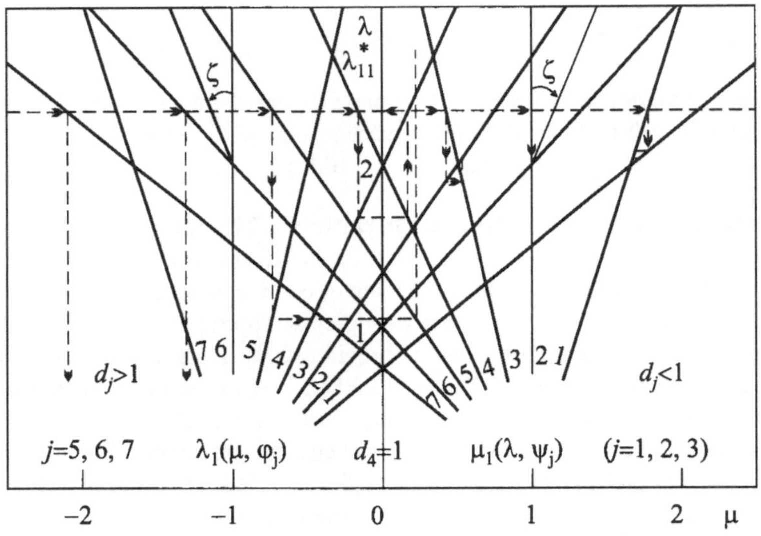

Fig. 13.2

13.4. Some Generalizations

Consider some more general classes of problems admitting an effective solution by the above method.

13.4.1. Perturbation of the surface density function. In applied problems, the additive representation (13.2.1) of the function $\rho = \rho_0(x, y)$ may happen to be unacceptable and the surface density $\rho(x, y)$ may have a more general form

$$\rho(x,y) = \rho_0(x,y) + \gamma r(x,y), \qquad 0 < \gamma \ll 1. \tag{13.4.1}$$

Here, $r(x,y)$ is an arbitrary integrable (in particular, continuous) function, γ is a small parameter characterizing the perturbation of the surface density $\rho_0(x,y)$. By the standard procedure of the perturbation method we obtain an approximate expression for $\lambda_{nm}(\gamma)$,

$$\begin{aligned} \lambda_{nm}(\gamma) &= \lambda_{nm}(0) + \gamma \Lambda_{nm} + O(\gamma^2), \\ \Lambda_{nm} &= -\frac{\lambda_{nm}(0)}{\|u_{nm}\|_0^2} \int_0^a \int_0^b r(x,y) u_{nm}^2(x,y)\,dx\,dy. \end{aligned} \tag{13.4.2}$$

Here, $\lambda_{nm}(0)$ is an eigenvalue of the unperturbed problem (13.1.4) with $\gamma = 0$; $u_{nm}(x,y)$ is the corresponding eigenfunction (see (13.2.3), (13.2.4)); $\|u_{nm}\|_0^2$ is the squared norm with the weight $\rho_0(x,y)$. It is assumed that the solution of the unperturbed problem (13.1.4) has been constructed with good precision and its error does not exceed $O(\gamma^2)$. Note that further refinement of the solution (13.4.2), with higher powers of γ taken into account, is very difficult, because it requires the construction of the Green function of the unperturbed problem. This requirement is connected with the problem of finding the èntire countable sets of unperturbed eigenvalues $\{\lambda_{nm}(0)\}$ and eigenfunctions $\{u_{nm}(x,y)\}$. In the general case, the above approach can be used for constructing approximations of the Green function.

13.4.2. Nonuniform membrane tension. Consider an equation of membrane motion which is more general than (13.1.1) and takes into account nonuniform and variable character of surface tension: $\sigma = \operatorname{diag}\,(\sigma_x^0(x),\, \sigma_y^0(y))$. Thus, we come to an eigenvalue problem more general than (13.1.4) [22]:

$$\frac{\partial}{\partial x}\left(\sigma_x^0(x)\frac{\partial u}{\partial x}\right) + \frac{\partial}{\partial y}\left(\sigma_y^0(y)\frac{\partial u}{\partial y}\right) + \lambda\rho_0(x,y)u = 0; \qquad u\big|_{x=0,a} = u\big|_{y=0,b} = 0. \tag{13.4.3}$$

where $0 < \sigma_{x,y}^- \le \sigma_{x,y}^0 \le \sigma_{x,y}^+ < \infty$.

Similarly to (13.2.1), the eigenvalue problem (13.4.3) admits separation of the variables,

$$\begin{aligned} &(\sigma_x^0(x)X')' + (\lambda\varphi(x) - \mu)X = 0, \qquad X(0) = X(a) = 0; \\ &(\sigma_y^0(y)Y')' + (\lambda\psi(y) + \mu)Y = 0, \qquad Y(0) = Y(b) = 0. \end{aligned} \tag{13.4.4}$$

Now, the method of accelerated convergence of Chapter 2 can be used for each family of one-dimensional problems (13.4.4). Highly precise approximate solutions of these problems are obtained by a procedure almost identical to that of Section 13.3. Of course, the form of the equations becomes different ((13.4.4) instead of (13.3.4)) and the refinement formulas (13.3.6) are replaced by

$$\begin{aligned} \lambda_1^{(k+1)}(\mu) &= \lambda_1^{(k)}(\mu) - \varepsilon^{(k)}(\mu)a\sigma_x^0(a)\frac{(U^{(k)\prime}(a,\mu))^2}{\Phi[U^{(k)}]}, \\ \mu_1^{(l+1)}(\lambda) &= \mu_1^{(l)}(\lambda) - \delta^{(l)}(\lambda)b\sigma_y^0(b)\frac{(V^{(l)\prime}(b,\lambda))^2}{\Psi[V^{(l)}]}. \end{aligned} \tag{13.4.5}$$

13.4.3. The presence of elastic environment. Note that apart from nonuniform and variable surface tension, one can take into account the influence of an external elastic medium (Winkler's base) whose coefficient of elasticity $\chi^0(x,y) = \chi_x(x) + \chi_y(y)$ is an additive function similar to $\rho_0(x,y)$. The construction of an approximate solution of this problem follows the above scheme, but in equations of the form (13.4.4), the coefficients of X and Y become $\lambda\varphi - \mu - \chi_x$ and $\lambda\psi + \mu - \chi_y$, respectively.

13.4.4. Taking into account perturbations of general form. By analogy with Subsection 13.4.1, it is possible to take into account more general perturbations than those in (13.4.1). Such perturbations may contain terms with mixed derivatives and first derivatives [22]. Suppose that the tensor of surface tension σ has the form

$$\sigma = \begin{pmatrix} \sigma_{xx} & \sigma_{xy} \\ \sigma_{yx} & \sigma_{yy} \end{pmatrix}, \qquad \sigma_{xx} = \sigma_x^0 + \gamma\omega_x(x,y), \quad \sigma_{yy} = \sigma_y^0 + \gamma\omega_y(x,y),$$
$$\sigma_{xy} = \sigma_{yx} = \gamma\omega_{xy}(x,y), \qquad \chi(x,y) = \chi^0(x,y) + \gamma\kappa(x,y), \tag{13.4.6}$$

where γ is a small parameter and the functions σ_x^0, σ_y^0, ω_x, ω_y, ω_{xy}, and κ are sufficiently smooth. An approximate expression for $\lambda_{nm}(\gamma)$ has the form (13.4.2), and the coefficient Λ_{nm} is expressed by

$$\Lambda_{nm} = \frac{1}{\|u_{nm}\|_0^2} \int_0^a \int_0^b \Big\{ [\kappa(x,y) - \lambda_{nm}(0) r(x,y)]\, u_{nm}^2(x,y)$$
$$+ 2\omega_{xy}(x,y) \frac{\partial u_{nm}}{\partial x} \frac{\partial u_{nm}}{\partial y} + \omega_x(x,y) \left(\frac{\partial u_{nm}}{\partial x}\right)^2 + \omega_y(x,y) \left(\frac{\partial u_{nm}}{\partial y}\right)^2 \Big\}\, dx\, dy. \tag{13.4.7}$$

Formula (13.4.7) requires highly precise integration of numerically defined functions of two arguments. It should also be mentioned that for the sake of simplicity, in (13.4.1), (13.4.6) we consider perturbations of the same order with respect to the small parameter γ.

Thus, the method of accelerated convergence developed in Section 13.3 is an effective tool for the construction of approximate solutions of membrane eigenvalue problems with boundary conditions of the first kind (rigid fixation). Together with these solutions, one calculates an approximate solution (13.1.5) of the original initial boundary value problem (13.1.1)–(13.1.3), provided that the functions f, $h^{0,a}$, $g^{0,b}$, w^0, w^1 admit fairly precise approximations in terms of a few lower vibration modes $u_{nm}(x,y)$, $n, m = 1, 2, \ldots$ It should be said that the determination of frequencies and forms of free vibrations is in itself an interesting problem from the standpoint of theory and applications.

A similar, but much more cumbersome procedure, can be proposed for obtaining approximate solutions of the boundary value problem with the boundary conditions of the third kind (elastic fixation). A highly effective algorithm of accelerated convergence can be constructed on the basis of the results from [6] and the approach described in Section 13.3, with subsequent generalizations similar to those considered in Section 13.4.

To illustrate the application of our method, let us consider some examples.

13.5. Examples

Let us examine free vibrations of an inhomogeneous uniformly stretched square membrane. To demonstrate the efficiency of the algorithm based on the method of accelerated convergence, we consider problems with some specific $\rho_0(x,y)$, in particular, problems admitting analytical solutions in terms of elementary functions.

13.5.1. Inhomogeneity with respect to one coordinate. Suppose that the functions ρ_0 and χ depend only on the coordinate x: $\rho_0(x,y) = \varphi(x)$, $\chi(x,y) = \chi_x(x)$, and thus, $\psi(y) = \chi_y(y) \equiv 0$. Following Section 13.2, we get

$$X'' + (\lambda\varphi(x) - \pi^2 m^2 - \chi_x(x))X = 0, \qquad X(0) = X(1) = 0; \qquad 0 \le x \le 1,$$
$$Y = Y_m(y) = c_Y \sin(\pi m y), \qquad \mu = \mu_m = \pi^2 m^2 \quad (m = 1, 2, \ldots), \quad 0 \le y \le 1. \tag{13.5.1}$$

Here and in what follows, the variables and the parameters have been normalized by a ($b = a$). Fixing m in (13.5.1), consider the Sturm–Liouville problem for λ and X. For some specific $\varphi(x)$, $\chi_x(x)$, this problem can be effectively solved by the method of accelerated convergence of Chapter 2, and two-sided estimates of λ are obtained in [5].

For simplicity, take $\varphi(x) = (1+x)^{-2}$, $\chi_x(x) \equiv -\pi^2 m^2$. In this case, equation (13.5.1) for X is of the Euler type and admits an exact analytical solution in terms of elementary functions. The sought quantities λ_n, $X_n(x)$ have the form (see Section 1.2)

$$\lambda_n = \frac{1}{4} + \left(\frac{\pi n}{\ln 2}\right)^2, \quad X_n(x) = \sqrt{1+x}\,\sin\left(\frac{\pi n}{\ln 2}\ln(1+x)\right), \quad n \geq 1,$$
$$\lambda_1 = 20.79229, \quad X_1(x) = \sqrt{1+x}\,\sin\left(\frac{\pi}{\ln 2}\ln(1+x)\right), \tag{13.5.2}$$
$$X_1'(x^*) = 0, \quad x^* = \exp\left(\frac{\ln 2}{\pi}\left(\pi - \arctan\frac{2\pi}{\ln 2}\right)\right) - 1 \approx 0.44892,$$

where, x^* is a point of maximum of the eigenfunction $X_1(x)$, $X_1(x^*) \approx 1.19645$.

Let us calculate λ_1, using the algorithm of Section 13.3 and taking the test function $p = \sin \pi x$; see (13.3.3). As the initial approximation we take the upper bound $\lambda_1^0 = \lambda_1^* = 22.22421$. The first root and the closeness parameter (see (13.3.5)) are $\xi = 0.95459$ and $\varepsilon = 4.5 \cdot 10^{-2}$, respectively. In view of the results of Section 2.2, λ_1 is estimated from below by $\lambda_{1*} = \xi^2 \lambda_1^* = 20.25164$. The refinement constructed by (13.3.6) with the error $O(\varepsilon^2)$ is $\lambda_1^{(1)} = 20.80330$. This value differs from the exact one by a quantity of the order $5 \cdot 10^{-4} = O(\varepsilon^2)$ (relative error), see (13.5.2), and the error is $|X_1^{(1)} - X_1| \sim 10^{-4}$.

Consider another problem which admits an analytical solution. Let $\varphi(x) = (1+x^2)^{-2}$. Then the first eigenvalue λ_1, the function $X_1(x)$, the point of minimum x^*, and the maximal value X_1 are the following:

$$\lambda_1 = 15, \quad X_1(x) = \sqrt{1+x^2}\,\sin(4\arctan x) = 4x(1-x^2)(1+x^2)^{-3/2},$$
$$x^* = \tfrac{1}{\sqrt{5}} \approx 0.447214, \quad X_1(x^*) = 1.08866. \tag{13.5.3}$$

Using (13.3.3) with the test function $p = \sin \pi x$, we obtain the upper bound $\lambda_1^* = 15.33728$. Formulas (13.3.4), (13.3.5) yield the values $\xi = 0.98383$, $\varepsilon = 1.617 \cdot 10^{-2}$. The lower bound turns out equal to $\lambda_{1*} = \xi^2 \lambda_1^* = 14.83533$, and the refined value is $\lambda_1^1 = 14.99719$ (see (13.3.6)). Thus, the relative error is $2 \cdot 10^{-4}$ and its order is ε^2 (see (13.5.3)). We see that our calculations ensure great precision after only one iteration in spite of a fairly rough choice of the test function, and the error is $|X_1^{(1)} - X_1| \sim 10^{-4}$.

13.5.2. Symmetric Inhomogeneity. Let us calculate the first eigenvalue λ_{11} in the case of $\rho_0(x,y) \equiv \rho_0(y,x) = 1 + x + y$. According to the scheme of Section 13.2, we take $\varphi(x) = 1 + x$, $\psi(y) = y$. The range S_{11} of (λ, μ) is determined by (13.2.5) with $\varphi^+ = 2$, $\varphi^- = 1$, $\psi^+ = 1$, $\psi^- = 0$. Let us restrict this region by the upper bound $\lambda_{11}^* = \pi^2$ obtained by the Rayleigh principle with the test function $U = p(x)q(y) = \sin \pi x \sin \pi y$. Let us construct the function $\mu_1(\lambda)$ in a neighborhood of λ_{11}^*, mainly for smaller values of λ, by solving the Sturm–Liouville problem (13.2.1) for (μ, Y), using the method of Section 13.3. Having calculated $\mu_1(\lambda_{11}^*)$, we construct $\lambda_1(\mu)$ in a neighborhood of this value, mainly for larger values of μ. As a result, we obtain two curves which cross at the point $\lambda_{11} \approx 9.7641$, $\mu_{11} \approx 4.8827$ (see Fig. 13.3). Using linear interpolation, we easily find approximate coordinates of the crossing point, $\lambda_{11} = 9.764255$, $\mu_{11} = 4.882141$. On the sixth step of the recurrent procedure described in Section 13.3, we obtain the values $\lambda_{11}^{(6)} = 9.764544$, $\mu_{11}^{(6)} = 4.882624$.

Note that the problem is symmetric, since $\rho_0(x,y) = \rho_0(y,x)$, $a = b = 1$, and therefore, the method of separation of variables yields λ_{11} equal to the eigenvalue λ_1 of the one-dimensional

Sturm–Liouville problem for the equation $X'' + \lambda(1/2 + x)X = 0$ with the boundary conditions $X(0) = X(1) = 0$. By the method of accelerated convergence of Chapter 2, we find that $\lambda_1 = 9.763874$. This value differs from the exact eigenvalue by a relative quantity of the order $4 \cdot 10^{-5}$. Thus, the approximate eigenvalue λ_{11} calculated according to the general scheme with the use of linear interpolation has the relative error $4 \cdot 10^{-5}$. The relative calculation error for the eigenfunctions $X_{11}^{(1)}(x)$, $Y_{11}^{(1)}(y)$ is determined by the discrepancies at the points $x = 1$, $y = 1$ and is of the order $\sim 10^{-5}$, in view of (13.3.9).

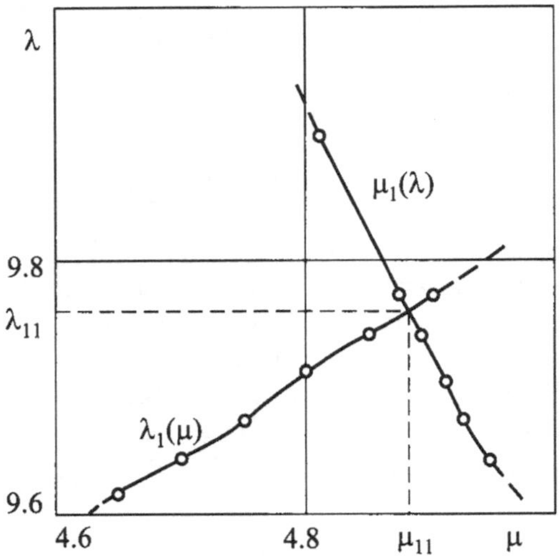

Fig. 13.3

13.5.3. Multi-coordinate approximation. Let us use the algorithm described in Chapter 9 to obtain an approximate solution of the original eigenvalue problem (13.1.4) for a partial differential equation with the Dirichlet boundary conditions. Consider free vibrations of a uniformly stretched inhomogeneous membrane of rectangular shape with clamped edges in an elastic medium. Upon separation of the time and the spatial variables, we come to the following eigenvalue problem in a rectangular domain:

$$u_{xx} + u_{yy} + [\lambda\rho(x, y) - q(x, y)]u = 0,$$
$$u|_{x=0} = u|_{x=a} = u|_{y=0} = u|_{b=0} = 0. \tag{13.5.4}$$

Here, $u = u(x, y, \lambda)$, $\rho(x, y) > 0$ is the surface density, $q(x, y) \geq 0$ is the coefficient of uniformly distributed elastic forces. Suppose that the dependence of these functions on one of the Euler variables, say y, is weak in the sense that for some scaling factor $\beta = B^{-1}$, with $B \gg b$, we have $\rho = \rho(x, \beta y)$, $q = q(x, \beta y)$. In the limit, as $\beta b \to 0$ $(b \sim a)$, the functions do not depend on y, and this allows us to separate the variables x and y in (13.5.4). We get

$$u_{nm}(x, y) = X_{nm}(x) Y_m^{(0)}(y), \quad Y_m^{(0)}(y) = \sin \frac{\pi m y}{b}, \quad n, m = 1, 2, \ldots,$$
$$X'' + \left[\lambda r(x, 0) - q(x, 0) - \left(\frac{\pi m}{b}\right)^2\right] X = 0, \qquad X(0) = X(a) = 0. \tag{13.5.5}$$

Here, X is an unknown scalar function corresponding to a given known function $Y_m^{(0)}(y)$. The eigenvalues λ_{nm} and the eigenfunctions $X_{nm}(x)$ of the Sturm–Liouville problem (13.5.5) can be effectively constructed by the method of accelerated convergence described in Chapter 2 (see also [4, 5]).

For $\rho(x,\beta y)$, $q(x,\beta y)$ with a relatively slow variation in y, it is natural to seek $u(x,y)$ in the form

$$u(x,y) = u_{(N)}(x,y) = (X, Y^{(0)}_{(y)}) \equiv \sum_{i=1}^{N} X_i Y_i^{(0)}(y), \quad N \geq 1,$$
$$PX'' + [\lambda R(x) - Q(x) - I]X = 0, \qquad X(0) = X(a) = 0. \tag{13.5.6}$$

Here, $X(x)$, $Y^0(y)$ are N-vectors, N is sufficiently large; the components of the vector-valued function $Y^0(y)$ are defined by (13.5.5). Symmetric matrices P, $R(x)$, $Q(x)$, I are obtained by substituting $u_{(N)}$ into (13.5.6), (13.5.4), multiplying equation (13.5.4) by $Y^{(0)}(y)$ and integrating in y, $0 \leq y \leq b$. Thus, we have

$$P = \operatorname{diag}\left(\tfrac{b}{2}, \ldots, \tfrac{b}{2}\right), \qquad R(x) = \int_0^b \rho(x,\beta y) Y^{(0)}(y) Y^{(0)T}(y)\, dy,$$
$$Q(x) = \int_0^b q(x,\beta y) Y^{(0)}(y) Y^{(0)T}(y)\, dy, \quad I = \operatorname{diag}\left(\frac{b}{2}\left(\frac{\pi}{b} i\right)^2\right), \quad i = 1, \ldots, N. \tag{13.5.7}$$

The properties of the functions $\rho(x,\beta y)$, $q(x,\beta y)$ ensure that the matrices P, $R(x)$, I are positive-definite, and the matrix $Q(x)$ nonnegative-definite.

Thus, we have obtained a vector Sturm–Liouville problem (13.5.6), (13.5.7) of a special type. A highly precise approximate solution $\lambda^{[1]}$, $X^{[1]}(x)$ of this problem is constructed by the method of accelerated convergence (see Chapter 2). This solution is arbitrarily close to the solution of the original problem (13.5.4), if $\beta b/N$ is sufficiently small. As shown above, for $\beta b \ll 1$, it suffices to consider the scalar variable $Y_m^{(0)} = \sin(\pi m y/b)$ (see (13.5.5)). It turns out that in the general case, if the functions $r(x,y)$, $q(x,y)$ admit separation of the variables, it suffices to have the representation (13.5.5) in which the functions $X_{nm}(x)$ and $Y_{nm}(y)$ constitute a solution of a coupled boundary value problem. Note also that weak dependence of the functions r and q on y may be of a more general nature, $r = r_0(x) + \beta r_1(x,y)$, $q = q_0(x) + \beta q_1(x,y)$, where β is small and r_1, q_1 are arbitrary sufficiently smooth functions of $(x,y) \in \Pi$.

If the dependence of $r(x,y)$, $q(x,y)$ on both variables cannot be claimed weak, the above procedure can be modified. The solution $\lambda^{[1]}$, $X^{[1]}(x)$ constructed above is used for the construction of a Sturm–Liouville problem similar to (13.5.6), in order to refine the approximations of λ, $Y(y)$. The matrices corresponding to P, $R(y)$, $Q(y)$, I are calculated by formulas similar to (13.5.7) after substituting the expression $u^{[1]}_{(N)} = (X^{[1]}, Y)$ into (13.5.4), multiplying the result by $X^{[1]}(x)$, and integrating in x, $0 \leq x \leq a$. The method of accelerated convergence is applied to the eigenvalue problem for λ and Y. As a result, one obtains a refined value $\lambda^{[2]}$ and a refined function $u^{[2]}_{(N)} = (X^{[1]}(x), Y^{[1]}(y))$. If necessary, this procedure can be recurrently continued with respect to λ, X on the basis of the refined $\lambda^{[2]}$, $Y^{[1]}(y)$. As an indirect criterion of closeness of the exact solution to its approximation it is natural to take the value of the functional corresponding to problem (13.5.4).

Computational experience shows that the finite-mode approach described here is very effective for solving the boundary value problem (13.5.4) with partial derivatives. Thus, for $N = 1$, a single iterative procedure in x or y yields practically the same value as the 4-mode approximation by the Rayleigh–Ritz method for problems with substantial dependence of r, q on x, y. The method based on a combination of the finite-mode approach and the algorithm of accelerated convergence requires further analytical justification in connection with partial differential equations with essentially variable coefficients.

13.6. Exercises

Exercise 1. Consider the functions $u_{nm}(x, y)$ and $u_{n'm'}(x, y)$ defined by (13.2.2)–(13.2.4). Prove their orthogonality with the weight $\rho_0(x, y)$ (13.2.1) for $(n, m) \neq (n', m')$.

Exercise 2. Calculate the squared norm of the functions $u_{nm}(x, y)$ with the weight $\rho_0(x, y)$ on the basis of properties (13.2.4) of the functions $X_{nm}(x)$ and $Y_{nm}(y)$.

Chapter 14

Free Vibrations of a Rectangular Membrane with Sharply Varying Surface Density

We describe the numerical-analytical method of accelerated convergence and the corresponding computational algorithm meant for constructing approximations of frequencies and shapes of free vibrations of a rectangular membrane with clamped boundary. The mass density of the membrane and its surface tension are described by functions which may change sharply and have a large variation. For definiteness, we describe calculations for the lowest vibration mode of a square membrane with constant surface tension and inhomogeneity of special type. The inhomogeneity is modelled by two orthogonal strips forming a cross or its modifications (shifted or asymmetrical cross, angle, T-shaped figure). The values of the density function (characterizing, in particular, the width of the strips and the position of their intersection) vary within a wide range. The membrane vibration characteristics are studied numerically and some interesting mechanical effects are detected and discussed.

14.1. Statement of the Problem of Free Vibrations of an Inhomogeneous Rectangular Membrane

14.1.1. Preliminary remarks. Problems of transverse vibrations of a plane membrane formulated in different terms are of great interest for theory and applications. Investigations of these problems are usually based on constructive solutions of the corresponding eigenvalue problems, which are used for the construction of finite-dimensional dynamical models. Of special interest is the problem of highly precise approximation of frequencies and shapes of lowest vibration modes depending on the mechanical characteristics of membranes. The lowest modes determine operational qualities of systems whose working surface consists of a stretched membrane. Higher vibration modes usually have small amplitudes and decay much faster than the lowest. The classical results for homogeneous uniformly stretched membranes of various shapes (rectangular, circular, annular, sectorial, elliptic, triangular, etc.) can be found in [22, 24, 28, 37, 43, 46, 61, 62, 64, 65].

For membranes of variable surface density and tension, the study and the calculation of frequencies and vibration shapes encounter great difficulties. To obtain rough estimates of these vibration characteristics numerical methods are widely used, for instance, the Rayleigh–Ritz method [22, 28] and the method of finite elements [18, 30, 60]. There are some cases for which there exist solutions in terms of special functions (for instance, for circular membranes with piecewise constant or some other simple radial inhomogeneity).

In Chapter 13, we have proposed an effective numerical-analytical method for calculating frequencies and shapes of free vibrations of an inhomogeneous rectangular membrane with clamped boundary. This method presumes that the surface density can be approximated by a sum of two functions, each depending on a single coordinate, and that the difference between this sum and the real density is relatively small and can be taken into account by the perturbation method. The class of membranes considered in Chapter 13 is fairly wide and meaningful (covers many cases occurring in practice), and the investigation of these problems has certain methodical value and is interesting for applications. The efficiency of this approach has been demonstrated by calculations for model examples in Section 13.5.

In this chapter, we consider vibrations of an inhomogeneous membrane. The density of the membrane changes sharply and to a great extent in each coordinate, and its inhomogeneity is cross-shaped. This model has engineering applications (space sails, large antennas, etc.). Such inhomogeneities formed by rectangular strips may be used as stiffening or juncture elements.

14.1.2. Statement of the boundary value problem. Consider the problem for frequencies and shapes of free vibrations of a rectangular uniformly stretched membrane with clamped boundary (see Chapter 13)

$$\begin{gathered} u_{xx} + u_{yy} + \lambda r(x_1, x_2) u = 0, \qquad 0 < x_1 < l_1, \quad 0 < x_2 < l_2; \\ u|_{x_1=0} = u|_{x_1=l_1} = u|_{x_2=0} = u|_{x_2=l_2} = 0. \end{gathered} \tag{14.1.1}$$

Here, l_1, l_2 are the side-lengths of the rectangle; $\lambda = \sigma\omega^2/T$ is the spectral parameter, σ is the characteristic value of surface density, T is surface tension, ω is the frequency, ω^2 is the constant which appears when the time variable and the spatial variables are separated. In what follows, it is assumed that equation (14.1.1) and the boundary conditions are written in dimensionless variables: the unknown function u is normalized by any characteristic quantity, and the variables x_1, x_2 and the parameters are normalized by a suitable length l (for instance, l_1 or l_2, $\sqrt{l_1^2 + l_2^2}$, or $|l_1| + |l_2|$), then, $\sigma \to \sigma l^2$.

The function r describes the variable surface density of the membrane and is assumed piecewise smooth and satisfying the conditions (see Subsection 13.4.1)

$$\begin{gathered} r(x_1, x_2) = r_0(x_1, x_2) + \rho(x_1, x_2), \qquad 0 < r^- \le r(x_1, x_2) \le r^+ < \infty, \\ r_0(x_1, x_2) = r_1(x_1) + r_2(x_2), \qquad 0 < r_0^- \le r_0(x_1, x_2) \le r_0^+ < \infty, \\ M_0 = \iint_\Pi r_0(x_1, x_2)\, dx_1\, dx_2 \gg \iint_\Pi |\rho(x_1, x_2)|\, dx_1\, dx_2 \equiv \delta M, \end{gathered} \tag{14.1.2}$$

where $\Pi = \{0 \le x_1 \le l_1,\ 0 \le x_2 \le l_2\}$.

Relations (14.1.2) have a clear mechanical meaning: the surface density function r is close (with respect to the integral measure or mass M) to an additive function r_0. The function ρ, which describes deviations from the said distribution, may have absolute values comparable with r_0; however, the distribution of the membrane mass M is mainly described by r_0. Without the loss of generality, it may be assumed that the integral of ρ over Π is equal to zero, i.e., the function ρ has zero mean value, $\langle\rho\rangle = 0$, $M = M_0$. If $\langle\rho\rangle \ne 0$, then one can introduce the function $\rho_* = \rho - \langle\rho\rangle$ and add $\langle\rho\rangle$ to r_0. Note that it is possible that $r_1 \equiv 0$ or $r_2 \equiv 0$, and the sign of r_1 or r_2 need not be fixed, but it is essential that r_0, r be strictly positive as in (14.1.2).

Our aim is to find frequencies and shapes of free vibrations of the membrane, i.e., eigenvalues λ and eigenfunctions $u(x_1, x_2, \lambda)$ of problem (14.1.1), (14.1.2). Our main attention is focused on effective highly precise calculations of these characteristics for lower vibration modes, which are of special interest for theory and applications. Such approximations can be constructed with arbitrary accuracy (with relative error of the order $(\delta M/M)^2$) for sufficiently small δM on the basis of a system of eigenvalues λ^0 and eigenfunctions $u^0(x_1, x_2, \lambda^0)$ of the generating (unperturbed) problem with $\rho \equiv 0$. The unperturbed eigenvalues and eigenfunctions are constructed by the method of accelerated convergence of Chapter 2 on the basis of rough estimates obtained by the Rayleigh–Ritz method, the iterative refinement algorithm (having quadratic convergence with respect to the initial error), and the procedure of continuation in the parameters of the problem. As mentioned above, our highly precise calculations are performed for a special class of functions $r_{1,2}(x_{1,2})$ which are nearly piecewise constant.

Note that the general theory of eigenvalue problems of the type (14.1.1) has been developed in great detail in classical works (the Schmidt theory [22, 24, 25, 33, 54, 66]). It has been shown that there exists a countable system of positive eigenvalues λ_k, some of which may be degenerate (multiple), and that the corresponding eigenfunctions form a basis.

14.2. Construction of the Generating Solution

14.2.1. Separation of the variables in the unperturbed problem. The unperturbed problem (14.1.1), (14.1.2) (with $\rho \equiv 0$) admits separation of the variables x_1, x_2 by means of the standard replacement $u = u_1(x_1)u_2(x_2)$,

$$\begin{aligned} &u_1'' + (\lambda r_1(x_1) - \mu)u_1 = 0, \qquad u_1(0) = u_1(l_1) = 0; \\ &u_2'' + (\mu + \lambda r_2(x_2))u_2 = 0, \qquad u_2(0) = u_2(l_2) = 0. \end{aligned} \tag{14.2.1}$$

Here, μ is the separation parameter which, together with λ ($\lambda > 0$) and u_1, u_2, is to be found from the system of two coupled eigenvalue problems. For problems in which the independent variables cannot be separated completely, there are no standard methods of solution, because of some difficulties of principal and computational character. To identify these difficulties we proceed as follows. Consider the special case $r_2(x_2) = r_2 = \text{const}$. Then, from the second problem (14.2.1) we establish a linear relation between λ and μ. Taking into account that relation, we obtain

$$\begin{aligned} &u_1'' + (\lambda(r_1(x_1) + r_2) - \nu_{2m}^2)u_1 = 0, \qquad u_1(0) = u_1(l_1) = 0; \\ &\mu + \lambda r_2 = \nu_{2m}^2 = \left(\frac{\pi m}{l_2}\right)^2 \qquad (m = 1, 2, \ldots). \end{aligned} \tag{14.2.2}$$

Thus, for each m, we have the standard Sturm-Liouville problem for $\lambda_{nm} = \lambda_n(\nu_{2m})$ and $u_1 = u_{1nm}(x_1) = u_{1n}(x_1, \nu_{2m})$ ($n = 1, 2, \ldots$). This problem can be solved by the usual methods (see Chapter 1). Especially effective is the method of accelerated convergence described in Chapter 2. Similar constructions may be considered in the case of $r_1(x_1) = r_1 = \text{const}$. Thus, in the situation under consideration, we have complete separation of the boundary value problems, since after obtaining λ_{nm}, we find, according to (14.2.2), that $\mu_{nm} = \nu_{2m}^2 - \lambda_{nm} r_2$. The eigenfunctions $u_{1nm}(x_1)$ can be easily found by the integration of the Cauchy problem, and the functions $u_2 = u_{2nm}(x_2) = c_{2m} \sin \nu_{2m} x_2$ are independent of the index n.

The situation is quite different for the general system (14.2.1). Arguing as above, we can find the family of solutions $\mu_m(\lambda)$, $u_2(x_2, \mu_m(\lambda), \lambda))$ of the second problem depending on the parameter $\lambda > 0$. Then, substituting $\mu_m(\lambda)$ into the first problem (14.2.1), we obtain the generalized Sturm-Liouville problem (see Chapter 4)

$$u_1'' + [\lambda r_1(x_1) - \mu_m(\lambda)]u_1 = 0, \qquad u_1(0) = u_1(l_1) = 0. \tag{14.2.3}$$

The construction of solutions of this problem encounters certain difficulties.

14.2.2. A scheme for the construction of the generating solution. In general, $\mu_m(\lambda)$ is a nonlinear function (it is linear only if $r_2 = \text{const}$; see above), which can be constructed only by analytical or numerical approximation. Thus, we have a generalized Sturm-Liouville problem. In order to construct its effective solution, a special approach is needed [8]; this approach is based on the method of accelerated convergence (see Chapter 4). It is natural to take this approach if the functions $\mu_m(\lambda)$ are determined in a relatively simple manner. Otherwise, one has to construct the solutions of two families of problems and find a common point (λ_{mn}, μ_{nm}) from the following relations:

$$\lambda = \lambda_n(\mu), \quad \mu = \mu_m(\lambda); \quad \lambda = \lambda_n(\mu_m(\lambda)), \quad (\lambda, \mu) \in S_{nm},$$
$$\lambda = \lambda_{nm} = \mathrm{Arg}_\lambda[\lambda - \lambda_n(\mu_m(\lambda))], \qquad \mu_{nm} = \mu_m(\lambda_{nm}). \tag{14.2.4}$$

The amount of calculations needed for the realization of this approach essentially depends on the region S_{nm} of the parameters μ and λ in which one constructs the functions $\lambda_n(\mu)$ and $\mu_m(\lambda)$ by (14.2.4). Rough estimates of the rectangular region S_{nm} with vertices $A_{1,2,3,4}$ are obtained in terms of the maximal and minimal values $r^{\pm}_{1,2}$ of the functions $r_{1,2}(x_{1,2})$ for $0 \le x_{1,2} \le l_{1,2}$ (see Section 13.2). Recall that $r_1^- > r_2^+$ by the construction of system (14.2.1). The characteristic shape of the region S_{nm} and a graphical representation of relations (14.2.4) are given in Fig. 14.1. Note that for $r_1 = \text{const}$ or $r_2 = \text{const}$, this region degenerates into a linear segment (A_1A_2 or A_2A_3). Practical methods for the construction of the functions $\lambda_n(\mu)$, $\mu_m(\lambda)$ and the solutions of system (14.2.1) rely on the procedure of continuation in the parameters μ and λ, combined with extrapolation of different orders (usually linear), in order to calculate approximate coordinates of the intersection point (λ_{nm}, μ_{nm}). As the initial approximation $\lambda^{(0)}_{nm}$, it is natural to use an upper bound λ^*_{nm} obtained by the Rayleigh–Ritz method. These approaches and their modifications are realized below in calculations of lower vibration characteristics.

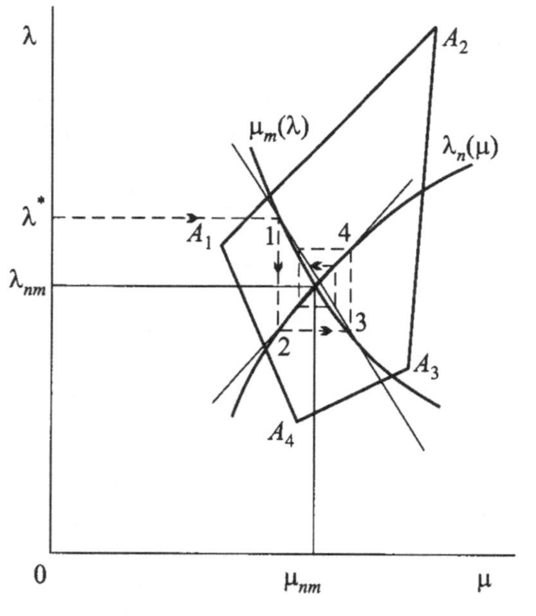

Fig. 14.1

14.3. Membrane Model with Sharply Changing Surface Density

14.3.1. Approximation of the density function. If the density function $r(x_1, x_2)$ has smooth and relatively small variation, then acceptable results can be obtained by the well-known methods, for instance, the Rayleigh–Ritz method, the finite element method, etc. If the functions $r_{1,2}(x_{1,2})$ vary sharply and to a great extent, these methods are of little use. In this case, effective highly precise mass calculations can be performed on the basis of the approach described in Section 14.2 (see [4–6, 7–9]). For definiteness, consider the function $r(x_1, x_2)$ of the form (14.1.2) with $\rho \equiv 0$. This function is described by the following relations with 14 arbitrary parameters:

$$\begin{gathered} r(x_1, x_2) = r_1(x_1) + r_2(x_2), \qquad 0 \le x_{1,2} \le l_{1,2}, \\ r_i(x_i) = \rho_i + h_i \left(\tanh \frac{x_i - a_i}{\delta_i} - k_i \tanh \frac{x_i - b_i}{\delta_i} \right) \qquad (i = 1, 2). \end{gathered} \tag{14.3.1}$$

Here, l_i, ρ_i, h_i, k_i, a_i, b_i, δ_i are constant parameters such that the function r is positive for $(x_1, x_2) \in \Pi$. It is assumed that all parameters are of the order of unity, except δ_i, $0 < \delta_i \ll 1$. As $\delta_i \to +0$, the functions r_i become close to step-functions (piecewise constant functions) which take from one to three constant values. The function $r(x_1, x_2)$ takes constant values on rectangles whose number may vary from 1 to 9 (more precisely, 1, 2, 3, 4, 6, 9). In particular, for $0 < a_i < b_i < l_i$, $k_i = 1$, and sufficiently small δ_i, we have $r = \rho_1 + \rho_2 = \rho$ outside the cross-shaped region $a_i \le x_i \le b_i$, $(i = 1, 2)$, and inside that region, we have $r = \rho + 2h_i$ on the nonintersecting pieces, and $r = \rho + 2h$, $h = h_1 + h_2$ on the intersection ($a_i \le x_i \le b_i$; $i = 1, 2$). It is convenient to represent the values of $r(x_1, x_2)$ in Π for arbitrary k_i as in Fig. 14.2, where $h_i^{\pm} = h_i(1 \pm k_i)$.

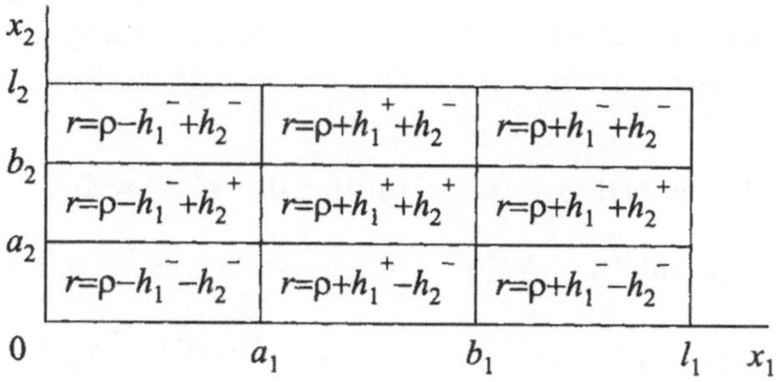

Fig. 14.2

Note that $h_i^- = h^- = 0$ for $k_i = 1$ (see above). For definiteness, we will calculate frequencies and shapes of free vibrations of the membrane with the parameters $k_i = 1$ and relatively small $(b_i - a_i)$, $l_i = 1$, and $\delta_i = 10^{-2}$, which corresponds to narrow strips forming a cross. Without the loss of generality, we can take $\rho = \rho_1 = 1$ and $\rho_2 = 0$, while the parameters $h_1 = h_2$ have a wide range such that $r(x_1, x_2)$ is positive. These strips may be located in any place of the rectangle: $0 \le a_i \le b_i$, $(b_i + a_i)/2 \le 1$. More detailed numerical data will be given and discussed below, together with the algorithm, the calculation scheme, and the calculation results for the eigenvalue λ_{11} and the eigenfunction $u_{11}(x_1, x_2)$ for a wide range of the determining parameters.

14.3.2. Brief description of the algorithm. Let us briefly describe the recurrent algorithm for the construction of approximations of the solution of problem (14.2.1), (14.3.1) with any given accuracy on the basis of the method of accelerated convergence and the procedure of continuation in parameters. For definiteness, consider the lowest vibration mode λ_{11}, $u_{11}(x_1, x_2)$ and, for brevity, drop the indices $n = m = 1$.

On the preliminary stage, it is assumed that an upper bound $\lambda^{(0)}$ has been calculated for problem (14.1.1), (14.3.1) with a fixed vector of the parameters. By the Rayleigh principle, we obtain the upper estimate

$$0 < \lambda < \lambda^{(0)} = \lambda^* = \iint_{\Pi} (\nabla \Psi)^2 \, dx_1 \, dx_2 \left[\iint_{\Pi} r(x_1, x_2) \Psi^2 \, dx_1 \, dx_2 \right]^{-1}, \tag{14.3.2}$$
$$\Psi\big|_{x_1=0} = \Psi\big|_{x_1=l_1} = \Psi\big|_{x_2=0} = \Psi\big|_{x_2=l_2} = 0.$$

Here, $\Psi = \Psi(x_1, x_2)$ is a continuously differentiable test function satisfying the boundary conditions, as well as some additional conditions of convexity (concavity), symmetry, and the like. This function is chosen from considerations of convenience and general physical notions, for instance, we can take $\Psi = \sin(\pi x_1/l_1)\sin(\pi x_2/l_2)$, $\Psi = x_1 x_2 (l_1 - x_1)(l_2 - x_2)$ or a similar function which takes into account the mass distribution $r(x_1, x_2)$. Note that for $h_{1,2} = 0$, the above trigonometric test function coincides with the exact eigenfunction ($r \equiv 1$), and from (14.1.1), (14.3.1) or (14.3.2), we obtain the exact eigenvalue $\lambda = \pi^2(l_1^{-2} + l_2^{-2})$. For $|h_{1,2}| \ll 1$, this value is an upper or a lower bound, depending on the behavior of the function r for $h_{1,2} \neq 0$.

Let us insert the value $\lambda^{(0)}$ (14.3.2) into the second equation in (14.2.1) and construct a solution of the boundary value problem for μ, u_2. By the Rayleigh principle, we obtain the upper bound $\mu^* = \mu^{(0)}(\lambda^{(0)})$ (for $h_2 = 0$, it coincides with the exact eigenvalue)

$$\mu(\lambda^{(0)}) < \mu^* = \mu^{(0)}(\lambda^{(0)}) = \int_0^{l_2} [\Psi_2'^2 - \lambda^{(0)} r_2(x_2) \Psi_2^2(x_2)] \, dx_2 \left[\int_0^{l_2} \Psi_2^2(x_2) \, dx_2 \right]^{-1}, \tag{14.3.3}$$
$$\Psi_2(0) = \Psi_2(l_2) = 0 \qquad (\mu(\lambda^{(0)}) = (\pi/l_2)^2, \quad h_2 = 0, \quad r_2 \equiv 0).$$

Then the bounds $\mu^{(0)}(\lambda^{(0)})$ are refined by means of the recurrent algorithm of accelerated convergence. This algorithm consists of the integration of the Cauchy problem with $\mu^{(j)}(\lambda^{(0)})$ found on the previous step, the determination of the first root $\xi_2^{(j)}$ of the function $v_2(x_2, \mu^{(j)})$, and the refinement of μ:

$$\begin{gathered}
v_2'' + \left[\mu^{(j)} + \lambda^{(0)} r_2(x_2)\right] v_2 = 0, \qquad v_2(0) = 0, \quad v_2'(0) = 1; \\
\xi_2^{(j)} = \arg_{x_2} v_2(x_2, \mu^{(j)}), \quad \varepsilon_2^{(j)}(\lambda^{(0)}) = l_2 - \xi_2^{(j)}(\lambda^{(0)}); \quad j = 1, 2, \ldots, \\
\mu^{(j+1)}(\lambda^{(0)}) = \mu^{(j)}(\lambda^{(0)}) - \varepsilon_2^{(j)}(\lambda^{(0)}) v_2'^2(\xi_2^{(j)}, \mu^{(j)}(\lambda^{(0)})) \left[N_2^{(j)}\right]^{-1}, \\
N_2^{(j)} = N_2(\xi_2^{(j)}, \mu^{(j)}), \quad N_2' = v_2^2(x_2, \mu^{(j)}), \quad N_2(0, \mu^{(j)}) = 0.
\end{gathered} \tag{14.3.4}$$

The quantity N_2 can be regarded as a squared norm of the function v_2. The algorithm (14.3.4) possesses accelerated (quadratic) convergence with respect to the small parameter $\varepsilon_2 = l_2 - \xi_2^{(0)}(\lambda^{(0)})$; in particular, for $h_2 = 0$, we have $\varepsilon_2 = 0$; and for $|h_2| \ll 1$, we have $\varepsilon_2 \ll 1$. In the general case ($h_2 \neq 0$), the parameter ε_2 can be made small by a suitable choice of the test function $\Psi_2(x_2)$ in the estimate (14.3.3). For $\varepsilon_2 \sim 0.1$, only a few iterations (usually 2 or 3) yield the values $\mu(\lambda^{(0)})$, $u_2(x_2, \lambda^{(0)})$ with the error of the order $10^{-4} \div 10^{-8}$. Further iterations are useless, because they would require all calculations to be performed with the precision hardly possible for modern computers.

Let us insert $\mu(\lambda^{(0)})$ constructed above into the first equation in (14.2.1) and construct the solution λ, u_1 of the boundary value problem by the method of accelerated convergence, as in the case of (14.3.4). We have

$$\begin{gathered}
v_1'' + \left[\lambda^{(j)} r_1(x_1) - \mu(\lambda^{(0)})\right] v_1 = 0, \qquad v_1(0) = 0, \quad v_1'(0) = 1; \\
\xi_1^{(j)} = \arg_{x_1} v_1(x_1, \lambda^{(j)}), \quad \varepsilon_1^{(j)}(\lambda^{(0)}) = l_1 - \varepsilon_2^{(j)}, \quad j = 1, 2, \ldots, \\
\lambda^{(j+1)} = \lambda^{(j)} - \varepsilon_1^{(j)} v_1'^2(\xi_1^{(j)}, \lambda^{(j)}) \left[N_1^{(j)}\right]^{-1}, \quad N_1^{(j)} = N_1(\xi_1^{(j)}, \lambda^{(j)}), \\
N_1' = r_1 v_1^2(x_1, \lambda^{(j)}), \quad N_1(0, \lambda^{(j)}) = 0.
\end{gathered} \tag{14.3.5}$$

The constant N_1 is the squared norm of the function v_1 with the weight $r_1(x_1)$. A practically exact solution is obtained after several iterations, and we thereby obtain $\lambda(\mu(\lambda^{(0)}))$, $u_1(x_1, \mu)$.

Thus, according to (14.3.4), (14.3.5), the first iteration step of the iteration procedure for finding λ, μ has been completed: first, we have found the point $\lambda^{(0)} = \lambda^*$, $\mu = \mu(\lambda^{(0)})$ on the curve $\mu(\lambda)$, and then the corresponding point $\lambda^{(1)} = \lambda(\mu^{(1)})$, $\mu^{(1)}$ on the curve $\lambda(\mu)$ (see Fig. 14.1, points 1 and 2, respectively). The following steps consist in the construction of the point $\lambda^{(1)}$, $\mu^{(1)}(\lambda^{(1)})$ on the curve $\mu(\lambda)$ and the point $\lambda^{(2)} = \lambda(\mu^{(2)})$, $\mu^{(2)}$ on the curve $\lambda(\mu)$ (see Fig. 14.1, points 3, 4). This process is continued until the differences $\lambda^{(k+1)} - \lambda^{(k)}$, $\mu^{(k+1)} - \mu^{(k)}$ $(k = 1, 2, \ldots)$ become less than a given small value. This process is uniformly convergent to the exact solution as $k \to \infty$, and the convergence rate is that of geometrical progression, in view of the construction of system (14.2.1) (the condition $r_1(x_1) > |r_2(x_2)|$, $(x_1, x_2) \in \Pi$, see Chapter 13). The convergence is represented by the dashed line in Fig. 14.1 resembling a trajectory near a focus or a node, if the tangents to the curves $\mu_m(\lambda)$, $\lambda_n(\mu)$ at their crossing point are both to the right of the vertical line. This can be ensured by adding a suitably large constant to r_1 and subtracting that constant from r_2. Obviously, this procedure practically attains the limit after a single step, if the curve $\mu_m(\lambda)$ near the intersection point is nearly vertical, i.e., $\mu_m(\lambda)$ weakly depends on λ (see Section 14.2). A faster convergence with respect to the parameters $\varepsilon_{1,2}$ can be ensured, if we use linear extrapolation of the function $\mu(\lambda)$ over two points 2 and 4 and of the function $\lambda(\mu)$ over two points 1 and 3; one can also use a higher-order extrapolation. In order to be ceratin that the process is convergent, it is necessary to integrate the Cauchy problems (14.3.4), (14.3.5) with $\lambda^{(k)}$, $\mu^{(k)}$ found previously and to find the discrepancies connected with the boundary conditions at the right end-points.

Each step of the above recurrent procedure contains the recurrent algorithm of accelerated convergence. As a result of that procedure we obtain the sought solution λ, μ, $u_1(x_1)$, $u_2(x_2)$ of the system of boundary value problems (14.2.1), and therefore, a solution λ, $u(x_1, x_2) = u_1(x_1)u_2(x_2)$ of the original eigenvalue problem (14.1.1), (14.1.2) for $\rho \equiv 0$ and fixed values of the parameters. Then, one should use the procedure of continuation in the parameters.

For sufficiently small variations of the parameter vector, the constructed solutions are used as estimates of the initial approximations (without the utilization of the Rayleigh principle or the Rayleigh–Ritz method). The solution is constructed according to the above scheme. If this solution is known for two close parameter vectors, then the initial approximation for the next variation of the parameters can be taken in the form of linear extrapolation, and for subsequent variations one takes quadratic, cubic, etc., extrapolation. This approach was realized for solving problem (14.2.1), (14.3.1) with the parameter h $(h_{1,2} = h)$ varying within a fairly wide range, while the other parameters had been fixed and we had chosen their numerical values so as to be able to evaluate their influence (see below).

14.3.3. Software. The software created for the realization of the above algorithm is based on the Maple computer algebra. Our choice of this system of symbolic programming was dictated by the following reasons: simplicity of the method of accelerated convergence and the high speed of modern computers, the wish to create a working tool with the simplest possible interface. The software includes the procedures of integration of ordinary differential equations, computation of frequencies and shapes of free vibrations, construction of their graphs depending on various parameters, extrapolation in parameters, and some other procedures. The Maple, Mathematica, MATLAB, and other commonly used software packages have their own procedures for solving ordinary differential equations, but these procedures are not very efficient for the integration of higher-dimensional systems of equations. This fact was discovered when solving some problems of celestial mechanics. Because of its universality, the procedure based on the Maple computer algebra takes no account of the properties of specific equations of the investigated models. Moreover, we need a procedure of integration which stops as soon as a ceratin condition is satisfied. The integration algorithm is realized in terms of the procedure intRK (realization of the Runge-

Kutta method and an algorithm for choosing the integration step) with the following tree formal parameters: *eqlist* is the list containing variable identifiers and the right-hand sides of integrated equations; *inilist* is the list of equalities whose left-hand side is the name of the integrated function, and its right-hand side is its initial value. The third parameter is the finite value of the integration variable or some functional (zero value of that parameter terminates the integration process). The parameters like the initial value of the integration variable, the integration step, the precision, etc., are automatically determined within the procedure, but may be changed through the corresponding global variables with specific names. For instance, in order to assign the value 0.1 to the iteration step, the operation *step:*=0.1 should be performed.

Calculations of frequencies and shapes of free vibrations are performed by the procedures *Iter* and *valU*. In order to prevent cycling, the number of iterations for μ and λ has been limited to three. This proved sufficient for obtaining highly precise results (with the relative error of the order 10^{-4}). For a third-order polynomial extrapolation in a parameter, it suffices to perform a single iteration with step $\Delta h = 0.25$.

The program for obtaining approximations of eigenvalues was implemented for the following three cases of parameter values.

1. *Symmetric cross.* The intersection of the strips is at the center of the rectangle (in particular, the square, for $l_1 = l_2 = 1$): $b_i = l_i - a_i$ $(i = 1, 2)$. The effect of the strip density (the parameter $h = h_{1,2}$) and their width $d = d_i = b_i - a_i$ is examined.

2. *Shifted cross.* One of the strips remains in the center (in particular, $b_2 = l_2 - a_2$) and the other is shifted to the edge $a_1 \to 0$. One examines the effect of its position relative to the center for different values of the density h and fixed widths of the strips, d_i.

3. *Nonsymmetric cross.* Both strips are shifted from the center to the boundary, $a_i \to 0$; the density parameter h is varied within a wide range; the strip widths d_i are fixed.

The corresponding numerical data and the graphs are given in Section 14.4, where we also discuss the results and the discovered mechanical effects.

14.4. Calculation Results and Conclusions

Figures 14.3–14.11 give the graphs of our calculation results for the eigenvalues λ, μ and the normalized eigenfunctions $u_i(x_i)$ $(i = 1, 2)$ for the above three cases of parameter values for a square membrane, $l_{1,2} = 1$. In all these cases, we take $\rho_1 = 1$, $\rho_2 = 0$, $h_1 = h_2 = h$, $k_1 = k_2 = 1$, $\delta_1 = \delta_2 = 0.01$. Recall that the functions r, $r_{1,2}$ (14.3.1) change sharply and to a great extent near $x_i = a_i$, $b_{i,j}$, and their derivatives for these x_i have the form $r'_{1,2} = 2h/\delta \sim 10^3$. The calculations were performed with the relative error of the order $10^{-3} \div 10^{-4}$ (the discrepancy in the boundary conditions).

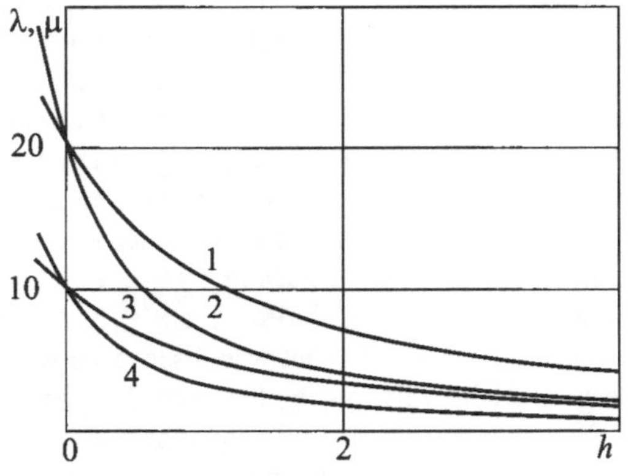

Fig. 14.3

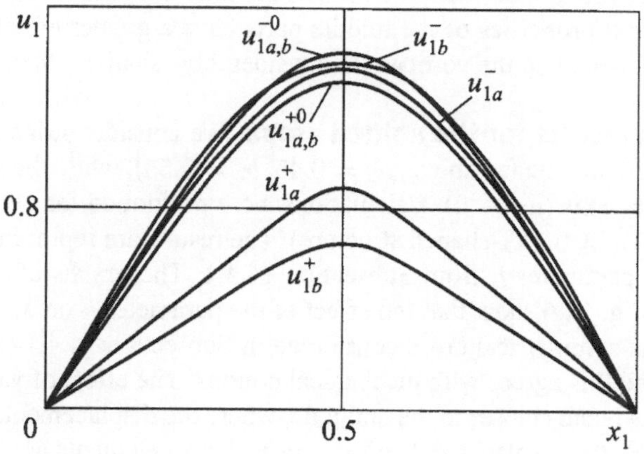

Fig. 14.4

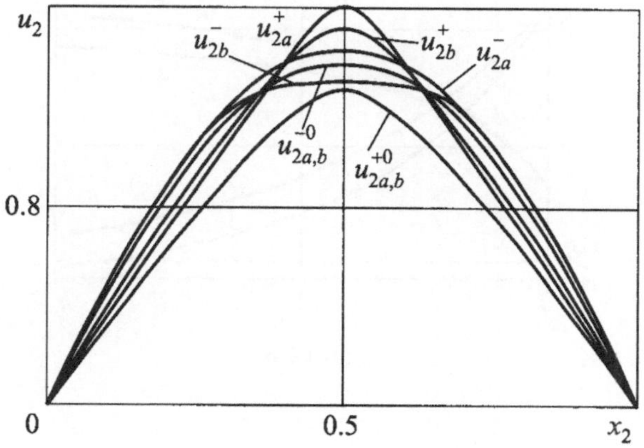

Fig. 14.5

14.4.1. Calculation results for the symmetrical cross. The calculation results for case 1 are represented in Figs. 14.3–14.5. The graphs in Fig. 14.3 give the dependence of the eigenvalues λ, μ of problem (14.2.1) (curves 1, 2 correspond to $\lambda_{a,b}$, and curves 3, 4 to $\mu_{a,b}$) for different $h \in [-0.225; 4.5]$ and two sets of a_i, b_i characterizing the strip widths $d_i = b_i - a_i$, namely, (a) $a_{1,2} = 0.45$, $b_{1,2} = 0.55$; (b) $a_{1,2} = 0.4$, $b_{1,2} = 0.6$. Thus, the strip widths in case (a) is equal to $d = 0.1$, and in case (b) the strip width is equal to $d = 0.2$. The density function r is such that outside the strips $r = 1$; on the strips outside the region of their intersection, $r \in [0.45; 9]$; inside the region of intersection, $r \in [0.1; 18]$. Note that with the decrease of h, a fast growth of $\lambda_{a,b}$ occurs, so that $\lambda_b > \lambda_a$ for $h < 0$ and $\lambda_b < \lambda_a$ for $h > 0$. With the growth of h, the values $\lambda_{a,b}$ monotonically decrease, which is natural. Figures 14.4, 14.5 represent the families of normalized "eigenfunctions" $u_1(x_1)$, $u_2(x_2)$ for $d = 0.1$, 0.2 (marked by the indices a, b) and different $h = -0.225$; -0.0225; 0.225; 4.5 (marked above by the indices $-$, -0, $+0$, $+$, respectively). Thus, we have eight functions $u^{\pm}_{1a,b}(x_1)$, $u^{\pm 0}_{1a,b}(x_1)$ in Fig 14.4 and eight functions $u^{\pm}_{2a,b}(x_2)$, $u^{\pm 0}_{2a,b}(x_2)$ in Fig. 14.5. Note that the functions $u^{\pm 0}_{1a,b}(x_1)$ are nearly identical and are therefore represented by the same curve. In a similar way we represent the "vibration shapes" $u^{-0}_{1a,b}(x_1)$, $u^{\pm 0}_{2a,b}(x_2)$. These graphs show that for $h = -0.225$ the curves flatten in the middle and their convexity is more pronounced near the ends of the intervals (small inertia of the central part). For $h = 4.5$, the inertial properties of the middle parts have a greater effect: the lateral pieces of the curves are nearly linear, and the vertices are considerably "sharper" than for $h \ll 1$ and $h < 0$.

14.4.2. Calculation results for the shifted cross. We consider strips of width $h = 0.1$, the horizontal strip is in the middle (as above, $a_2 = 0.45$, $b_2 = 0.55$), while the vertical strip is shifted to the left along the x_2-axis ($a_1 \to 0$). Calculations were performed for two cases: (a) $a_1 = 0.2$; $b_1 = 0.3$; (b) $a_1 = 0$; $b_1 = 0.1$ (T-shaped structure). The results are represented in Figs. 14.6–14.8 for the values of the parameter h from Subsection 14.4.1. The graphs of $\lambda_{a,b}$ (curves 1, 2) and $\mu_{a,b}$ (curves 3, 4) in Fig. 14.6 show that the effect of the parameter h on $\lambda_{a,b}$ is somewhat weaker than in the case of the symmetrical cross considered in Subsection 14.4.1 (Fig. 14.3). Its effect is greater in case (a), and this agrees with mechanical notions. The effect of variation of h increases, if the density variation occurs closer to the antinode, where the displacements and the accelerations are greater. In case (b), the displacements of the vertical strip elements are insignificant.

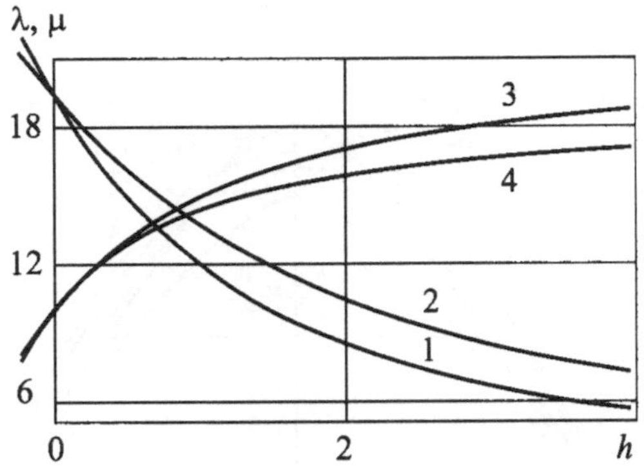

Fig. 14.6

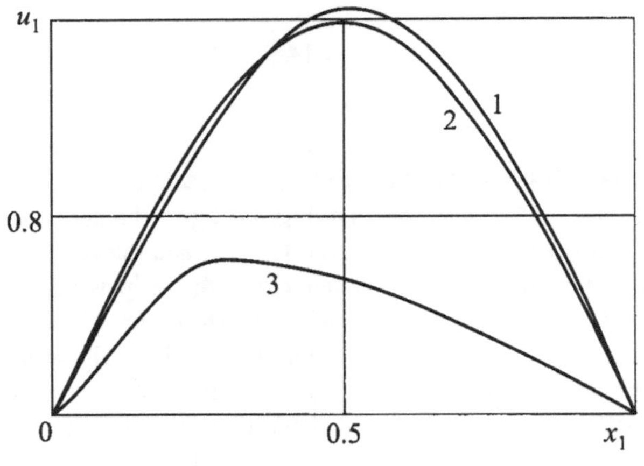

Fig. 14.7

The "eigenfunction" $u_{1a,b}(x_1)$ becomes nonsymmetric with respect to the middle (Fig. 14.7), which seems natural. In case (a) for the maximal $h = 4.5$, the "shape" u_1 changes substantially. Curve 1 for u_{1b}^{+} and curve 2 for $u_{1a,b}^{-}$ are relatively close (curve 3 corresponds to u_{1a}^{+}); and these curves are also very close to those representing $u_{1a,b}^{\pm 0}(x_1)$ (omitted in Fig. 14.7 to avoid too many graphs). It is interesting to observe a subtle property of the curves $u^{\pm 0}_{1a,b}$, namely, that their maximal points are shifted to the right of the middle point $x_1 = 1/2$. The family of the curves $u^{\pm}_{2a,b}$ in Fig. 14.8 (curves 1, 2, 3 correspond to u^{+}_{2a}, $u^{-}_{2a,b}$, u^{+}_{2b}) are symmetrical with respect to the middle; the graphs of $u^{+}_{2a,b}$ are similar to the corresponding graphs in Figs. 14.4, 14.5. The curves $u^{-}_{2a,b}$ practically coincide, the curves of $u^{\pm 0}_{2a,b}$ are omitted, being very close to $u^{-}_{2a,b}$ and to $\sqrt{2}\sin \pi x_2$ (see Figs. 14.4, 14.5.).

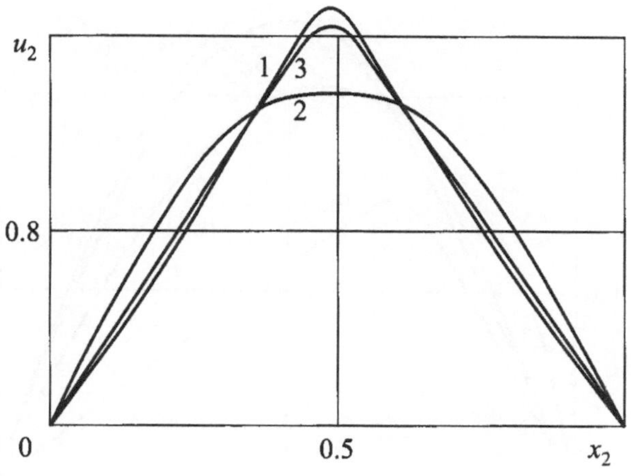

Fig. 14.8

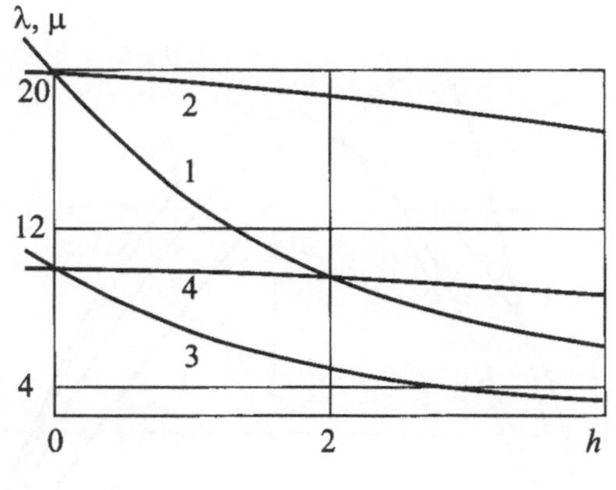

Fig. 14.9

14.4.3. Calculation results for the nonsymmetric cross. Figures 14.9–14.11 graphically represent the results obtained for the eigenvalues λ, μ and the "shapes" $u_1(x_1)$, $u_2(x_2)$ for the nonsymmetric cross. The strips of width $d = 0.1$ were jointly shifted: (a) $a_{1,2} = 0.2$; $b_{1,2} = 0.3$; (b) $a_{1,2} = 0$, $b_{1,2} = 0.1$ (angular structure). The graphs of $\lambda_{a,b}$, $\mu_{a,b}$ (curves 1–4 in Fig. 14.9) show that their dependence on h becomes even weaker. The curves λ_b, μ_b are close to linear functions of a small slope, which can be explained by the small effect of density of the strips located near the fixed boundary. The effect of h is much greater on the curves λ_a, μ_a corresponding to shifted strips.

The family of "eigenfunctions" $u^{\pm 0}_{1a,b}(x_1)$ is represented in Fig. 14.10. These curves correspond to mechanical notions. Observe that the graphs of $u^{-}_{1a,b}$ are relatively close to one another and to the curves $u^{\pm 0}_{1a,b}$ (omitted here), and also to the sine graph mentioned above. It should again be said that the points of maximum of $u^{-}_{1a,b}$ are also shifted to the right of the middle line. The behavior of the curves $u^{\pm}_{2a,b}(x_2)$ is similar, but the influence of $r_2(x_2)$ on their shape is more pronounced (see Fig. 14.11).

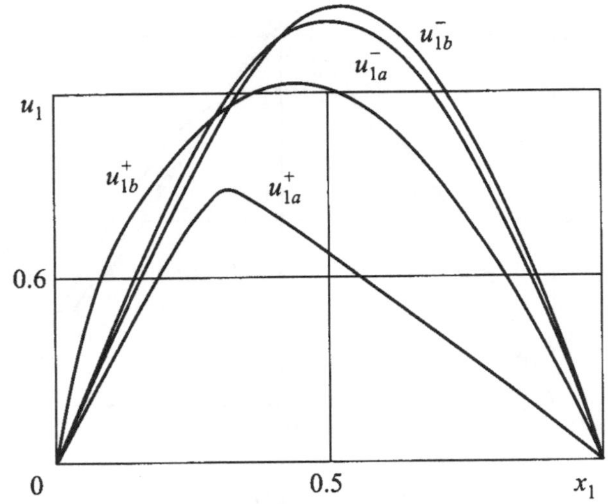

Fig. 14.10

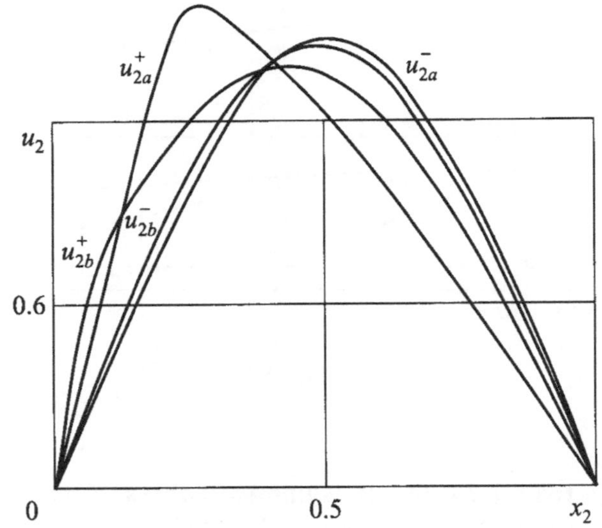

Fig. 14.11

14.4.4. Conclusions. We have described a constructive and fairly detailed investigation of frequencies and shapes of free vibrations of the fundamental vibration mode of a square membrane with cross-shaped inhomogeneities. Calculations, performed for a wide range of determining parameters, allowed us to identify and study some interesting mechanical effects.

Our rapidly convergent algorithm can be used for calculating approximations of higher vibration modes, λ_{nm}, $n+m \geq 3$, and this does not entail any difficulties of principal or computational character. The difference from the procedure of Section 14.3 consists in the choice of the corresponding test function $\Psi(x_1, x_2)$ in the calculation of the bound $\lambda_{nm}^{(0)}$ (in particular, for $h = 0$, one has to take $\lambda_{nm} = \pi^2(n^2 + m^2)$, which is the exact value) when determining the m th positive root of the function $v_2(x_2, \mu^{(j)})$ by (14.3.4), and the n th positive root of the function $v_1(x_1, \lambda^{(j)})$ by (14.3.5).

Perturbations $\rho(x_1, x_2)$ (14.1.2) which are small in the integral sense can be taken into account quite easily on the basis of the perturbation method [9] (see Chapter 13).

The above approach can be used for highly precise calculations of frequencies and shapes of free vibrations of an inhomogeneous rectangular membrane with variable surface tension and an inhomogeneous Winkler's base. The above method of accelerated convergence, combined with the procedure of continuation in parameters, is an effective tool for the construction of highly precise two-sided estimates of frequencies and shapes of free vibrations of inhomogeneous membranes for some classes of domains and boundary conditions of the second or the third kind.

Chapter 15

Free Vibrations of Elastic Systems in Elliptic Domains

Vibrations of distributed systems in domains with elliptic (in particular, circular) boundary is an object of numerous investigations in the theory of elasticity (plates and membranes), hydrodynamics, acoustics, electrodynamics (resonators), etc. In spite of the fact that such problems are formulated very simply and admit partial separation of spatial variables in elliptic coordinates, their numerical solution for large eccentricity values is difficult. For this purpose, new classes of special functions have been introduced. Great achievements in this field should be credited to Mathieu, Rayleigh, Ince, Strutt, McLachlan, Sretensky, and others.

In this chapter, we describe some approaches to the construction of highly precise two-sided estimates for eigenvalues (frequencies) and eigenfunctions (vibration shapes) of boundary value problems representing mechanical models of elliptic plates and membranes with arbitrary eccentricity values. These approaches are based on the rapidly convergent procedure of the method of accelerated convergence.

15.1. Free Vibrations of a Homogeneous Elliptic Membrane

15.1.1. Preliminary remarks regarding the present state of the investigations. The problem of vibrations of a homogeneous uniformly stretched membrane is of great importance for theory and applications. Its investigation attracted the attention of numerous specialists in mechanics and mathematics. One of the first works in this field dates back to 1868 and is due to Mathieu, who introduced special functions (now bearing his name) for solving the related boundary value problems of Strum–Liouville type [24, 33, 43, 56, 61, 62, 67]. These functions were tabulated and thoroughly studied in many analytical and numerical aspects. Still, the studies of the problem of free vibrations of elliptic membranes are far from being complete, because there are no fairly simple means for the calculation of the Mathieu functions for different values of their argument and their parameter. It is important to note that in order to solve this two-dimensional boundary value problem of the second order, one has to find countably many roots of a system of two numerically defined transcendental equations, because the original independent variables cannot be completely separated [10, 43, 61] (the separation parameters enter both equations, the common and the modified Mathieu equations).

Modern computers and software utilizing the Rayleigh–Ritz method, the finite element method, etc., allow us to construct numerical solutions for elliptic membranes with some specific parameters, in particular, with the eccentricity not too large, and the results obtained in this way are not always sufficiently precise. It is important to obtain highly precise numerical-analytical approximations of frequencies and shapes of free vibrations of elliptic membranes with arbitrary semi-axes under various additional conditions, in particular, the condition of fixed area or a fixed semi-axis. No universal curves similar to the Ince–Strutt diagrams [61] and characterizing the dependence of the eigenvalues (frequencies) and eigenfunctions (vibration shapes) on the eccentricity have been constructed so far. In this chapter, we construct such curves for lower vibration modes and analyze their behavior (see [10]).

15.1.2. Setting of the problem. The problem of frequencies and shapes of free vibrations of an ideal uniformly stretched homogeneous elliptic membrane with clamped boundary can be formulated as the following self-adjoint eigenvalue problem [14, 42]:

$$\frac{\partial^2 u}{\partial x^2} + \frac{\partial^2 u}{\partial y^2} + \lambda u = 0, \qquad u|_{\partial G} = 0. \tag{15.1.1}$$

Here, $u = u(x, y)$ is a smooth unknown function characterizing the membrane vibration shapes; $u \in G$, where $G = \{x^2 a^{-2} + y^2 b^{-2} \le 1\}$ is the ellipse with the boundary ∂G; $\lambda = \omega^2 \rho T^{-1}$ is the spectral parameter to be found; the surface density ρ, the surface tension T, and the frequency of free vibrations ω are assumed constant. For definiteness, we assume that a is the larger semi-axis of the ellipse and b is its the smaller semi-axis.

The traditional approach to problem (15.1.1) is based on the introduction of elliptic coordinates α, φ and passing from the variables (x, y) to (α, φ) in equation (15.1.1) [46],

$$\begin{gathered} x = d\cosh\alpha\cos\varphi, \quad y = d\sinh\alpha\sin\varphi, \quad d = ae, \quad e = (1 - b^2 a^{-2})^{1/2}, \\ 0 \le \alpha \le \alpha^* = \operatorname{arcosh} e^{-1}, \quad 0 \le \varphi \le 2\pi \pmod{2\pi}, \quad 0 \le e < 1; \\ \frac{\partial^2 u}{\partial \alpha^2} + \frac{\partial^2 u}{\partial \varphi^2} + \mu(\cosh^2\alpha - \cos^2\varphi)u = 0, \qquad \mu = \lambda d^2 > 0. \end{gathered} \tag{15.1.2}$$

Here, e is the eccentricity of the ellipse G; the parameter d is such that the boundary ∂G for $\alpha = \alpha^*$, $0 \le \varphi \le 2\pi$ coincides with one of the coordinate ellipses. Equation (15.1.2) admits separation of the variables α and φ, i.e., its solutions may be sought in the form of monomials [10, 43, 46, 61]

$$u = A(\alpha)\Phi(\varphi), \qquad 0 \le \alpha \le \alpha^*, \qquad 0 \le \varphi \le 2\pi, \tag{15.1.3}$$

where A, Φ are unknown functions, $\Phi(\varphi)$ is periodic with period π or 2π. Substituting (15.1.3) into (15.1.2) and introducing a separation constant χ for separating the variables α and φ, we obtain the system of equations

$$\begin{aligned} &A'' + (\mu\cosh^2\alpha - \chi)A = 0, \qquad && 0 \le \alpha \le \alpha^*(e), \\ &\Phi'' + (\chi - \mu\cos^2\varphi)\Phi = 0, && 0 \le \varphi \le 2\pi. \end{aligned} \tag{15.1.4}$$

This system contains two unknown parameters μ, χ to be determined jointly. The structure of equations (15.1.4) allows us to write the boundary conditions corresponding to the original ones in (15.1.1) as follows [43, 46, 61]:

$$A(0) = A(\alpha^*) = 0, \qquad \Phi(0) = \Phi(\pi) = 0; \tag{15.1.5}$$

$$A'(0) = A(\alpha^*) = 0, \quad \Phi'(0) = \Phi'(\pi) = 0. \tag{15.1.6}$$

Conditions (15.1.5) determine the so-called odd solutions of the Mathieu equation for Φ (15.1.4) and the corresponding solution of the modified Mathieu equation for A. Similarly, conditions (15.1.6) determine even solutions of the problem. Thus, the original eigenvalue problem (15.1.1) for a partial differential equation in an elliptic domain has been reduced, by the transformations (15.1.2), (15.1.3), to boundary value problems of two types (15.1.4)–(15.1.6) for a system of two ordinary differential equations coupled by the parameters μ, χ.

The above relations can be used for the construction of numerical and analytical solutions of the original problem. It should be observed that the transformation (15.1.2) is degenerate for $e = 0$ (for a circular domain G, ∂G is the periphery of the circle), since $\alpha^* = \infty$ in this case. For $e \to 0$, one has to be careful when passing to the limit and take into account that $d \cosh\alpha \to \alpha r$, $d \sinh\alpha \to ar$, where r is the polar radius ($0 \le r \le 1$). Equation (15.1.4) for the function A turns into the Bessel equation, and in the Mathieu equation for Φ, one should take $\mu = 0$ [43, 46, 61]. Thus, the problem for the unknown Φ has been completely separated (χ is the separation parameter) and can be easily solved in terms of trigonometric functions. Now, in order to solve the problem for a circle, one can use well-known arguments [24, 64, 65]. Moreover, (15.1.2) implies that for $e \to 1$, the replacement transformation also becomes degenerate, since $\alpha^* \to 0$, i.e., the interval of the argument α becomes infinitely small. Therefore, the case $e \to 1$, too, requires careful asymptotic analysis when solving the boundary value problem (see Subsection 15.1.5).

It is required to develop efficient numerical-analytical methods for the construction of highly precise approximations of joint solutions of the eigenvalue problems (15.1.4)–(15.1.6) and for their analysis. Interpolation procedures based on tabulated Mathieu functions require much effort and are often unsuitable for highly precise reliable calculations. The approaches based on formal separation of the problems for the variables A and Φ, with one of the parameters χ or μ assumed known from some general considerations, have no justification and have resulted in gross computational errors [39, 51].

15.1.3. Variational approach and the construction of highly precise estimates. The eigenvalue problem (15.1.1) is equivalent to the following variational problem [24, 46]: *find the minimum of the functional $J[u]$ on the class of continuously differentiable functions $\{u\}$ satisfying zero boundary conditions and the conditions of normalization,*

$$\begin{gathered} J[u] = \iint_G (\nabla u)^2 \, dx \, dy \to \min_u, \qquad u|_{\partial G} = 0, \\ I[u] = \iint_G u^2 \, dx \, dy = I_0 = \text{const}. \end{gathered} \tag{15.1.7}$$

Here, I_0 is a fixed positive constant, say, $I_0 = 1$. If a function $u_1(x, y)$ realizing this minimum has been found, then the smallest eigenvalue λ_1 characterizing the lowest vibration mode has the form $\lambda_1 = J[u_1]/I[u_1]$. Subsequent eigenfunctions $u_k(x, y)$ and eigenvalues $\lambda_k (k \ge 2)$ are determined by relations (15.1.7) with the additional orthogonality conditions

$$\begin{gathered} I_l[u] = \iint_G u u_l \, dx \, dy = 0, \quad \lambda_k = \frac{J[u_k]}{I[u_k]}; \\ l = 1, 2, \ldots, k-1, \; k = 1, 2, \ldots \end{gathered} \tag{15.1.8}$$

Relations (15.1.7), (15.1.8) form the basis of the Rayleigh–Ritz method for the construction of effective upper bounds λ_k^* of eigenvalues of problem (15.1.1), and also for the construction of the corresponding eigenfunctions $u_k^*(x,y)$. The precision of these bounds may depend, sometimes to a great extent, on the choice of the test functions which should take into account specific properties of the problem and the vibration mode under consideration. Experience shows that disregard of these factors may result in gross computational errors.

Below, we give some examples with aptly chosen test functions allowing us to obtain highly precise approximations of the first and subsequent eigenvalues on the basis of the Rayleigh principle. For this purpose, we pass from the coordinates (x,y) to the generalized polar coordinates (r,φ) by

$$x = ar\cos\varphi, \quad y = br\sin\varphi, \quad 0 \le r \le 1, \quad 0 \le \varphi \le 2\pi,$$
$$r = (x^2a^{-2} + y^2b^{-2})^{1/2}, \qquad dS = abr\,dr\,d\varphi. \tag{15.1.9}$$

Here, dS is the area element of the ellipse. Note that the coordinate system (r,φ) is nonorthogonal. For $r = \text{const}$, we have the homothetic ellipses $e = (1-b^2a^{-2})^{1/2} = \text{const}$. It seems that the variables (r,φ) of the form (15.1.9) have never been used in previous studies of vibrations of elliptic plates and membranes.

First, consider the case in which the test functions in the generalized polar coordinates (r,φ) do not depend on the angle φ: $u = u(r)$. Then the variational problem (15.1.7) for the unknown $u(r)$, after the integration in φ, reduces to the problem

$$J[u] = \pi ab(a^{-2}+b^{-2})\int_0^1 u'^2 r\,dr \to \min_u, \quad u = u(r),$$
$$I[u] = 2\pi ab\int_0^1 u^2 r\,dr = 1, \quad u(1) = 0, \quad |u(0)| < \infty. \tag{15.1.10}$$

Let us write the Euler–Lagrange equation corresponding to problem (15.1.10). Thus, we obtain the classical Sturm–Liouville problem for the Bessel equation of zero order [32, 33, 46, 67]:

$$r^{-1}(ru')' + \nu u = 0, \qquad r(1) = 0, \quad |r(0)| < \infty; \qquad \nu = 2\lambda(a^{-2}+b^{-2})^{-1}. \tag{15.1.11}$$

The solution of problem (15.1.1) has the form $u_{n0}(r) = c_{n0}J_0(\gamma_{n0}r)$, where γ_{n0} are zeroes of the function $J_0(\gamma)$. Consider the first vibration mode $(n = 1)$

$$u_{10}(r) = c_{10}J_0(\gamma_{10}r), \quad \gamma_{10} = 2.4048, \quad \lambda_{10}^* = \tfrac{1}{2}\gamma_{10}^2(a^{-2}+b^{-2}),$$
$$\lambda_{10}^*(e) = (\gamma_{10}^2a^{-2})(1-\tfrac{1}{2}e^2)(1-e^2)^{-1} = \gamma_{10}^2a^{-2}(1+\tfrac{1}{2}e^2+O(e^4)), \quad e \ll 1. \tag{15.1.12}$$

Here, $J_0(z)$ is the Bessel function of zero order, γ_{10} is the first root of the function $J_0(\gamma)$, c_{10} is a constant chosen from the normalization condition, $c_{10}^2 = 1/2J_1^2(\gamma_{10})$ (J_1 is the Bessel function of the first order). Thus, (15.1.2) gives us an upper bound for the first eigenvalue of problem (15.1.1) and the corresponding approximation of the first eigenfunction, since the functional class $\{u(r)\}$ is more narrow than the class $\{u(r,\varphi)\}$. Note that for the first eigenvalue λ_{10}, which determines the fundamental frequency of free vibrations, we have obtained the two-sided estimates [62]:

$$\gamma_{10}^2(ab)^{-1} \le \lambda_{10} \le \tfrac{1}{2}\gamma_{10}^2(a^{-2}+b^{-2}) = \lambda_{10}^*. \tag{15.1.13}$$

For $a = b$, the bounds in (15.1.3) are precise in the sense that they are attained for a circular membrane, $\lambda_{10} = \gamma_{10}^2a^{-2}$. From (15.1.13), it follows that among elliptic membranes of the fixed area $S = \pi ab$, the circular membrane has the smallest fundamental frequency.

Calculations show that the upper bound λ_{10}^* (15.1.12) is fairly precise for a wide range of eccentricity values, e (15.1.2). Let us fix the half-axis a, say, take $a = 1$. Then, $b = (1-e^2)^{1/2}$. Let us vary e on the interval $0 \le e \le 0.9$ and compare the upper bound $\lambda_{10}^*(e)$ (15.1.12) with the highly precise lower bound $\lambda_{10*}(e)$ obtained by the numerical-analytical method of accelerated convergence [8–10] for large values of e with the error of the order 10^{-5}. The calculation results are listed in Table 15.1.

Table 15.1

e	λ_{10}^*	λ_{10}	$\Delta\lambda/\lambda$
0.5	6.74691	6.74653	10^{-4}
0.6	7.40955	7.38894	$1.4 \cdot 10^{-3}$
0.7	8.56120	8.52048	$5 \cdot 10^{-3}$
0.8	10.92356	10.84783	10^{-2}
0.9	18.11012	17.93740	10^{-2}

Table 15.1 demonstrates the precision of the bound $\lambda_{10}^*(e)$ obtained by (15.1.12) for reasonable values of the parameter e. Note that for $e \le 0.5$, the value $\lambda_{10}^*(e)$ is even closer to the exact value $\lambda_{10}(e)$. Moreover, if we consider the class of elliptic membranes of the fixed area S for different a and b such that $S = \pi ab$, then the bound $\lambda_{10}^*(e)$ takes the form

$$\lambda_{10}^* = \tfrac{1}{2}\pi\gamma_{10}^2 S^{-1}(2-e^2)(1-e^2)^{-1/2}, \qquad \lambda_{10}(e) \le \lambda_{10}^*(e).$$

This bound is also fairly precise for the said values of e. By analogy with radially symmetric vibrations of a circular membrane, the vibration shape $u_{10}(r)$ obtained above may be called "elliptically symmetric" (see (15.1.9)). However, this is only an approximate solution, although fairly precise for e not too large ($e \le 0.5$). It can be expected that calculations of the upper bounds $\lambda_{n0}^*(e)$ ($n = 1, 2, \ldots$) for the eigenvalues $\lambda_{n0}(e)$ on the basis of the roots γ_{n0} (similarly to (15.1.12)) would result in larger absolute errors, which may be admissible for small enough $e > 0$ (the relative error will be substantially smaller).

Now, consider a vibration shape depending on the angular variable φ and containing a nodal line along the y-axis. For this purpose, we take a test function of the form $u(r,\varphi) = v(r)\cos\varphi$, where $v(r)$ is an unknown admissible function. Substituting u into (15.1.7) and integrating in φ, we obtain the following variational problem for $v(r)$ (cf. (15.1.10)):

$$J[v] = \pi ab \int_0^1 (A^c r v'^2 + B^c v v' + A^c r^{-1} v^2)\,dr \to \min_v,$$

$$I[v] = \pi ab \int_0^1 v^2 r\,dr = 1, \qquad v(1) = 0, \quad |v(0)| < \infty, \tag{15.1.14}$$

$$A^c = \tfrac{3}{4}a^{-2} + \tfrac{1}{4}b^{-2}, \qquad B^c = a^{-2} - b^{-2}.$$

The problem for eigenvalues and eigenfunctions corresponding to (15.1.14) can be written as the first-order Bessel equation with the corresponding boundary conditions. Thus, we have

$$r^{-1}(rv')' - r^{-2}v + \nu v = 0, \qquad v(1) = 0, \quad |v(0)| < \infty;$$

$$\nu = \frac{\lambda}{A^c}, \quad \nu_{n1} = \gamma_{n1}^2, \quad J_1(\gamma) = 0, \quad v_{n1}(r) = c_{n1} J_1(\gamma_{n1} r). \tag{15.1.15}$$

Note that problem (15.1.15) does not contain B^c, since $vv' = \frac{1}{2}v^{2'}$ in (15.1.14). The solution of the Sturm–Liouville problem (15.1.15) is constructed similarly to (15.1.11), (15.1.12). The smallest eigenvalue λ_{11}^{*c} provides a good upper bound for the original problem,

$$\lambda_{11}^c \le \lambda_{11}^{*c} = A^c\gamma_{11}^2, \quad \gamma_{11} = 3.8317, \quad A^c = a^{-2}(1 - \tfrac{3}{4}e^2)(1 - e^2)^{-1}. \tag{15.1.16}$$

In the limit case, $a = b$, we have $A^c = a^{-2}$ and the bound λ_{11}^{*c} becomes equal to the exact value $\lambda_{11}^c = \gamma_{11}^2 a^{-2}$. Since the function $v_{11}(r)\cos\varphi$ is orthogonal to $u_{10}(r)$, we have $\lambda_{11}^{*c} > \lambda_{10}^*$.

Now, let us find the error of the bound $\lambda_{11}^{*c}(e)$ for different e and $a = 1$, $b = (1 - e^2)^{1/2}$ (see above). These data are given in Table 15.2. The relative error of the bound $\lambda_{11}^{*c}(e)$ for $e \le 0.6$ is less than 10^{-4} and is therefore omitted. Thus, Table 15.2 shows that the vibration shape for which the nodal line coincides with the smaller semi-axis of the ellipse is adequately described by the function $v_{11}(r)\cos\varphi$, and the frequency of these vibrations can be determined with high precision by (15.1.16) for a wide range of eccentricity values. Naturally, the absolute error grows for larger n.

Table 15.2

e	λ_{11}^{*c}	λ_{11}^c	$\Delta\lambda/\lambda$
0.5	15.9054	–	10^{-4}
0.6	16.7466	–	10^{-4}
0.7	18.2085	18.1321	$5\cdot10^{-3}$
0.8	21.2073	20.9461	$5\cdot10^{-3}$
0.9	30.3330	29.6271	$2\cdot10^{-2}$

Now, consider the test function $u = w(r)\sin\varphi$ for which the nodal line coincides with the larger semi-axis of the ellipse. Calculations similar to those described above bring us to a problem of the type (15.1.15), and the bound λ_{11}^{*s} has the form

$$\begin{gathered} \lambda_{11}^{*s} = A^s\gamma_{11}^2, \quad A^s(a,b) = A^c(b,a) = a^{-2}(1 - \tfrac{1}{4}e^2)(1 - e^2)^{-1}, \\ w_{11}(r) = c_{n1}J_1(\gamma_{n1}r), \quad \lambda_{11}^{*c,s}(a,a) = \gamma_{11}^2 a^{-2} \end{gathered} \tag{15.1.17}$$

From (15.1.17), it follows that $\lambda_{11}^{*s} \ge \lambda_{11}^{*c} > \lambda_{10}^*$ (this agrees with intuitive physical notions; see below), and the equality (degeneration) takes place for $a = b$, i.e., for a circular membrane, which is obvious. For $e > 0$, the degeneration disappears and the eigenvalues split. For $e \to 1$, we have $\lambda_{11}^{*s}/\lambda_{11}^{*c} \to 3$. The results of calculations based on the estimate (15.1.17) and the numerical method of accelerated convergence are given in Table 15.3. The estimate (15.1.17) is less precise, as expected. Note that $A^s/A^c \approx 2$ for $e = 0.9$ (see the last rows in Tables 15.2 and 15.3). From (15.1.16), (15.1.17), it also follows that (see Subsection 15.1.4)

$$\lambda_{11}^{*c}(e) = \gamma_{11}^2 a^{-2}(1+\tfrac{1}{4}e^2+O(e^4)), \quad \lambda_{11}^{*s}(e) = \gamma_{11}^2 a^{-2}(1+\tfrac{3}{4}e^2+O(e^4)), \quad e \ll 1. \tag{15.1.18}$$

Table 15.3

e	λ_{11}^{*s}	λ_{11}^{s}	$\Delta\lambda/\lambda$
0.4	16.7875	16.7722	10^{-3}
0.5	18.3614	18.3314	$1.2 \cdot 10^{-3}$
0.6	20.8861	20.8385	$2.2 \cdot 10^{-3}$
0.7	25.2739	25.0891	$8 \cdot 10^{-3}$
0.8	34.2746	33.7240	$1.6 \cdot 10^{-3}$
0.9	61.6536	60.6543	$2 \cdot 10^{-2}$

The results obtained for these upper bounds and the results of highly precise calculations of lower bounds by the method of accelerated convergence are represented in graphical form in Fig. 15.1. Calculations of lower bounds (practically precise values; their relative error is less than 10^{-5}) for relatively large values of the eccentricity e (marked by crosses) were performed with the aim of testing the precision of the aforementioned highly precise upper bounds obtained by the Rayleigh principle with suitably chosen test functions. The dashed lines represent the exact eigenvalues λ_{10}^{0}, λ_{11}^{0} for circular membranes whose area is equal to that of the ellipses $S = \pi(1-e^2)^{1/2}$ with the given eccentricity e: $\lambda_{1l}^{0} = \gamma_{1l}^{2}(1-e^2)^{-1/2}$, $l = 0, 1$. It should be observed that $\lambda_{10}^{*} > \lambda_{10}^{0}$, $\lambda_{11}^{*s}(e) > \lambda_{11}^{0}(e)$, but the curve $\lambda_{11}^{*c}(e)$ for smaller values of e lies below the curve $\lambda_{11}^{0}(e)$, and for large enough e ($e \geq 0.94$, i.e., $b/a \leq 1/3$) it lies above that curve. As mentioned above, this behavior of eigenvalues (and frequencies) is suggested by mechanical intuition. Obviously, the circle is a "more compact" figure than the ellipse of the same area, and the half-circle may be "more or less" compact than the half-ellipse with the variation of its eccentricity.

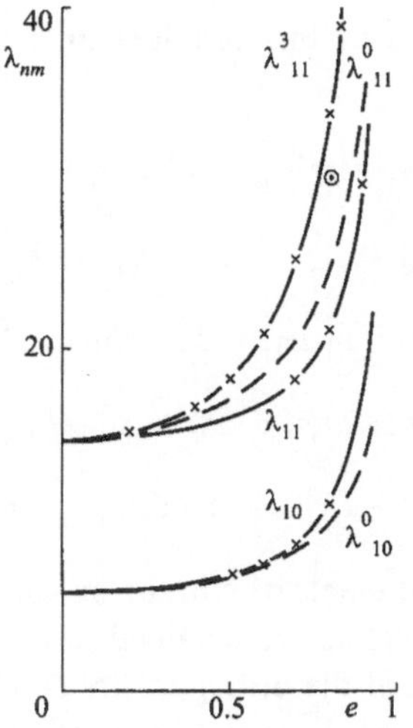

Fig. 15.1

Thus, our main result consists in the construction of universal curves describing highly precise approximations of fundamental frequencies $\omega_{1l}^{*c,s} = (\lambda_{1l}^{*c,s}T/(a^2\rho))^{1/2}$, $l = 0, 1$, and approximations of the corresponding vibration shapes for a wide range of eccentricity values. These frequencies may be provisionally viewed as a result of a linear transformation (depending on the eccentricity e) of the eigenfrequencies of a circular membrane. For $e > 0$, the frequencies of double multiplicity split and their divergence tends to infinity as $e \to 1$.

In order to perform stable calculations on the basis of variational estimates and numerical methods, it is necessary to obtain approximate analytical expressions of eigenvalues and eigenfunctions depending on the eccentricity e. These expressions could be used for testing the precision of calculations and could also be taken as initial approximations of the sought quantities in the procedures based on the recurrent numerical-analytical algorithm of accelerated convergence (see Subsection 15.1.6). Approximate analytical expressions of eigenvalues and eigenfunctions can be effectively constructed for $e \ll 1$ by the method of regular perturbations, and also for $1 - e = \delta \ll 1$ by the method of asymptotic expansions. Next, we briefly describe the derivation of these expressions.

15.1.4. Construction of approximate analytical expressions for eigenvalues of elliptic membranes with small eccentricity. Regular expansions of the sought eigenvalues and eigenfunctions in terms of the small parameter $\varepsilon = e^2 \ll 1$ can be obtained on the basis of the variational statement of the problem (15.1.7), (15.1.8). Let us fix the larger semi-axis a, taking $a = 1$, $b(\varepsilon) = (1-\varepsilon)^{1/2}$. Changing the variables $(x, y) \to (r, \varphi)$, we obtain the problem

$$J[u,\varepsilon] = b \int_0^1 \int_0^{2\pi} [u_r'^2(\cos^2\varphi + b^{-2}\sin^2\varphi)$$
$$+ r^{-2}u_\varphi'^2(b^{-2}\cos^2\varphi + \sin^2\varphi)]r\,dr\,d\varphi \to \min_u, \qquad (15.1.19)$$

$$I[u,\varepsilon] = b \int_0^1 \int_0^{2\pi} u^2 r dr d\varphi = 1, \qquad I_l[u,\varepsilon] = b \int_0^1 \int_0^{2\pi} u_l u r dr d\varphi = 0.$$

We seek an approximate solution of this problem in the form of expansions in powers ε^j $(\varepsilon = e^2)$. For $j = 0, 1$, we have

$$J = J_0 + \varepsilon J_1 + \varepsilon^2 J_2 + \cdots \to \min_u, \qquad I = I_0 + \varepsilon I_1 + \varepsilon^2 J_2 + \cdots = 1,$$
$$u = u_0 + \varepsilon u_1 + \varepsilon^2 u_2 + \cdots, \qquad \lambda = \lambda_0 + \varepsilon\lambda_1 + \varepsilon^2\lambda_2 + \cdots,$$
$$J_0[u_0] \to \min_{u_0}, \qquad I_0[u_0] = 1, \qquad \lambda_0 = \lambda^0_{nm} = \gamma^2_{nm}, \qquad (15.1.20)$$
$$u_0^c = c_{nm} J_m(\gamma_{nm} r)\cos m\varphi, \qquad u_0^s = s_{nm} J_m(\gamma_{nm} r)\sin m\varphi,$$
$$J_1[u_1] = -\tfrac{1}{2} J_0[u_0] + F_1[u_0] + L_1[u_1] \to \min_{u_1}, \qquad I_1[u_1] = -\tfrac{1}{2} + H_1[u_1].$$

Here, F_1, L_1, I_1, H_1 are the coefficients of ε in the expansions of J, I obtained after inserting the expansion of u. These coefficients are obtained in a standard way and their derivation is omitted. Note that no expansions of the orthogonality conditions are needed for what follows. The coefficient $L_1[u_1]$ is a linear integral operator applied to $u'_{1r}u'_{0r}$ and $u'_{1\varphi}u'_{0\varphi}$. This coefficient is calculated by integration by parts, with the boundary conditions for $r = 0, 1$, $\varphi = 0, 2\pi$ taken into account. Then, we use the condition $I_1 = 0$ (15.1.20), where H_1 is a linear integral operator applied to $u_1 u_0$. Thus, we finally obtain the formula $L_1[u_1] = \frac{1}{2}\lambda_0$. Let us calculate $F_1[u_0]$. After elementary trigonometric transformations, we obtain the explicit expressions:

$$
\begin{aligned}
&F_1[u_0] = \frac{1}{2}\lambda_0 + \frac{1}{2}\int_0^1\int_0^{2\pi}(-u'^2_{0r} + r^{-2}u'^2_{0\varphi})\cos 2\varphi r\,dr\,d\varphi;\\
&F_1[u_0^{(c)}] = 0, \quad F_1[u_0^{(s)}] = 0 \quad \text{if} \quad m = 0, 2, 3, \dots \quad (m \neq 1);\\
&F_1[u_0^{(c)}] = \tfrac{1}{4}\lambda_{n1}^0, \quad F_1[u_0^{(s)}] = \tfrac{3}{4}\lambda_{n1}^0 \quad \text{if} \quad m = 1, \quad n = 1, 2, \dots
\end{aligned}
\tag{15.1.21}
$$

The upper minus in (15.1.21) corresponds to u_0^c, and the plus corresponds to u_0^s (see (15.1.20)). Substituting the above finite expressions of L_1, F_1 into $J_1[u_1]$, we obtain approximate values $\lambda_{nm}^{c,s}$, which can be regarded as generalizations of the Rayleigh formulas [62],

$$
\begin{aligned}
&\lambda_{nm}^{c,s}(e) = \gamma_{nm}^2 + \tfrac{1}{2}\gamma_{nm}^2 e^2 + O(e^4), \qquad n = 1, 2, \dots, \qquad m \neq 1;\\
&\lambda_{n1}^{c}(e) = \gamma_{n1}^2 + \tfrac{1}{4}\gamma_{n1}^2 e^2 + O(e^4), \quad \lambda_{n1}^{s}(e) = \gamma_{n1}^2 + \tfrac{3}{4}\gamma_{n1}^2 e^2 + O(e^4).
\end{aligned}
\tag{15.1.22}
$$

From (15.1.22), it follows that for $e > 0$ in the first approximation in $\varepsilon = e^2$, the eigenvalues $\lambda_{n1}^{c,s}(e)$ split, as in the case of the Mathieu equation [43, 46, 61, 62]. The values $\lambda_{n1}^0 = \gamma_{n1}^2$ have double multiplicity, which is well known. For $n = 1$, the expressions (15.1.22) coincide with those obtained in Subsection 15.1.3 on the basis of variational estimates. The values $\lambda_{nm}^{c,s}(e)$ for $m \geq 2$ do not split in the first approximation in ε; the splitting order with respect to ε will be higher, and there is no splitting of $\lambda_{n0}(e)$.

Note that the calculations based on (15.1.22) are in complete agreement with the variational bounds given in Tables 15.1–15.3 and the graphs in Fig. 15.1 for $0 \leq e \lesssim 0.5$. By the perturbation method, one can construct subsequent approximations of the sought quantities. However, the resulting expressions are very cumbersome and the calculations based on these yield poor results. A more efficient approach to highly precise mass calculations of eigenvalues $\lambda_{nm}^{c,s}(e)$ and eigenfunctions $u_{nm}^{c,s}(r, \varphi, e)$ is based on the numerical-analytical method of accelerated convergence [8–10, 13], which is equivalent to the Newton method of tangential lines combined with the procedure of extension in a parameter (see Subsection 15.1.6.).

15.1.5. Asymptotic expansions of eigenvalues for large eccentricity values. For the construction of asymptotic approximations of eigenvalues and eigenfunctions it is convenient to use the mathematical techniques developed for boundary value problems (15.1.4)–(15.1.6). Performing elementary transformations of the functions $\cosh^2\alpha$, $\cos^2\varphi$ and changing notation, we obtain the following equations for A, Φ [43, 46, 61]:

$$
\begin{aligned}
&A'' + (\Lambda\cosh 2\alpha - N)A = 0, \qquad \Phi'' + (N - \Lambda\cos 2\varphi)\Phi = 0,\\
&\Lambda = \tfrac{1}{2}\mu, \qquad N = \chi - \tfrac{1}{2}\mu \qquad (\mu = \lambda d^2)
\end{aligned}
\tag{15.1.23}
$$

with the boundary conditions (15.1.5), (15.1.6). Approximate asymptotic expressions for $\Lambda(e)$, $N(e)$ will be constructed under the assumption that $\delta = 1 - e$ is small.

First, consider the asymptotic properties of Λ, N obtained from the boundary value problems for the modified Mathieu equation for A. As $\delta \to 0$, the interval $\alpha \in [0, \alpha^*]$ becomes infinitely small ($\alpha^* \to 0$), and therefore, equation (15.1.23) for A should be transformed by the replacement $\alpha = \alpha^* z$ to the following ($A = A(z)$):

$$
\begin{aligned}
&A'' + (\sigma\cosh 2\alpha^* z - \nu)A = 0, \qquad 0 \leq z \leq 1, \quad \sigma = \alpha^{*2}\Lambda, \quad \nu = \alpha^{*2}N,\\
&\alpha^* = \operatorname{arcosh} e^{-1} = -\ln(1-\delta) + \ln\bigl(1 + \sqrt{\delta(2-\delta)}\bigr) = \sqrt{2\delta} + \delta + O(\delta^{3/2}).
\end{aligned}
\tag{15.1.24}
$$

Boundary conditions (15.1.5), (15.1.6) should be transformed accordingly: instead of the conditions for $\alpha = \alpha^*$, we have to write the conditions for $z = 1$. Approximate solutions of problems (15.1.24), (15.1.5), (15.1.6) for sufficiently small α^* can be constructed analytically with arbitrary precision with respect to powers of the small parameter $\varepsilon = \alpha^*(e) > 0$ ($\varepsilon \approx \sqrt{2\delta}$, $\delta = 1 - e$). This can be done on the basis of the results from [14] (see Chapter 3). In particular, taking into account the terms of the order $O(e^3)$ (with the error of the order $O(e^4)$), we obtain the following relations between the parameters σ, ν:

$$\sigma = \nu + \gamma + \varepsilon^2 l(\nu + \gamma) + O(\varepsilon^4), \quad \gamma = \gamma_n^{c,s}, \quad l = l_n^{c,s}, \quad n = 1, 2, \ldots,$$

$$\gamma_n^c = \pi^2(n - \tfrac{1}{2})^2, \quad l_n^c = -\tfrac{2}{3} + 4\pi^{-2}(2n-1)^{-2}, \quad \gamma_n^s = (\pi n)^2, \quad l_n^s = -\tfrac{2}{3} + 2(\pi n)^{-2}. \tag{15.1.25}$$

Here, the superscripts c, s refer to even and odd solutions of the basic Mathieu equation, in view of (15.1.6), (15.1.5). Dividing (15.1.25) by ε^2 and thus going back to the parameters Λ, N introduced in (15.1.23), we obtain the desired relation

$$\Lambda - N = \Gamma + \varepsilon^2 l \Lambda + O(\varepsilon^2), \qquad \Gamma = \gamma\varepsilon^{-2} \quad (\Gamma \gg 1). \tag{15.1.26}$$

Relation (15.1.26), in view of (15.1.25), contains the index $n = 1, 2, \ldots$ characterizing the mode number with respect to α.

Now, let us use the known [43, 46] asymptotic relation between the parameters in the basic Mathieu equation. For N and $\Lambda \gg 1$, we obtain

$$N + \Lambda = p(2\Lambda)^{1/2} - \tfrac{1}{8}(p^2 + 1) - \tfrac{1}{64\sqrt{2}}(p^3 + p)\Lambda^{-1/2} + O(\Lambda^{-1}). \tag{15.1.27}$$

The integer p in (15.1.27) is determined by the mode index m and the symmetry properties of the solution (c or s; see below). Let us resolve system (15.1.26), (15.1.27) with respect to the sought quantities Λ, N by means of the recurrent procedure of successive approximations. We have

$$\begin{aligned} 2\Lambda &= \Gamma + O(\varepsilon^{-1}) = \Gamma + p\sqrt{\Gamma} + O(1) = \Gamma + p\sqrt{\Gamma} + \tfrac{1}{8}(3p^2 - 1) + \tfrac{1}{2}l\gamma + O(\varepsilon) \equiv \\ &\equiv 2\Lambda_{(2)} + O(\varepsilon) = \Gamma + p\sqrt{2\Lambda_{(2)}} + \tfrac{1}{2}l\gamma - \tfrac{1}{8}(p^2+1) - \tfrac{1}{2}\varepsilon l p\sqrt{\gamma} \\ &- \tfrac{1}{64}\varepsilon(p^3 + 3p)/\sqrt{\gamma} + O(\varepsilon^2), \qquad 2N = 2\Lambda - 2\Gamma - l(\gamma + \varepsilon p\sqrt{\gamma}) + O(\varepsilon^2). \end{aligned} \tag{15.1.28}$$

The final formulas (15.1.28) have absolute error of the order $O(\varepsilon^2) = O(\delta)$, and therefore, the relative error of calculations will be of the order $O(\varepsilon^4) = O(\delta^2)$. Formulas (15.1.25)–(15.1.28) may be refined by taking into account higher powers of the parameter ε. However, the expressions obtained in this way are lengthy and can hardly be used for calculations. Practically, in order to continue analytical and numerical calculations for $\varepsilon \geq 0.9$, it suffices to construct the second approximations $\Lambda_{(2)}$ and $N_{(2)}$ with the small relative error $O(\varepsilon^3)$ and the absolute error $O(\varepsilon)$ ensured by (15.1.28).

Let us turn to the approximation of the original parameters λ and χ defined by (15.1.23),

$$\lambda = 2d^{-2}\Lambda, \qquad \chi = N + \Lambda, \qquad d^2 = a^2e^2, \qquad e \approx 1. \tag{15.1.29}$$

Here, the parameters Λ, N as functions of $\varepsilon = \alpha^*(e)$ are defined by (15.1.28). It follows from (15.1.25)–(15.1.29) that $\Lambda \approx \frac{1}{2}\Gamma$, $N \approx -\frac{1}{2}\Gamma$, $\chi \approx p\sqrt{\Gamma}$, which means that the separation parameter χ for the variables α and φ is much smaller than λ: $\chi/\lambda \sim \varepsilon$, and this is in spite of the fact that $\chi \sim \varepsilon^{-1}$. The value $\chi = 0$ in the modified Mathieu equation (15.1.4) entails a large absolute error [43, 46, 61, 62] (see Subsection 15.1.6).

In order to find the approximations $\lambda_{nm}^{c,s}(e)$, $\chi_{nm}^{c,s}(e)$ one the basis of (15.1.25)–(15.1.28), we have to know the relation between the indices m and $p = 2i + 1$. The coefficients $\Gamma_n^{c,s}$, $\gamma_n^{c,s}$, $l_n^{c,s}$ and the index n are determined by (15.1.25), (15.1.26). The integers p, i depend on m (the index of vibration in φ) and the type c, s of the solution of the Mathieu equation. This dependence is expressed by the relations [46]

$$p = p_m^{c,s} = 2i_m^{c,s} + 1, \quad i_m^c = m, \quad i_m^s = m - 1, \quad m \geq 1, \quad p_0^c = 1, \quad p_0^s = 0. \qquad (15.1.30)$$

Thus, relations (15.1.27), (15.1.30) imply that we have the same asymptotic expansions for N_0^c and N_1^s, N_1^c and N_2^s, ..., N_m^c and N_{m+1}^s, since $p_m^c = p_{m+1}^s$, i.e., in (15.1.27) for p one takes $p = 2m + 1$ or $p = 2m - 1$ when calculating N_m^c or N_m^s, respectively. Note that in these calculations, the parameter $\Lambda > 0$ in (15.1.27) is assumed arbitrary and independent of m. However, from (15.1.25), (15.1.28), it follows that the asymptotic formulas for $\Lambda_{nm}^{c,s}$ with n fixed will not be the same in the above sense. This is due to the fact that the coefficient $\Gamma_n^{c,s}$ is different for even (c) and odd (s) solutions. Note also that the leading asymptotic term $O(\varepsilon^{-2})$ for $\lambda_{nm}^{c,s}$ is determined by the index n, while the next, smaller asymptotic term $O(\varepsilon^{-1})$ depends on the indices n, m. The intermediate separation parameter $\chi_{nm}^{c,s}$ has the same asymptotic behavior as the second term of the expansion for $\lambda_{nm}^{c,s}$. It is interesting to observe that the leading term λ_{nm} coincides with the leading term in the case of a narrow rectangular membrane.

The formulas obtained here can be used in practice for $\delta \sim 0.1$, in particular, for the calculations described in Subsection 15.1.3. Thus, we obtain highly precise numerical and numerical-analytical values λ_{10} and $\lambda_{11}^{c,s}$ for all $0 \leq e < 1$ (the eccentricity may be arbitrarily close to $e = 1$); see Subsections 15.1.3–15.1.5.

15.1.6. Finding eigenfrequencies and vibration shapes of an elliptic membrane by the method of accelerated convergence. In order to apply the highly precise numerical-analytical method of accelerated convergence, it is convenient to write boundary value problem (15.1.23), (15.1.5), (15.1.6) with both arguments varying on unit intervals, similarly to (15.1.24). Thus, we obtain two eigenvalue problems ($\alpha = \alpha^* z$, $\varphi = \pi\theta$):

$$\begin{aligned} &A'' + \alpha^{*2}(e)(\Lambda \cosh 2\alpha^* z - N)A = 0, \qquad A = A(z, e), \\ &\Phi'' + \pi^2(N - \Lambda \cos 2\pi\theta)\Phi = 0, \qquad \Phi = \Phi(\theta, e); \end{aligned} \qquad (15.1.31)$$

$$A(0) = A(1) = 0, \quad \Phi(0) = \Phi(1) = 0 \quad \text{(1st problem)}; \qquad (15.1.32)$$

$$A'(0) = A(1) = 0, \quad \Phi'(0) = \Phi'(1) = 0 \quad \text{(2nd problem)}. \qquad (15.1.33)$$

Our aim is to construct the eigenvalues $\Lambda_{nm}^{c,s}(e)$ and the functions $A_{nm}^{c,s}(z, e)$, $\Phi_{nm}^{c,s}(\theta, e)$ corresponding to even (15.1.33) and odd (15.1.32) solutions of the Mathieu equation for a possibly wide interval $0 < e < 1$ ($n = 1, 2, ..., m = 0, 1, 2, ...$). The limit solutions corresponding to $e \to 0$ or $e \to 1$ have been constructed with any given accuracy in Subsections 15.1.3–15.1.5. Note that the problem for Φ (15.1.31)–(15.1.33) does not explicitly depend on e. Therefore, it is reasonable to use the method of consecutive determination of the parameters N and Λ depending on e and the indices n, m for both boundary value problems. This method is based on the utilization of the functions $N_m^{c,s}(\Lambda)$ from the Mathieu equations. The analytical and the numerical properties of these functions have been studied in detail (the Ince–Strutt diagrams can be found in [61]). Suppose that we know an analytical expression of $N_m^{c,s}(\Lambda)$ with the desired precision, say, from the finite Hill's determinant or a segment of a continued fraction [39, 56], for a sufficiently wide range of $\Lambda \geq 0$. Substituting this expression into the Mathieu equation for A and using (15.1.31)–(15.1.33), we obtain two families of boundary value problems (e is the parameter of each family)

$$A'' + \alpha^{*2}(e)(\Lambda \cosh 2\alpha^* z - N_m^{c,s}(\Lambda))A = 0, \qquad 0 \leq z \leq 1,$$
$$A^s(0) = A^s(1) = 0, \qquad A^{c'}(0) = A^c(1) = 0. \tag{15.1.34}$$

These problems of Sturm–Liouville type belong to the class of generalized problems, because the parameter Λ, whose eigenvalues (critical values) $\Lambda_{nm}^{c,s}(e)$ are sought, is involved in a nonlinear manner through the functions $N_m^{c,s}(\Lambda)$. A regular numerical-analytical method for solving generalized problems of the type (15.1.34) is described in [8] (see Chapter 4). This method consists in obtaining a preliminary rough bound by some variational approach (for a fixed e) and utilizing the recurrent numerical-analytical procedure of refining that bound by the numerical-analytical method of accelerated convergence. For other values of e, the procedure of continuation in the parameter is used, either a simple procedure or that with extrapolation (linear, quadratic, etc.).

For relatively small Λ, the function $N_m^{c,s}(\Lambda)$ is weakly dependent on Λ [43, 46, 61]. Therefore, just as in Subsection 15.1.3, one can successfully apply the modified method based on the solution of the classical problem. This method uses Λ obtained on the preceding iteration step in the calculation of $N_m^{c,s}$. As mentioned above, this approach presumes that Λ (i.e., e) is not too large. When used in Subsection 15.1.3 for $0 < e \leq 0.9$, this approach resulted in relative calculation errors of the order $10^{-4} \div 10^{-5}$ and did not require considerable computational resources.

If Λ is large and the utilization of the functions $N_m^{c,s}(\Lambda)$ is problematic, then highly precise solutions of problems (15.1.31)–(15.1.35) can be constructed by the method of accelerated convergence used in this case for the joint refinement of the sought values $\Lambda_{nm}^{c,s}(e)$, $N_{nm}^{c,s}(e)$. This method consists in the construction of the functions $N_m^{c,s}(\Lambda)$, $\Lambda_n^{c,s}(e)$ with subsequent determination of the crossing point of the corresponding curves with the help of a numerical-graphical solution (obtained by linear or quadratic interpolation) of the equation

$$\Lambda = \Lambda_n^{c,s}(N_m^{c,s}(\Lambda), e), \qquad \Lambda = \Lambda_{nm}^{c,s}(e), \qquad N = N_{nm}^{c,s}(e) = N_m^{c,s}(\Lambda_{nm}^{c,s}(e)). \tag{15.1.35}$$

The utilization of these relations for solving vibration problems requires special software and modern computers.

15.1.7. Conclusions. A survey of fundamental works dedicated to calculations of free vibration frequencies of elliptic membranes and plates [43, 61] shows that the intermediate constant of separation of the variables is often neglected (taken equal to zero), even for relatively large eccentricity values (for instance, $e = 0.8$ [43]). This leads to fairly gross errors of the approximate values obtained for eigenfrequencies of elliptic membranes (the value λ_{11}^c obtained by the method of [43] corresponds to the small circle in Fig. 15.1). The separation constant is also neglected in calculations of eigenfrequencies of elliptic plates [51], which again results in gross errors. Note that such assumptions are acceptable only on the first step of the iteration procedure which consists in the refinement of the said separation constant from the Mathieu equation on the basis of a rough bound for the eigenvalue, which is then refined from the modified Mathieu equation. The authors of classical works noticed this fact, but their followers copied their results and used them without due critical analysis.

15.2. Free Vibrations of an Elliptic Plate with Clamped Edge

In this section, a modified version of the Rayleigh–Ritz method is used to obtain highly precise analytical approximations of frequencies and shapes of free vibrations of an elliptic plate clamped at the edge. We establish a relation between the vibration spectra of an elliptic and a circular plate. The estimates obtained here are compared with known numerical results and experimental data.

15.2.1. Preliminary remarks. The study of free vibrations of two-dimensional systems with elliptic boundary is of undeniable interest for theory and applications (see [43, 56, 61] et al.) For practical problems, it is important to find frequencies and vibration shapes (primarily, for lower modes) of stretched membranes and elastic plates, heavy fluids in a reservoir, electromagnetic and acoustic waves in wave-guides and resonators, and many other objects. The theoretical interest is due to the fact that the corresponding boundary value problems in elliptic domains represent a natural generalization of the thoroughly studied problems for circular domains. An attractive feature of such problems in elliptic domains is that in elliptic coordinates, the variables can be separated and one can apply numerical-analytical methods to the resulting coupled one-dimensional boundary value problems of Sturm–Liouville type.

Traditionally, eigenvalues and vibration shapes are calculated by variational methods, the Rayleigh–Ritz method, the Bubnov–Galiorkin method, the finite element method, grid methods, etc. [25, 28, 29, 33, 37, 39, 41, 50–52]. In what follows, a version of the Rayleigh–Ritz method is used for obtaining highly precise approximations of frequencies and shapes of lower vibration modes of an elliptic plate. Our method is based on the introduction of generalized polar coordinates and the transverse displacement (deflection) is assigned a given oscillation structure with a fixed number of nodal lines (just as for a circular plate). The dependence of the deflection on the radial coordinate is not prescribed but is determined from the solution of the corresponding Euler–Lagrange equation, as in Section 15.1 and [10] for elliptic membranes.

15.2.2. Setting of the problem. Frequencies and shapes of free vibrations of an elliptic plate with clamped boundary are determined by the following boundary value problem [56, 61]:

$$\Delta\Delta w - \lambda w = 0, \tag{15.2.1}$$

$$w\big|_{\partial G} = \frac{\partial w}{\partial n}\bigg|_{\partial G} = 0, \tag{15.2.2}$$

Here, $w = w(x, y)$ is the deflection of the plate; $w \in G$, where $G = \{x^2a^{-2} + y^2b^{-2} \le 1\}$ is the ellipse with the boundary ∂G; n is the normal to the elliptic contour on the middle surface of the plate; a is the larger and b is the smaller semiaxis of the ellipse; Δ is the Laplace operator, $\Delta w \equiv w_{xx} + w_{yy}$. The sought quantities in problem (15.2.1), (15.2.2) are the eigenvalues λ_n and the corresponding eigenfunctions $w_n(x, y)$ $(n = 1, 2, \ldots)$ for arbitrary $a > b > 0$. The eigenvalues λ_n in (15.2.1) are related to the eigenfrequencies ω_n by

$$\omega_n = h\sqrt{\frac{E\lambda_n}{3\rho(1-\nu^2)}}, \qquad n = 1, 2, \ldots, \tag{15.2.3}$$

where h is the half-thickness of the plate, ρ is the density, E is the Young modulus, and ν is the Poisson ratio of its material. The boundary value problem (15.2.1), (15.2.2) is equivalent to the problem of minimizing the following functional:

$$J[w] = \iint_G (\Delta w)^2 \, dx \, dy \tag{15.2.4}$$

with the isoperimetric condition

$$\Phi[w] = \iint_G w^2 \, dx \, dy = 1 \tag{15.2.5}$$

on the class of four times differentiable functions $w(x, y)$ satisfying the boundary conditions (15.2.2).

15.2.3. Estimates for the frequency of the lowest vibration mode with the help of an elliptically symmetrical test function. As in Section 15.1 and [10], we introduce generalized polar coordinates r and φ by

$$x = ar\cos\varphi, \qquad y = br\sin\varphi, \qquad r \geq 0, \qquad 0 \leq \varphi \leq 2\pi \tag{15.2.6}$$

($r = 1$ on the boundary of the ellipse).

Assume first that the sought function on which the absolute minimum of the functional in (15.2.4) is attained depends only on r, i.e., $w = w(r)$. Passing to the variables r, φ in (15.2.4), (15.2.5) for $w(r)$ and integrating in φ, we get

$$J[w] = \frac{\pi ab}{4} \int_0^1 \left\{ A\left[\left(\frac{d^2w}{dr^2}\right)^2 + \frac{1}{r^2}\left(\frac{dw}{dr}\right)^2\right] + \frac{2}{r}B\frac{dw}{dr}\frac{d^2w}{dr^2} \right\} r \, dr, \tag{15.2.7}$$

$$\Phi[w] = 2\pi ab \int_0^1 w^2 r \, dr = 1, \tag{15.2.8}$$

$$A = \frac{3}{a^4} + \frac{3}{b^4} + \frac{2}{a^2b^2}, \qquad B = \frac{1}{a^4} + \frac{1}{b^4} + \frac{6}{a^2b^2}.$$

Using the Lagrange method for solving the variational problem (15.2.7), (15.2.8) and writing the corresponding Euler–Lagrange equation [24], we obtain the linear fourth-order differential equation

$$\frac{1}{r}\frac{d}{dr}\left\{ r\frac{d}{dr}\left[\frac{1}{r}\frac{d}{dr}\left(r\frac{dw}{dr}\right)\right]\right\} - \mu^4 w = 0, \qquad \mu^4 = \frac{8\lambda}{A}. \tag{15.2.9}$$

This equation should be integrated under the boundary conditions which follow from (15.2.2) and the condition that $w(r)$ and its derivative at $r = 0$ are bounded:

$$w(1) = \frac{dw(1)}{dr} = 0, \qquad |w(0)| \leq M, \qquad \left|\frac{dw(0)}{dr}\right| \leq M. \tag{15.2.10}$$

The solution satisfying all these conditions has the form

$$w(r) = c_1 J_0(\mu r) + c_2 I_0(\mu r). \tag{15.2.11}$$

Here $J_0(\mu r)$ is the zero-order Bessel function of the first kind, $I_0(\mu r)$ is the modified Bessel function of zero-order [32, 67].

From the boundary conditions (15.2.10), we obtain the following system of algebraic equations for the constants c_1, c_2:

$$c_1 J_0(\mu) + c_2 I_0(\mu) = 0, \qquad c_1 J_1(\mu) - c_2 I_1(\mu) = 0. \tag{15.2.12}$$

Equating the determinant of this system to zero, we obtain the known frequency equation for circular plates [24, 28, 61]:

$$J_0(\mu)I_1(\mu) - I_0(\mu)J_1(\mu) = 0. \tag{15.2.13}$$

In our case, this equation determines approximate values (upper bounds) of eigenfrequencies of the elliptic plate. The smallest root of equation (15.2.13) is (see [32])

$$\mu_{10} = 3.1961. \tag{15.2.14}$$

In view of (15.2.14), the expression

$$\lambda_{10} = \frac{\mu_{10}^4}{8}\left(\frac{3}{a^4} + \frac{3}{b^4} + \frac{2}{a^2b^2}\right) \tag{15.2.15}$$

provides an upper bound for the smallest eigenvalue, and

$$f_{10} = \frac{h}{2\pi}\sqrt{\frac{E\lambda_{10}}{3\rho(1-\nu^2)}} \tag{15.2.16}$$

is an upper bound for the smallest eigenfrequency. Below, the expression (15.2.16) is compared with experimental results.

Consider the family of ellipses of a fixed area, $\pi ab = \pi R^2$, where R is the radius of a circle of that area. Then, using (15.2.15), one can prove an isoperimetric theorem similar to the known theorem for elliptic membranes [10, 62]. Namely, *among all elliptic plates of equal area, the circular plate has the smallest fundamental frequency*. Indeed, letting $b = R^2/a$, we get

$$\lambda_{10} = \frac{\mu_{10}^4}{8}\left(\frac{3}{a^4} + \frac{3a^4}{R^8} + \frac{2}{R^4}\right),$$

and from the relation

$$\frac{d\lambda_{10}}{da} = \frac{\mu_{10}^4}{8}\left(\frac{12a^3}{R^8} - \frac{12}{a^5}\right) = 0,$$

we find that $a = R$, which proves our statement. It seems possible that such a statement can be made about arbitrary plates of equal area.

Calculations based on (15.2.15) were performed for the dimensionless frequency parameter $\omega_{10}^* = \sqrt{\lambda_{10}}a^2$, with the length b of the smaller semi-axis expressed through the length of the larger semi-axis a and the eccentricity e: $b = a\sqrt{1-e^2}$, $0 \le e < 1$. Table 15.4 gives the values of ω_{10}^* corresponding to the first three frequencies of vibrations of the type under consideration for different values of the eccentricity e (these vibrations may be called *elliptically symmetric* by analogy with axially symmetric vibrations of a circular plate). For the sake of comparison, we also give the values ω_{10}^T obtained from the exact solution of problem (15.2.1), (15.2.2) in terms of the Mathieu functions [56]. Empty cells mean that the corresponding data are absent. For the eccentricity $e = 0.866$ (the ratio of the semi-axes $a/b = 2$, the plate thickness $2h = 1$ mm, the larger semi-axis $a = 100$ mm, the material is aluminum) the fundamental frequency was measured experimentally. Its value is equal to $f_{10} = 610$ Hz ($\omega_{10} = 27.281$), which is very close to the value obtained from the exact solution ($\omega_{10}^T = 27.468$). Calculations by (15.2.15), (15.2.16) yield $f_{10}^* = 618$ Hz, which differs from the experimental value by less than 1.5%.

Table 15.4

e	ω_{10}^*	ω_{10}^T	e	ω_{10}^*	ω_{10}^T
0	10.212	10.212	0.6	13.245	13.167
0.1	10.267	–	0.7	15.515	–
0.2	10.429	–	0.8	20.335	–
0.3	10.726	–	0.866	27.737	27.468
0.4	11.209	11.183	0.9	35.502	–
0.5	11.978	–	0.943	58.502	56.820
			0.95	66.505	–

Table 15.4 shows that the relative error for the first frequency calculated by (15.2.15), (15.2.16) does not exceed 0.6% for $e \le 0.6$, 1% for $e \le 0.866$, and 3.6% for $e \le 0.943$. Thus, formulas (15.2.15), (15.2.16) can indeed be claimed highly precise, since similar estimates obtained by the Rayleigh–Ritz method with the test function $w = c(1 - x^2/a^2 - y^2/b^2)^2$ yield errors from 2% to 5% for the above range of eccentricity values.

15.2.4. Estimates for the second vibration modes. The next eigenvalues (frequencies) correspond to vibration shapes with the nodal lines coinciding with the smaller semi-axis (the second eigenvalue) and the larger semi-axis (the third eigenvalue). To obtain approximations of the second vibration shape, consider admissible test functions of the form

$$w(r, \varphi) = V(r) \cos \varphi. \tag{15.2.17}$$

Substituting such functions into the integrals (15.1.4), (15.1.5), we come to the following Euler–Lagrange equation:

$$\begin{gathered}\frac{d^4V}{dr^4} + \frac{2}{r}\frac{d^3V}{dr^3} - \frac{3}{r^2}\frac{d^2V}{dr^2} + \frac{3}{r^3}\frac{dV}{dr} - \left(\gamma^4 + \frac{3}{r^4}\right) V = 0,\\ \gamma^4 = \frac{8\lambda a^4 b^4}{a^4 + 2a^2b^2 + 5b^4}.\end{gathered} \tag{15.2.18}$$

It can be shown that this equation admits the solutions $J_1(\gamma r)$, $Y_1(\gamma r)$, $I_1(\gamma r)$ and $K_1(\gamma r)$ [32]. Therefore, taking into account the condition of boundedness at the origin, the general solution of equation (15.2.18) can be written in the form

$$V(r) = c_1 J_1(\gamma r) + c_2 I_1(\gamma r). \tag{15.2.19}$$

Substituting (15.2.19) into the boundary conditions similar to (15.2.10), we obtain the frequency equation

$$J_1(\gamma) I_1'(\gamma) - J_1'(\gamma) I_1(\gamma) = 0, \tag{15.2.20}$$

whose smallest root is $\gamma_{11}^c = 4.611$ (see [32]). Note that the eigenvalue γ_{11} for $e = 0$ becomes doubly degenerate. Thus, we obtain an upper bound for the second eigenvalue (frequency) for the elliptic membrane with a nodal line coinciding with the smaller semi-axis. We find that

$$\lambda_{11}^c = \frac{\gamma_{11}^{c4}}{8}\left(\frac{5}{a^4} + \frac{1}{b^4} + \frac{2}{a^2b^2}\right). \tag{15.2.21}$$

Taking $w(r, \varphi) = V(r) \sin \varphi$, we can obtain an approximation of the next (third) vibration mode. Arguing as above, we find that

$$\lambda_{11}^s = \frac{\gamma_{11}^{s4}}{8}\left(\frac{1}{a^4} + \frac{5}{b^4} + \frac{2}{a^2 b^2}\right), \qquad \gamma_{11}^s = \gamma_{11}^c = 4.611. \tag{15.2.22}$$

It should be observed that the values λ_{11}^c and λ_{11}^s are non-degenerate for an elliptic plate, whereas $\lambda_{11}^c = \lambda_{11}^s$ for a circular plate with $a = b$,

Table 15.5 gives upper bounds for the first eigenfrequencies of an elliptic plate with a single nodal line, $\omega_{10}^c = \sqrt{\lambda_{11}^c}a^2$ and $\omega_{10}^s = \sqrt{\lambda_{11}^s}a^2$, depending on the eccentricity e.

Table 15.5

e	ω_{10}^c	ω_{10}^s	e	ω_{10}^c	ω_{10}^s
0	21.261	21.261	0.6	24.435	30.378
0.1	21.315	21.423	0.7	26.858	36.937
0.2	21.484	21.927	0.8	32.136	50.502
0.3	21.793	22.844	0.866	40.477	70.904
0.4	22.297	24.320	0.9	49.422	92.074
0.5	23.101	26.636	0.95	85.940	175.886

For eccentricity values $e \le 0.6$, the bounds for eigenfrequencies obtained by formulas (15.2.21), (15.2.22) differ from the exact values by less than 1%. For $e = 0.866$, the experimental values of the eigenfrequencies are $f_{11}^c = 902$, $f_{11}^s = 1390$, which corresponds to $\omega_{10}^c = 38.86$, $\omega_{10}^s = 62.40$. The divergence of the first experimental value from the theoretical bound is 4.2% and of the second 14%. It should be said that these experimental values are somewhat smaller than the precise theoretical values of eigenfrequencies. This may be an indication of the fact that as the frequency grows, its approximations have a greater error, if they are obtained by calculations with finitely many terms preserved in the expansion of the deflection in terms of the Mathieu functions. Another reason for the observed difference may be that for higher frequencies, the conditions of clamped edge cannot be perfectly realized in experiment and become more "soft", while the measured frequencies become smaller than the theoretically expected values.

15.2.5. Estimates of eigenfrequencies for higher vibration modes. Let us briefly describe an algorithm for obtaining approximations of higher eigenfrequencies ($n \ge 2$). The test functions are represented in the form

$$w(r, \varphi) = V(r)\begin{cases} \cos n\varphi, \\ \sin n\varphi, \end{cases} \qquad n = 2, 3, \ldots \tag{15.2.23}$$

Reasoning as above for $n = 2$, upon the insertion of (15.2.23) into (15.2.7), (15.2.8), we obtain the Euler–Lagrange equation

$$\frac{d^4V}{dr^4} + \frac{2}{r}\frac{d^3V}{dr^3} - \frac{9}{r^2}\frac{d^2V}{dr^2} + \frac{9}{r^3}\frac{dV}{dr} - \gamma^4 V = 0, \qquad \gamma^4 = \frac{16\lambda a^4 b^4}{7a^4 + 2a^2b^2 + 7b^4}. \tag{15.2.24}$$

Equation (15.2.24) can be integrated in terms of the Bessel functions and the modified Bessel functions of the second order. The general solution which is bounded at the origin can be written in the form [32]

$$V(r) = c_1 J_2(\gamma_2 r) +_2 I_2(\gamma_2 r). \tag{15.2.25}$$

Now, by the standard arguments we construct the frequency equation on the basis of (15.2.25). In a similar way, one can construct and integrate the Euler–Lagrange equations for $n > 2$.

It should be said that the upper bounds obtained for $n \geq 2$ regarded as approximations of the exact frequencies have a greater error for larger n. However, for applied problems, these bounds yield acceptable results, if the eccentricity values are not too large.

15.2.6. Conclusions.

1. A modification of the Rayleigh–Ritz method has been proposed for the construction of upper bounds for free vibration frequencies of elliptic plates with clamped edge.
2. An analytical relation has been established between approximations of frequencies and shapes of free vibrations of an elliptic plate and frequencies and vibration shapes of a circular plate with similar boundary conditions.
3. The approximations of eigenfrequencies are described by simple analytical formulas, which can be used for the calculation of eigenfrequencies of elliptic plates with a wide range of eccentricity values, and the precision of these calculations is sufficient for practical purposes.
4. A survey of publications shows that the relations between the frequency spectra of elliptic and circular plates obtained so far are valid only for small eccentricity values, $e \ll 1$, i.e., when the perturbation theory is applicable. It seems that the transformation (15.2.6) has never been previously used for the construction of test functions and expressions like (15.2.15), (15.2.21), (15.2.22).

15.3. Exercises

Exercise 1. Consider a triangular membrane with two lateral sides of length a and the angle α between them. Suppose that the membrane has constant density ρ and tension T. Replacing the membrane by the inscribed and the circumscribed sectors, find analytically the upper and the lower bounds for the first eigenvalue. Find the asymptotics of the first eigenvalue as $\alpha \to 0$.

Exercise 2. Using the generalized spherical coordinates $x = ar\cos\varphi\sin\theta$, $y = br\sin\varphi\sin\theta$, $z = cr\cos\theta$ and the Rayleigh principle, obtain an upper bound for the first eigenvalue of the three-dimensional boundary value problem (for a three-axial ellipsoid)

$$\Delta u + \lambda u = 0, \qquad u|_{r=1} = 0.$$

Exercise 3. It is known that the first eigenvalue of an ellipsoid is larger than that of a ball of equal volume. Find a lower bound for the first eigenvalue from Exercise 2.

Exercise 4. Consider an elliptic plate with eccentricity $e = \sqrt{1 - b^2/a^2} \ll 1$ (see the problem stated in Subsection 15.2.3). Using formulas (15.2.15) and (15.2.16), construct the expansion of the smallest eigenvalue λ_{10} and the lowest frequency f_{10} in powers of e up to e^4.

Exercise 5. Using formulas (15.2.21) and (15.2.22), construct the expansions of the eigenvalues λ_{11}^c and λ_{11}^s in powers of e up to e^4 (see the problem for Subsection 15.2.4).

References

1. **Abramowitz, M. and Stegun, I. A. (Editors),** *Handbook of Mathematical Functions with Formulas, Graphs and Mathematical Tables*, National Bureau of Standards Applied Mathematics, Washington, 1964.
2. **Akulenko, L. D.,** High-frequency natural oscillations of mechanical systems. *J. Appl. Math. Mech.,* Vol. 64, No. 5, pp. 783–796, 2000.
3. **Akulenko, L. D., Kostin, G. V., and Nesterov, S. V.,** Numerical-analytical method for studying natural vibrations of nonuniform rods. *Mech. Solids*, Vol. 30, No. 5, pp. 173–182, 1995.
4. **Akulenko, L. D. and Nesterov, S. V.,** Accelerated convergence method in the Sturm-Liouville problem, *Rus. J. Math. Phys.*, Vol. 3, No. 4, pp. 517–521, 1995.
5. **Akulenko, L. D. and Nesterov, S. V.,** An effective method of investigating oscillations of substantially inhomogeneous distributed systems, *J. Appl. Maths. Mechs.*, Vol. 61, No. 3, pp. 451–461, 1997.
6. **Akulenko, L. D. and Nesterov, S. V.,** Determination of frequencies and forms of oscillations of non-uniform distributed systems with boundary conditions of the third kind, *J. Appl. Maths. Mechs.*, Vol. 61, No. 4, pp. 531–538, 1997.
7. **Akulenko, L. D. and Nesterov, S. V.,** Oscillations of interacting systems with distributed parameters, *Mech. Solids*, Vol. 34, No. 2, pp. 11–19, 1999.
8. **Akulenko, L. D. and Nesterov, S. V.,** Natural oscillations of distributed inhomogeneous systems described by generalized boundary-value problems, *J. Appl. Maths Mechs*, Vol. 63, No. 4, pp. 617–625, 1999.
9. **Akulenko, L. D. and Nesterov, S. V.,** Vibration of nonhomogeneous membrane, *Mech. Solids*, Vol. 34, No. 6, pp. 112–121, 1999.
10. **Akulenko, L. D. and Nesterov, S. V.,** Free vibrations of a homogeneous elliptic membrane, *Mech. Solids*, Vol. 35, No. 1, pp. 153–162, 2000.
11. **Akulenko, L. D. and Nesterov, S. V.,** Solution of self-adjoint boundary value problems of order four, *Russ. J. Math. Phys.*, Vol. 10, No. 2, pp. 121–133, 2003.
12. **Akulenko, L. D. and Nesterov, S. V.,** A frequency-parametric analysis of natural oscillations of non-uniform rods, *J. Appl. Math. Mechs*, Vol. 67, No. 4, pp. 525–537, 2003.
13. **Akulenko, L. D., Nesterov, S. V., and Shmatkov, A. M.,** Generalized parametric oscillations of mechanical systems, *J. Appl. Maths. Mechs.*, Vol. 63, No. 5, pp. 705–713, 1999.
14. **Akulenko, L. D. and Shamayev, A. S.,** Approximate solutions of some perturbed boundary value problems, *J. Appl. Maths. Mechs.*, Vol. 50, No. 2, pp. 146–154, 1986.
15. **Anastassiou, G.,** *Handbook of Analytic Computational Methods in Applied Mathematics*, CRC Press, Boca Raton, FL, 2000.
16. **Arscott, F.,** *Periodic Differential Equations*, Macmillan (Pergamon), New York, 1964.
17. **Banach, S.,** *Théorie des opérations lineáires. Monografje Matematyczne*, Warsaw, 1932.
18. **Bathe, K. J and Wilson, E., L.,** *Numerical Methods in Finite Element Analysis*, Prentice-Hall, Englewood Cliffs, NJ, 1976.

19. **Blaquière, A.,** *Nonlinear Systems Analysis*, Academic Press, New York, 1966.
20. **Boyce, W. E. and DiPrima, R. C.,** *Elementary Differential Equations and Boundary Value Problems, 8th Edition*, Wiley, New York, 2004.
21. **Chapman, S. and Lindzen, R. S.,** *Atmospheric Tides (Termal and Gravitational)*, D. Reided Publ., Dordrecht, 1970.
22. **Collatz, L.,** *Eigenwertaufgaben mit Thechnischen Auwendungen,* Akad. Verlagsgesellschaft, Leipzig, 1963.
23. **Collatz, L., Albrecht, J., and Velte, W.,** *Numerical Treatment of Eigenvalue Problems*, Springer Verlag, Berlin, 1987.
24. **Courant, R. and Hilbert, D.,** *Methods of Mathematical Physics, Vols. 1 and 2*, Wiley–Interscience Publ., New York, 1989.
25. **Dunford, N. and Schwartz, J. T.,** *Linear Operators. In 3 parts,* Interscience Publ., New York. *P. 1. General Theory*, 1958; *P. 2. Spectral Theory. Self Adjoint Operators in Hilbert Space*, 1963; *P. 3. Spectral Operators*, 1971.
26. **Eckart, C.,** *Hydrodynamics of Oceans and Atmospheres*, Oxford, 1960.
27. **Friedrichs, K. O.,** *Perturbation of Spectra in Hilbert Space*, Amer. Math. Soc., Providence, RI, 1965.
28. **Gould, S. H.,** *Variational Methods for Eigenvalue Problems: An Introduction to the Methods of Rayleigh, Ritz, Weinstein, and Aronszajn*, Dover Publ., New York, 1995.
29. **Hartman, P.,** *Ordinary Differential Equations*, John Wiley & Sons, New York, 1964.
30. **Hinton, D. and Schaefer, P. W. (Editors),** *Spectral Theory & Computational Methods of Sturm-Liouville Problems*, Marcel Dekker, New York, 1997.
31. **Ince, E. L.,** *Ordinary Differential Equations*, Dover Publ., New York, 1964.
32. **Janke, E., Emde, F., and Lösch, F.,** *Tafeln Höherer Funktionen*, B. G. Teubner Verlagsgesellschaft, Stuttgart, 1960.
33. **Kamke, E.,** *Differentialgleichungen: Lösungsmethoden und Lösungen, I, Gewöhnliche Differentialgleichungen*, B. G. Teubner, Leipzig, 1977.
34. **Kevorkian, J. and Cole, J. D.,** *Perturbation Methods in Applied Mathematics*, Springer-Verlag, New York, 1981.
35. **Kevorkian, J. and Cole, J. D.,** *Multiple Scale and Singular Perturbation Methods*, Springer-Verlag, New York, 1996.
36. **Korn, G. A. and Korn, T. M.,** *Mathematical Handbook for Scientists and Engineers, 2nd Edition*, Dover, New York, 2000.
37. **Krylov, N. M.,** *Selected Papers, Vols 1, 2, and 3* [in Russian], Izd-vo AN UkrSSR, Kiev, 1961.
38. **Lee, H. J. and Schiesser, W. E.,** *Ordinary and Partial Differential Equation Routines in C, C++, Fortran, Java, Maple, and MATLAB*, Chapman & Hall/CRC Press, Boca Raton, FL, 2004.
39. **Leis, R.,** *Initial-Boundary Value Problems in Mathematical Physics*, John Wiley & Sons, Chichester, 1986.
40. **Love, A. E. H.,** *A Treatise on the Mathematical Theory of Elasticity*, Cambridge Univ. Press, Cambridge, 1927.
41. **Mackie, A. G.,** *Boundary Value Problems*, Scottish Academic Press, Edinburgh, 1989.
42. **Malkin, I. G.,** *Some Problems in Nonlinear Oscillation Theory* [in Russian], Gostekhizdat, Moscow, 1956.
43. **McLachlan, N. W.,** *Theory and Application of Mathien Functions*. Clarendon Press, Oxford, 1947.

44. **Mennicken, R. and Moller, M.,** *Non-Self-Adjoint Boundary Eigenvalue Problems*, North-Holland, Amsterdam, 2003.
45. **Mikhlin, S. G.,** *Variational Methods in Mathematical Physics* [in Russian], Nauka, Moscow, 1970.
46. **Morse, P. M. and Feshbach, H.,** *Methods of Theoretical Physics, Vols. 1 and 2*, McGraw-Hill, New York, 1953.
47. **Naimark, M. A.,** *Linear Differential Operators* [in Russian], Nauka, Moscow, 1969.
48. **Nayfeh, A. H.,** *Introduction to Perturbation Techniques*, J. Wiley & Sons, New York, 1981.
49. **Nayfeh, A. H.,** *Perturbation Methods*, John Wiley & Sons, New York, 1973.
50. **Parlett, B. N.,** *The Symmetric Eigenvalue Problem*, Prentice-Hall, Engewood Cliffs, NJ, 1980.
51. **Polyanin, A. D.,** *Handbook of Linear Partial Differential Equations for Engineers and Scientists*, Chapman & Hall/CRC Press, Boca Raton, FL, 2002.
52. **Polyanin, A. D. and Zaitsev, V. F.,** *Handbook of Exact Solutions for Ordinary Differential Equations, 2nd Edition*, Chapman & Hall/CRC Press, Boca Raton, FL, 2003.
53. **Pryce, J. D.,** *Numerical Solution of Sturm-Liouville Problems*, Clarendon Press, Oxford, 1994.
54. **Riesz, F. and Sz.-Nagy, B.,** *Lecons D'analyse Fonctiounelle*, Akadémiai Kiadó, Budapest, 1972.
55. **Sagan, H.,** *Boundary and Eigenvalue Problems in Mathematical Physics*, Dover Publ., New York, 1989.
56. **Shibaoka, Y.,** On the transverse vibration of an elliptic plate with clamped edge, *J. of the Phys. Soc. of Japan*, Vol. 11, No. 7, pp. 797-803, 1956.
57. **Sretenskii, L. N.,** *Dynamics Theory of Tides* [in Russian], Nauka, Moscow, 1987.
58. **Stakgold, I.,** *Boundary Value Problems of Mathematical Physics, Vols. 1 and 2*, SIAM, Philadelphia, 2000.
59. **Stepanov, V. V.,** *A Course of Differential Equations* [in Russian], Gostekhizdat, Moscow, 1953.
60. **Strang, G. and Fix, G. I.,** *An Analysis of the Finite Element Method,* Prentice-Hall, 1973.
61. **Strutt, M. J. O.,** *Lamésche-Mathiéusche und Verwandte Funktionen in Physik und Technik*, T. 1, Berlin, 1932.
62. **Strutt, J. W. (Baron Rayleich),** *The Theory of Sound, Vol. 2,* MacMillan, London, 1926.
63. **Timoshenko, S. P. and Gere, T. M.,** *Theory of Elastic Stability, 2nd Edition,* McGraw-Hill, NJ, 1961.
64. **Timoshenko, S., Young, D. H., and Weaver, W., Jr.,** *Vibration Problems in Engineering, 4th Edition*, Wiley, New York, 1974.
65. **Tikhonov, A. N. and Samarskii, A. A.,** *Equations of Mathematical Physics*, Dover Publ., New York, 1990.
66. **Vulikh, B. Z.,** *An Introduction to Functional Analysis* [in Russian], Nauka, Moscow, 1967.
67. **Whittaker, E. T. and Watson, G. N.,** *A Course of Modern Analysis, Vols 1 and 2*, Cambridge Univ. Press, Cambridge, 1952.
68. **Yakubovich, V. A. and Starzhinskii, V. M.,** *Linear Differential Equations with Periodic Coefficients, Vols 1 and 2.* J. Wiley, New York, 1975.
69. **Zwillinger, D.,** *Handbook of Differential Equations*, Academic Press, San Diego, 1998.

Index

A

B

C

D

E

F

G

H

I

J

K

L

M

N

O

P

R

S

T

U

V

W